Withdrawn

PUMP

ALSO BY BILL SCHUTT

Dark Banquet: Blood and the Curious Lives of Blood-Feeding Creatures

Cannibalism: A Perfectly Natural History

ALSO BY BILL SCHUTT AND J. R. FINCH

Hell's Gate

The Himalayan Codex

The Darwin Strain

PUMP

A Natural History
of the Heart

BILL SCHUTT

illustrations by
PATRICIA J. WYNNE

ALGONQUIN BOOKS OF CHAPEL HILL 2021

Published by
ALGONQUIN BOOKS OF CHAPEL HILL
Post Office Box 2225
Chapel Hill, North Carolina 27515-2225

a division of
WORKMAN PUBLISHING
225 Varick Street
New York, New York 10014

Library of Congress Cataloging-in-Publication Data

Names: Schutt, Bill, author. | Wynne, Patricia, illustrator.
Title: Pump : a natural history of the heart / Bill Schutt ; illustrated by Patricia J. Wynne.
Description: Chapel Hill, North Carolina : Algonquin Books of Chapel Hill, 2021. |
Includes bibliographical references. | Summary: "Zoologist Bill Schutt delivers a look
at hearts from across the animal kingdom, from insects to whales to humans.
Illustrated with black-and-white line drawings"— Provided by publisher.
Identifiers: LCCN 2021019011 | ISBN 9781616208936 (hardcover) |
ISBN 9781643752143 (ebook)
Subjects: LCSH: Heart. | Heart—Anatomy.
Classification: LCC QP111.4 .S38 2021 | DDC 612.1/7—dc23
LC record available at https://lccn.loc.gov/2021019011

10 9 8 7 6 5 4 3 2 1
First Edition

For Elaine Markson (*Bill Schutt*)

and

Ted Riley (*Patricia J. Wynne*)

HEART (*n.*)

1 A hollow muscular organ that pumps the blood through the circulatory system by rhythmic contraction and dilation. In vertebrates, there may be up to four chambers (as in humans), with two atria and two ventricles.

2 Used to refer to a person's character, or the place within a person where feelings or emotions are considered to come from.

3 The firm central part of a vegetable, especially one with a lot of leaves.

4 Courage, determination, or hope.

5 A shape consisting of two half circles next to each other at the top and a V shape at the bottom, often colored pink or red and used to represent love.

6 One of the four suits in playing cards, represented by a red heart shape.

7 The central or most important part.

Hearts cannot be broken, they're small squishy things.
—JEFF HEISKELL

I am reminded of the advice of my neighbor,
"Never worry about your heart till it stops beating."
—WILLIAM STRUNK JR. and E. B. WHITE,
The Elements of Style

Contents

PUMP

Most things in life come as a surprise.
—LYKKE LI

A Small Town with a Big Heart

IN MID-APRIL 2014, a sharp-eyed resident of Trout River, Newfoundland, looked out into the Gulf of Saint Lawrence and saw something peculiar. What had first appeared as a small dot on the horizon was growing larger and larger. By the time the giant thing washed ashore, the media had descended, and so, too, had the ungodly stink, which someone described to me as "a sickly perfume smell combined with the reek of decaying flesh." And, indeed, this was more decaying flesh than anyone had ever seen before—around a hundred tons of it.

Soon the tiny fishing village was buzzing with reporters and gawkers as word of mouth gave rise to sensational headlines. The chatter between locals turned from bewilderment and disgust to health concerns, the potential for lost income, and even the threat of a horrific explosion. Stranger yet, something almost identical was taking place just up the coast, in the small town of Rocky Harbour.

Canadian winters are often frigid, but the winter of 2014 had been the coldest in memory. For the first time in decades, the Great Lakes had frozen over and their outlet to the Atlantic Ocean, the Gulf of Saint Lawrence, had a heavy buildup of sea ice. The high winds and currents had also piled up ice in the Cabot Strait, turning the gulf's widest channel

to the sea into a bottleneck. But if the inhabitants of Trout River and Rocky Harbour were struggling through the harsh weather conditions, a far more desperate struggle was taking place roughly two hundred miles to the south—in the Cabot Strait itself.*

In the late winter and early spring, blue whales (*Balaenoptera musculus*) typically begin to leave the Atlantic Ocean and enter the Gulf of Saint Lawrence to feed on tiny crustaceans called krill. The largest animal known to have lived on Earth,† a blue whale can reach one hundred feet in length and can weigh up to 163 tons. By way of comparison, this is equivalent

---

* Located between Nova Scotia and Newfoundland, the Cabot Strait is an important international shipping lane, named after Italian maritime navigator Giovanni Caboto. After exploring coastal North America in 1497, he was subsequently referred to as John Cabot by the English, who had commissioned his flag-planting exploits.

† The largest organism is a humongous fungus (*Armillaria ostoyae*) living in Oregon and covering an area of nearly four square miles.

to twenty African bull elephants or about sixteen hundred average-sized adult human males. Despite their enormous size, blue whales were not hunted for their oil-rich blubber until 1864. The reasons for this were related to the great speeds they can attain—up to thirty-one miles per hour—and their tendency to sink when slain. Whalers preferred the three species of *Eubalaena*, since their bodies have a higher blubber content and tend to float after death. Thus, they were christened "right wales"; they were the *right* whales to throw harpoons at. Things went horribly wrong for blue whale populations after faster, steam-driven whaling ships began using the newly invented harpoon cannon, and more than 380,000 blue whales were slain between 1866 and 1978. Most countries don't allow whale hunting anymore, but the blue whale's propensity to sink after death remains an inconvenience to those attempting to study its anatomy.

In March 2014, Mark Engstrom, senior curator and deputy director of collections and research at the Royal Ontario Museum (ROM) in Toronto, received a call from his friend Lois Harwood. Harwood, who worked for Canada's Department of Fisheries and Oceans (DFO), wondered if Engstrom had heard the news that nine blue whales feeding in the Cabot Strait had died. Apparently, she said, they were unable to escape a massive ice floe, had gotten trapped in the ice, and perished. This was tragic, especially because blue whales were critically endangered, and the loss of nine individuals meant the loss of something like 3 to 5 percent of the total North Atlantic population.

Harwood knew, though, that Engstrom was looking to obtain specimens of every whale species found in Canadian waters. She told him that three of the whales hadn't sunk, possibly because they had been buoyed by the thick ice. Engstrom became even more interested after Harwood put him in touch with Jack Lawson, a researcher with the DFO who had been tracking the dead whales by helicopter for the past month. He told Engstrom that he expected the trio of whales to wash up on the shore sooner or later—and in April, they did.

"The thing is, the whales drifted ashore in these three tiny villages,"
Engstrom told me during my visit to the ROM in 2018. "Trout River doesn't
really get the normal tourist traffic. It's sort of a struggling community.
The mayor told me one day he looked out and he could see the whale in
the water and he said, 'Oh, please, God, don't let that thing come ashore
here.' He said the next morning there it was, on the *only* stretch of beach
they have, and right underneath their *only* restaurant—this giant dead
blue whale, stinking to high heaven."

I asked Engstrom what happened next.

Engstrom laughed. "Then it started to bloat."

"That must have lightened things up," I offered.

"Not really," he said. "By then, they'd all seen YouTube videos of whales
exploding."

Videos of whales detonating from an accumulation of gases have
been making the rounds on the internet for years. At last count, they
numbered over two hundred and included one pitching "The Exploded
Whale Song." My personal favorite, though, depicts a fifty-six-foot,
sixty-ton sperm whale that beached in Taiwan in 2004. Local university-
scientist types quickly decided to take advantage of the unexpected
opportunity by carrying out an autopsy on the megacorpse. They also
decided that it would be best to do this at their labs, and so a massive
effort was undertaken to move the thing. Three cranes, fifty workers, and
thirteen hours later, the whale was driven off, strapped to the open bed of
a tractor-trailer. But on the way through the busy streets of Tainan City,
the putrefying giant exploded spontaneously. The blast spewed thousands
of pounds of rotten blood, blubber, and entrails onto cars, motor scooters,
and shops. It even soaked some unfortunate onlookers.*

"But blue whales don't do that," Engstrom assured me, just as he had

* The event left such an impression on me that I immediately posted a newspaper photo of the
postexplosion carnage outside my office door at LIU Post, labeling one car owner's particularly bad
parking spot.

previously tried to assure the freaked-out and unconvinced residents of Trout River. He told the townsfolk that unless people decided to jump up and down on the dead behemoth or cut it open, the tissue breakdown would likely allow the accumulating gases to escape slowly, like from an old balloon. "Which is what eventually happened," he said.

Engstrom explained that most of the questions he got from the reporters on the scene in Newfoundland were related to one of two topics: smell and size. "How big is the heart? We hear it's as big as a car." He and his team heard the heart-size question so many times that, finally, one of his technicians responded with a question of his own. "Why don't we try to save the sucker?"

Engstrom was immediately intrigued by the possibility, though he knew that his team had to move quickly. One of the three whales had drifted into an uninhabited cove and had broken up in the tide during a storm. The second specimen was currently impersonating the Goodyear Blimp for whale-bomb-wary crowds in Trout River, a situation that did not bode well for the preservation of its internal organs.

But Engstrom knew that the last and smallest whale (seventy-six feet), the one that had come ashore at Rocky Harbour, lay partially submerged in the cold water—potentially slowing down the process of organ decomposition. He asked ROM colleague Jacqueline Miller, a mammalogy technician assigned to the Rocky Harbour recovery team, if she could salvage the heart.

The expert anatomist responded immediately and enthusiastically, "Yeah, we can save it." Later, she confessed to me that she wasn't quite sure *what* they would find when they opened the whale up, or if it would be salvageable. But the prospect of preserving a blue whale heart was exciting enough that she was eager to try.

Miller and seven other intrepid researchers began by "flensing" the Rocky Harbour whale carcass—whaling-speak for removing the flesh and soft tissue from tail to head. Once the muscles surrounding the heart- and lung-containing thoracic cavity had been removed, members of the recovery team got their first look at the megapump—something no researchers had ever seen before. Instead of a typical mammalian heart, the specimen more closely resembled a four-hundred-pound flesh-colored soup dumpling. Undaunted by the heart's resemblance to a gargantuan Chinese appetizer, they took a further look through the gore and were completely thrilled to see that although the heart had collapsed into a six-foot-wide blob, it had *not* decomposed.

"It was still pink," Miller told me, although she also remembers some mildew and a bit of necrotic (i.e., dead) tissue. "It had a lot of elasticity, and it still held a lot of fluid."

Several years later, in 2017, Miller would be invited to necropsy a North Atlantic right whale (*Eubalaena glacialis*) after a mass mortality event during which seventeen whales had died mysteriously. Her hope was to recover a cetacean heart from an additional species.* But even though that particular whale had been dead for less time than the Newfoundland blues whales, it turned out to be the wrong right whale. The heart had already decayed into an unsalvageable mess. The episode, which took place during the summer, led Miller to realize how fortuitous it had been that the Rocky Harbour specimen had died in the winter and spent three months in ice water. "I think we were just lucky," she said.

Miller, whose graduate school studies had focused on mice and other pint-sized mammals, got in and got dirty. She and her rainwear-clad colleagues worked with flensing knives and machetes to sever the vena cava and the aorta, the great vessels leading to and from the blue whale's heart, respectively. Then they attempted to free the organ from the gigantic animal's body. But after positioning themselves inside the creature, Miller and three associates discovered that, try as they might,

---

* *Necropsy* comes from the Greek for "corpse" or "the dead" (*nekros*), while the -*opsy* part (also derived from Greek) means "sight," thus referring to the visual inspection of a dead body. *Autopsy* (from Greek, "seeing for oneself"), when used in the context of dead bodies, is reserved for postmortem examination of human bodies.

they could not maneuver the heart through the space they had cut for it between two of the ribs. Even after detaching the heart from the lungs by cutting through the pulmonary arteries and veins, it wouldn't budge. Eventually, after forcing several ribs apart, the four researchers were able to shove what turned out to be a 386-pound heart from its original home into a nylon mesh bag spacious enough to package a Volkswagen Beetle.

With the aid of a front-end loader, a forklift, and a dump truck, the blue whale heart was transferred to a refrigerated truck and shipped out to a facility where it was frozen at −20°C. It would remain on ice for an entire year before a team of experts could be assembled to carry out the next phase of the project: preservation.

This process, Engstrom explained, would include restoring the heart to its original shape. This was necessary because, unlike a human heart would do, the blue whale's heart had collapsed like a deflated beach ball after its great vessels were severed. Engstrom told me that, although no one was quite sure, this was likely an adaptation to the great pressure experienced by blue whales during deep dives.

Preservation efforts began with the specimen being placed into a tap water bath to thaw. The heart would need to be filled with preservative to halt decomposition, stiffen the muscles, and kill any bacteria that might have survived the trip to the freezer. First, though, the team searched for appropriately sized objects to plug the dozen or so severed blood vessels coming off the organ. The cork job was necessary so that they could fill the interior chambers of the heart with preservative without having it flow back out. It would also allow the researchers to reinflate the specimen, remedying the unsightly collapsed-balloon look that the mighty heart had assumed since its removal.

Ultimately, the items they chose as plugs ranged from soft drink bottles, for the smallest vessels, all the way up to a five-gallon bucket, which fit quite nicely into the giant caudal vena cava. This particular mega-vein was responsible for carrying oxygen-depleted blood from the whale's body

and tail to its right atrium, one of the heart's two "receiving chambers."*
The right atrium also received blood from the only slightly smaller cranial
venae cavae, which returned blood from the whale's massive head region.
In two-legged creatures like humans, the equivalent vessels are known as
the inferior vena cava and the superior vena cava, respectively. As in all
mammals, the venae cavae transport carbon dioxide–rich and oxygen-
poor blood back to the heart, which then pumps it to the lungs.

During the initial preservation effort, Jacqueline Miller and her team
used seven hundred gallons of everyone's favorite embalming agent,
formaldehyde. This tissue fixative has been known to be a carcinogen since
the early 1980s, and though most people remember its distinctive smell
from biology class, our most common exposure results from the chemical's
nearly undetectable inclusion in building materials like particleboard,
plywood, and fiberboard. Although the whale preservation crew diluted
their formaldehyde into a somewhat more biologically friendly solution
known as formalin (generally around 40 percent formaldehyde), the
liquid was still, in scientific parlance, some particularly nasty shit.

"The funny thing," Miller told me, "is that in a typical lab, you risk
getting splashed by formalin. Here, the risk was falling into a vat full of it."

The heart sat in formalin for five months, undergoing the process of
fixation, during which all tissue decay ceases. The formerly pink organ
also took on the beige color typical of similarly fixed specimens. But
although it could have remained in the same solution for decades, Mark
Engstrom and his colleagues decided that sticking it in the equivalent of a
giant bottle of poison would not do justice to the great heart. Instead, after
consultation with a pair of conservators versed in the art of preserving
large specimens, a decision was made to "plastinate" it. Plastination is a
unique process of specimen preservation invented in 1977 by the decidedly
weird German anatomist Gunther von Hagens. Known affectionately as

---

* *Atrium* is Latin for "entrance hall."

Dr. Death, von Hagens created the controversial *Body Worlds* exhibit, which consists of dozens of skinned and plastinated human bodies, posed in a variety of positions, each chosen to better illustrate a range of anatomical systems.*

Since the researchers at the ROM were not trained or equipped to carry out the complex procedure by themselves, they shipped the whale heart to the Plastinarium, a *Body Worlds* gallery and plastination facility in Guben, Germany. Otherwise known as Gubener Plastinate GmbH, the former cloth factory is staffed with von Hagens–trained experts, each eager to satisfy their customers' every plastination need. Though they were used to dealing with museum specimens of many shapes and sizes, the blue whale heart would be their largest undertaking ever.

During the initial steps of the process, all of the water and soluble fats are slowly drawn out of the specimen and replaced by acetone, an organic compound that is as toxic to humans as it is flammable. In the very definition of "Don't try this at home," the ROM's blue whale heart required a total of six thousand gallons of the stuff. The heart sat in acetone for eighty days at freezing temperatures, the cold expediting the loss of water from the cells and its replacement with the poisonous solvent.

The staff of the Plastinarium then put the heart through a process known as forced impregnation, during which the acetone was replaced by liquid plastic, specifically a silicone polymer. To achieve this, they placed the organ in a vacuum chamber and gradually lowered the air pressure. This environment caused the acetone to vaporize within the cells, drawing the polymer in behind it to fill the empty space. With most of the cell mass now occupied by liquid polymer, the process had literally transformed the formerly living tissue into plastic. The Plastinarium

---

* In January 2011, in a macabre chapter to a story many had considered more than a bit macabre to begin with, the then sixty-five-year-old von Hagens revealed publicly that he was terminally ill. He also expressed a desire to have his body skinned and plastinated after his death. The current plan is for the plastinate version of von Hagens to "greet" visitors as they enter one of the permanent *Body Worlds* exhibitions. Reportedly, Dr. Death will be wearing his trademark black fedora.

employees then employed a curing agent to harden the silicone, a step that took an additional three months.

Once fully firm, the blue whale heart was shipped back across the pond in May 2017, to become the centerpiece of an elaborate exhibit that was constructed at the Royal Ontario Museum to highlight the amazing specimen. For size comparison purposes, the heart was displayed alongside a Smart car, while from the adjacent ceiling stretched the fully articulated skeleton of the Trout River whale specimen. Now weighing in at 440 pounds, the plastinated blue whale heart would never decay or smell, and the enormous pump would be viewed by hundreds of thousands of museum visitors during its four-month-long star turn in Toronto.

PUMP IS A story about hearts and the circulatory systems associated with them. Big hearts, small hearts, cold hearts, and even nonexistent hearts. It is also the story of some of the notable structures, fluids, findings, and foul-ups associated with them. The history of our attempts to understand the function of the heart and circulatory system is long and, until relatively recently, riddled with errors. For example, among the medical communities of the seventeenth and eighteenth centuries, there was a belief that blood carried within it the essence of its owner's personality. Terms like "blue blood," "bloodthirsty," "cold-blooded," and "hot-blooded" are linguistic vestiges from a very different world. Armed with the knowledge of just *how* different that world was, it will be easier to understand why the history of cardiovascular medicine has no shortage of strange stories and bizarre treatments.

Pump is certainly not a textbook, nor is it my goal to cover every type of heart and every facet of every circulatory system. Instead, I will wander through these broad topics, making interesting stops along the way. For those of you who have gone exploring with me before, there will be quite a few of these side trips, most of them with a zoological or historical perspective. Some of these seemingly tangential stops will be necessary to better explain poorly understood or misunderstood concepts, while others will help explain how hearts and circulatory systems work—covering topics like diffusion, the blood-brain barrier, and Mothra.

Hearts and their related circuitry show a serious degree of variation in invertebrates like insects, crustaceans, and worms—and there are good reasons for that. There exist far fewer differences in creatures that come equipped with a backbone, whether fish, fowl, or farmer. But in addition to exploring some prime examples of cardiovascular diversity across the animal kingdom, we'll learn how some of these creatures are now saving lives and providing answers to difficult questions about cardiac health and the ailing human heart.

Pump is also the story about what happened when one relatively new species of mammal decided that the heart was something far more than an organ keeping everybody alive—that it was no less than the center of emotion and the seat of the soul. Where did *that* belief come from? Why does it cross so many cultural boundaries? Why does it persist? And just as importantly, is there any truth to the link between hearts and minds?

By the end of this journey, you will gain a new appreciation for the degree to which the heart plays a vital role in the natural and human world, both as the engine that drives the circulatory system *and* as the mysterious organ at the core of human culture and human nature itself. From a hollow cluster of cells with a unique ability to shorten its length to the golf cart–sized heart of the blue whale, from beliefs about the origin of love and the soul to early cardiac medicine, futuristic therapies, and beyond—my hope is that you will never think about these topics in quite the same way again.

In fact, it's my heart's desire.

PART 1

Wild at Heart

One size does not fit all.

—UNKNOWN (POSSIBLY FRANK ZAPPA)

[1]

Size Matters I

IN AUGUST 2018, I traveled to Toronto's Royal Ontario Museum with artist Patricia J. Wynne to examine the famous blue whale heart. Patricia and I have been friends and officemates at the American Museum of Natural History since the mid-1990s, and she has illustrated every paper, book chapter, and book (fiction and nonfiction) that I've ever written. Although the blue whale exhibit had already closed and the specimen was being stored at an off-site facility, researcher Bill Hodgkinson had uncrated the heart in preparation for our arrival. In a room the size of a small aircraft hangar, the preserved whale heart sat perched upon a two-inch-thick stainless-steel rod, giving it the appearance of having been skewered from below. The bottom end of the skewer was secured to a wooden floor stand while the business end had been connected to a metal armature, invisible to viewers, that served as the heart's permanent internal scaffold.

Because the specimen's official dimensions are forty-two inches from top to bottom by thirty-eight inches in width, I was quite surprised to find it looming over me at a height of what I estimated to be well over six feet. The explanation for the added height was the massive blood vessels situated atop the plastinated organ. The most prominent of these was the

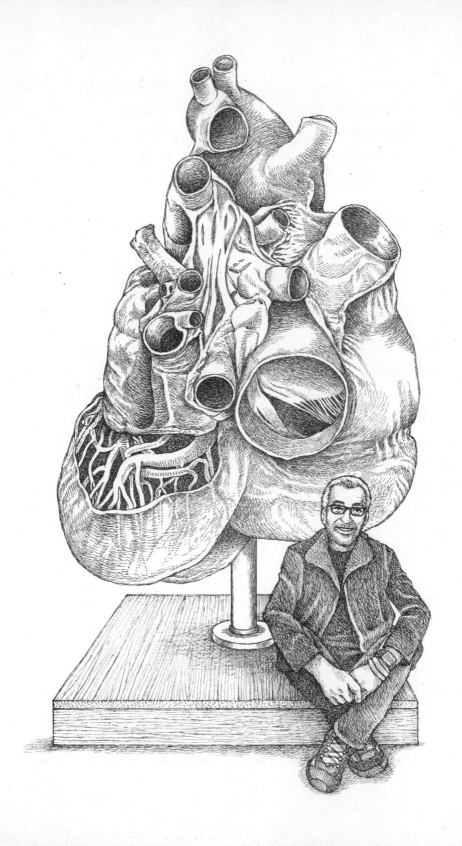

great arch of the aorta and its offshoots, a pair of carotid arteries that had once carried oxygenated blood from the left ventricle of the heart to the animal's head. If the previously mentioned atria can be envisioned as the heart's receiving chambers (the left atrium and the right atrium receiving blood from the lungs and the body, respectively), then the ventricles are the heart's pumping chambers—the right ventricle pumps oxygen-poor/ CO_2-rich blood to the lungs while the left pumps oxygenated blood out to supply the body.

During the blue whale heart's lengthy preparation period, a special type of colored silicone polymer had been injected into the blood vessels, and so veins and arteries could now be differentiated, because veins were blue and arteries were red. The multicolored heart was really quite beautiful, and I was immediately drawn to a porthole-shaped section that had been cut through the right ventricle by plastination expert Vladimir Chereminsky. The window allows viewers to peer inside the chamber, where, among other things, they can see the odd-looking arrangement of inch-thick muscle strands that line its walls. These strands are known as trabeculae carneae (meaty ridges) by anatomy types and medical professionals, and smaller versions can be seen in many mammals, including humans. The ridges increase the surface area of the ventricular walls as compared to a smooth wall, packing more muscle fibers into a limited space. This is important because the extra muscle translates to stronger ventricular contractions, which propel blood out of the heart. Additional functions of this odd-looking chamber surface remain to be explored.

The right and left atria of the whale heart also contract, and their thinner walls reflect the fact that their job is less difficult: pumping blood into their adjacent ventricles instead of out to the body. Located between the atria and ventricles are the aptly named atrioventricular (AV) valves. Through Chereminsky's porthole, museum visitors could see the blue whale's right AV valve, which appeared to have the diameter of a toddler's toy drum. In humans, the corresponding valve spans about three-quarters of a square inch, about the diameter of a marble, and is more commonly known as the tricuspid valve, due to its three flap-like valve cusps.*

The AV valves regulate blood flow from the atria to the ventricles, but equally important is their job preventing blood from reversing direction and heading *back* into the atria when the ventricles contract. Vital to this role, and clearly visible within the blue whale heart, are a dozen or so tough fibers known as chordae tendineae. Colloquially known as the heartstrings (since they resemble pieces of string), these cords are composed primarily of a structural protein called collagen.† With one end of the chordae tendineae firmly anchored to the floor of the ventricle and the other end attached to the valve cusp, the cusps are prevented from extending into the atria when the ventricles contract—effectively sealing off the two chambers.

To visualize this, picture a dog with its collar fastened to a long leash, with the nondog end staked to the ground. The dog (standing in for the valve cusps) can travel only so far before the leash (the chordae tendineae) pulls tight, preventing the dog from advancing past an open gate. In humans, the terms "ventricular prolapse" or "prolapsed valve" are used to describe medical conditions in which one or more of the AV valve cusps bulges into an atrium (think of the dog's leash that has been stretched from the pup's constant tugging, allowing it to advance beyond the gate). Since this prolapse breaks the seal separating the atrium and the ventricle, some of the ventricular blood "regurgitates" back into the atrium when the ventricle contracts, instead of leaving the heart, as it

---

* On the left side, the bicuspid valve is named for its two cusps. Confusing the issue, it is also known as the mitral valve, due to its *supposed* resemblance to a miter, the ceremonial headwear worn by bishops. Thankfully, there are no hat-derived alternative names for the tricuspid valve.

† Wound into fibers, collagen is the most abundant protein in mammals. It is commonly found in tendons, ligaments, and the skin. Collagen also give bones their varying degrees of flexibility.

would normally. These so-called "floppy" valves can result from previous heart attacks, infections like bacterial endocarditis (frequently found in intravenous drug users), or rheumatic fever, a now-rare consequence of untreated strep throat or scarlet fever. Mitral valve prolapse can also be congenital in nature.

Valve problems can also be a consequence of aging. As the heart valves stiffen and lose their flexibility, they lose their ability to efficiently seal off the heart chambers. With some of the blood moving backward into the atrium with each heartbeat, less blood is pumped out of the heart, and so it has to work harder (by increasing its rate or contracting harder) to compensate. The extra effort can put added stress on the heart, which can lead to serious problems. These become especially apparent if the heart reaches a point at which it can no longer provide sufficient oxygen- and nutrient-rich blood to the body.

Once blood passes through the AV valves, filling the right and left ventricles, it must next pass through the semilunar valves, named for their half-moon-shaped cusps. As the ventricles contract, blood rushes through them into two large arteries. On the right side is the pulmonary trunk, which sends deoxygenated blood to the lungs via the pulmonary arteries that branch from it. On the left side, ventricular contraction pumps oxygenated blood out through the aorta, whose branches distribute it to the rest of the body. Though their anatomy is different from the AV valves before them—no chordae tendineae here—pulmonary and aortic semilunar valves also prevent the backflow of blood, here from the pulmonary artery and aorta back into the ventricles.

In humans, slight valvular abnormalities are often symptom-free and don't require treatment. In more serious cases, a prolapsed valve can cause irregular heartbeat (arrhythmia), dizziness, fatigue, and shortness of breath, and surgery may be required to fix it. Until the early 2000s, valve repair or replacement required complicated open-heart surgery. Now, though, transcatheter valve replacements can be accomplished through

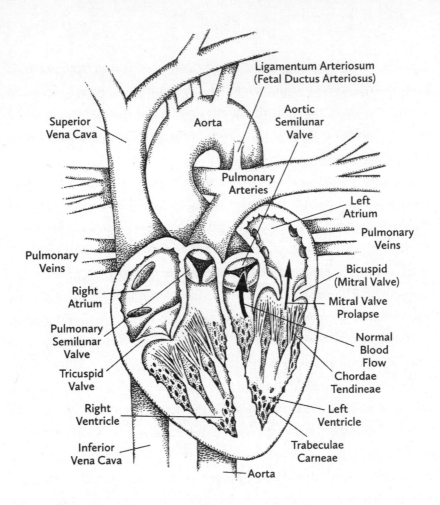

Ligamentum Arteriosum
(Fetal Ductus Arteriosus)

Superior
Vena Cava

Aorta

Aortic
Semilunar
Valve

Pulmonary
Arteries

Left
Atrium

Pulmonary
Veins

Pulmonary
Veins

Bicuspid
(Mitral Valve)

Right
Atrium

Mitral Valve
Prolapse

Pulmonary
Semilunar
Valve

Normal
Blood
Flow

Tricuspid
Valve

Chordae
Tendineae

Right
Ventricle

Left
Ventricle

Trabeculae
Carneae

Inferior
Vena Cava

Aorta

small incisions, or even no incisions at all, as a result of major advances in cardiac catheterization—a process whose history is as interesting as any fiction writer could have dreamed up. But more on that topic later.

To give viewers a look at the blue whale's heart just below its surface, plastination-meister Chereminsky also removed a section of the whale's visceral pericardium. This is the thin, protective layer of the heart that lies atop all that muscle. It's also the *inner* layer of the saclike pericardium, which lubricates and cushions the heart while holding it in place. To visualize the relationship between heart and the pericardium, picture a Ziploc storage bag containing a bit of water. Push your fist (the heart) into the side of the bag so that the bag wraps around your fist. The bag of water is the pericardium, and the part of the bag plastered against your fist is the visceral pericardium. The space inside the bag is the pericardial cavity, partially filled with its supply of pericardial fluid. To complete the metaphor, the part of the Ziploc bag farthest from your fist is the parietal pericardium, and it is attached to the surrounding walls of the chest cavity. This connection anchors the heart in place while cushioning it from external shocks. It's worth noting that the pericardium does not *contain* the heart, but rather is wrapped around it.

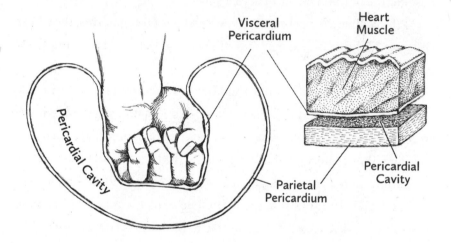

Having observed the plastinated whale heart, inside and out, I left my friend Patricia at the warehouse to sketch the specimen while I set off for the ROM to interview some of the people responsible for its recovery and preservation. But beyond the story of how this one-of-a-kind specimen came to be, I was most interested in what Jacqueline Miller, Mark Engstrom, and their colleagues learned from it that they had not known before.

I asked Miller about the plastinated heart's odd shape. Typically, the mammalian heart is conical, coming to a single point at the bottom or apex. I had been struck by the fact that in blue whales the apex of the heart is split. Miller explained that this bifurcation is a characteristic of rorquals, a name used to group the largest of the baleen whales.* Another unique characteristic, she told me, is that this particular heart is flatter and wider than most mammalian hearts.

"The typical terrestrial mammal has a spiral heart—a heart in which the connective tissue and muscle fibers are oriented so that they spiral around the left and right ventricles," added Engstrom. "When the heart contracts, the overall action is more like wringing out a towel."

But in rorquals, the fibers run straight from the top of the heart (the base) to the bottom, rather than in a spiral.

"I think what's happening is that when they do deep dives, their heart collapses," Engstrom told me.† "It's still beating, but it collapses due to the pressure."

Because of this, and as Miller and her team discovered back in Rocky Harbour, once the heart had been severed from its moorings and removed from the body, it had collapsed "like an enormous spongy bag," according to Miller, thus requiring reinflation during the preservation process.

Adding to the list of things the researchers at the ROM had learned about blue whales, Engstrom mentioned how many times over his career he had been asked about the actual size of the world's largest heart.

"I was getting tired of the question," he admitted. "And I really wanted to be able to say 'It's *that* big' and then point to it."

For decades, in both popular and scientific literature, it was written that a blue whale heart would be the size of a sedan and weigh at least a metric ton.‡ Miller told me that in preparing to extract the heart, she and her colleagues had read about how "you'd be able to swim down one of the greater vessels, presumably the caudal vena cava, which is the largest vessel on the blue whale heart."

---

* "Baleen" is the arrangement of bristles inside the mouths of certain whale species into a filter-feeding device. Composed of keratin (the stuff that makes up our nails and hair), it is used to trap krill after large gulps of water are taken in and then forced out of the mouth.

† While the dive record for a tagged blue whale is 315 meters (1,033.5 feet), a Cuvier's beaked whale (*Ziphius cavirostris*) holds the record for dive depth by a mammal, at 2,992 meters (or 1.86 miles)!

‡ 1 metric ton = 2,204.6 pounds.

As I looked over the impressive vasculature attached to the ROM specimen, it was easy to see that even the largest blood vessel wasn't wide enough for a human to swim through, though I figured an otter or a migrating salmon could make the journey with relative ease.

Indeed, Miller told me, once the heart had been preserved it was significantly *smaller* than they had thought it would be. And this wasn't an undersized blue whale by any stretch. So *why* was it so much smaller than anticipated?

The answer turned out to be that blue whale hearts are simply not as large as the hearts of most other mammals. While quite humongous by human standards, a blue whale's heart apparently makes up only around 0.3 percent of the animal's total body weight. For comparative purposes, the relative size of the heart in both mice and elephants has been calculated to be about 0.6 percent.

Interestingly, some of the world's smallest animals have disproportionately large hearts. For example, the masked shrew (*Sorex cinereus*) is one of the smallest mammals in the world, weighing in at around five grams,* but its heart makes up about 1.7 percent of its body weight, which is approximately three times larger than one would predict for a typical terrestrial mammal, and nearly six times the relative size of a blue whale heart. Birds, meanwhile, tend to have relatively larger hearts than mammalian hearts, due to the metabolic demands of flight. In hummingbirds, the smallest of which can weigh as little as two grams (less than a dime), the heart-to-body weight numbers are even more extreme, with the heart reaching 2.4 percent of body weight. Relatively speaking, this means that hummingbird hearts are *eight times* larger than those of a blue whale.

It is thought that the reason for possessing a relatively large heart relates to the lifestyles of the small and hyperactive. For example, hummingbirds

---

* The smallest mammal in the world is the Kitti's hog-nosed bat (*Craseonycteris thonglong-yai*) from Thailand and Myanmar. Also known as the bumblebee bat, it weighs in at barely two grams.

can beat their wings at eighty times per second, and shrews are such nonstop hunters that during my mammal-trapping days as a PhD student at Cornell University I was taught that they would starve to death if not removed from a live trap within an hour. The manic behavior of these tiny animals causes an extremely high cellular demand for both energy and oxygen. These metabolic requirements are met in part by increasing heart rate, thus also increasing the frequency at which oxygen-rich and nutrient-laden blood is pumped to the body. The resulting heart rate numbers are truly astonishing. Hummingbird heart rates can reach 1,260 beats per minute, while shrews hold the vertebrate record at 1,320 beats per minute—roughly seven times the maximum heart rate of a thirty-five-year-old human.

Though these are eye-popping numbers, the increase in beat frequency is not unlimited, and researchers believe that there *is* a maximum rate at which a heart can beat. For a shrew, one heartbeat lasts forty-three milliseconds—that's forty-three *thousandths* of a second. During this split second, the heart needs to fill with venous blood, contract and eject the arterial blood, and relax in preparation for the next filling cycle. All of that can occur only so fast—and if shrews aren't at the upper limit of heart rate, then they're awful damn close. So if the physical design of the heart limits it to something like a maximum of fourteen hundred beats per minute, then the only way to pump *more* blood is to increase the size of the heart. That way, the larger chambers are able to receive and pump a relatively greater measure of blood with each beat.* This explains the comparatively enormous heart size of creatures like shrews and hummingbirds. But as we'll soon see, increasing heart size among the ubersmall also has its limits.

Before leaving blue whale hearts, though, and whale hearts in general, it should be emphasized that there is much, much more to learn: How

---

* An average-sized man has about five liters of blood. At rest, cardiac output is approximately five liters/minute, so the average time it takes our blood to take a full circuit of the body (from heart to lungs, back to the heart, out to the body, and back to the heart) is approximately one minute.

exactly do these hearts collapse, and how can their owners survive when they do? Other diving mammals, like seals, reduce their heart rates and cut off blood flow to different regions of their body. Do blue whales possess the same oxygen-saving adaptations? Initial research indicates that this could be so, since a recent study by biologist Jeremy Goldbogen and his colleagues at Stanford University found that blue whale heart rates can drop to as low as two beats per minute.* On the anatomy side of things, other serious questions remain, some as simple as identifying the blood vessels in the confusing assemblage sprouting from the now-famous ROM specimen. Until more research can be done, much of the physiology of the rorqual heart will remain in the realm of hypothesis and conjecture.

---

* Goldbogen and his team used suction cups to attach a heart rate monitor to a single blue whale, and were able to monitor the animal's heart rate for nearly nine hours. They did not seek to determine if blood flow was redirected to specific regions of the body during the dramatic drop in heart rate that they recorded.

The Microbe is so very small
You cannot take him out at all.
—HILAIRE BELLOC

[2]

Size Matters II

FOR THOSE OF you who have a body of less than one millimeter across, nothing much in this book applies to you. Why's that, you ask? The answer is that much of what has come before in this book and much that follows is about the heart. By definition, a heart is a hollow muscular organ that receives circulatory fluid from the body before rhythmically pumping it back out again. Collectively, the pump, the fluid, and the vessels through which the fluid travels are referred to as a circulatory system . . . and you don't have one. Because of your minuscule size, nutrients and oxygen can be distributed to your cells (or cell, if you're small enough to have only one), and waste products can be removed from them, by a simple exchange with the external environment, which for most of you probably consists of water.

That exchange is known as diffusion, which is a vitally important process for all living things, whether they're microbes or blue whales. Basically, diffusion occurs when molecules—like oxygen, or nutrients, or waste products—exist at different concentrations on either side of a barrier. Imagine that you've just cleaned your room by cramming everything into your closet and forcing the door shut. There is a higher concentration of stuff inside the closet than outside, with the closet door acting as the barrier. If you were to cut a hole in the door, anything smaller than that hole would have the potential to escape and tumble out, always moving from an area of higher concentration (your closet) to an area of lower concentration (your room). So now, instead of bumming out whenever you open your closet door and stuff falls out, you can think of the mini avalanche as your belongings following their concentration gradient.

But what does your closet have to do with circulatory systems? As previously touched upon, the answer relates to one of the system's key functions, which is to deliver nutrients and oxygen from outside the body to the cells and tissues inside the body. Conversely, circulatory systems also function by helping transport potentially harmful stuff, like toxins, cellular waste products, and carbon dioxide, *out* of the body before it can cause problems.

Organisms less than a millimeter wide are generally composed of a single cell. In these microbes, both the good stuff moving in and the waste moving out pass through tiny pores in the cell membrane, a barrier that separates the inside of the cell from the outside. These gaps are the equivalent of the hole in our metaphorical closet door. Like junk from a closet, the movement of material follows its particular concentration gradient. If there is more oxygen outside the microbe than inside, then it diffuses *into* the organism. Nutrients, including carbohydrates and sugars, also diffuse in. And when waste products accumulate at a higher concentration inside the microbe than outside . . . Well, you get the picture.* Finally, as in the closet example, some substances are prevented from crossing the cell membrane. As a result, the cell membrane is said to be "semipermeable." This property explains why cell structures like

---

* The back-and-forth movement described above takes place with little or no energy expenditure by the cell, making it a "passive" process. Material can also be moved in either direction if engulfed by the cell (as is seen in organisms like amoebas) or packaged by it in tiny membrane-bound bags called vesicles, which can be ejected from the cell. Both of these "active" processes require an input of energy, as would moving a substance across a membrane *against* its concentration gradient.

organelles (the nucleus and mitochondria, for example) remain inside the cell: basically, because they can't fit through the pores.*

Now I know what a few of you are thinking—or *would* be thinking if you had a central nervous system. "Some of us are a whole lot wider than a millimeter, and we don't have any of that circulatory-system junk you just mentioned. So explain that one, Mr. Science."

Well, all right, but I'm going to make this quick.

It *is* true that some of you—flatworms (a.k.a. platyhelminths), for example—can form chains up to eighty feet long, and, yes, you're all doing ever so well without a circulatory system—too well if you ask me. But like every other living creature, the twenty thousand or so species belonging to Team Flatworm exist and thrive because they have adapted to the specific demands of their environments (so-called selection pressures). For some flatworms, this resulted in the evolution of folded bodies, or of long threadlike shapes. Just as a walnut has more surface area than a smooth ball of the same size, a flatworm with a folded body has more surface area for gas, nutrient, and waste exchange than a smooth flatworm of the same size and shape. Extending that concept to the closet door example, an accordion door would have more surface area than a flat door, allowing for more holes to be cut in it.

But there's more to the success of flatworms than just shape. Notably, there are no high-activity sprinters here. No speedy swimmers or fliers either. Instead, their lives are pretty much fulfilled once they hook their headlike scolex to the lining of someone's colon. Others while away the hours lying low in a streambed, or maybe in the shade of some moist leaf litter. It's a lazy existence, and as a result, these couch potatoes need less energy and oxygen to get them through the day.

---

* In addition to size constraints, some substances exhibit other physical properties that prevent their movement across the membrane. An example of this might be a molecule with an electrical charge that repels it if it gets too close to a similarly charged membrane.

But, hey, guys, don't take this the wrong way. Though you lack circulatory and respiratory systems, and many of you live parasitic lifestyles and infect three hundred million people per year and defecate out your mouths, please know that none of this is meant to make you feel bad.* It's just that this book isn't about you—so we'll talk later, okay?

All right. Are they gone? Cool.

Now for those of you who are a bit thicker around the middle than our minuscule friends and who might live somewhere other than in someone else's intestine or under a layer of pond scum, you should know that there were real problems during your evolutionary journey from single-celled organisms into dung beetles, leeches, and insurance salesmen. Perhaps the most serious issue was the fact that diffusion does not work well over large distances. In fact, it's a no-go for pretty much anything wider than a millimeter. As a consequence, diffusion alone is extremely ineffective for moving vital substances and waste products in creatures with beefy three-dimensional bodies, composed of layers hundreds and even thousands of cells thick.

You might ask, how then did organisms evolve to become as large as they are?

That's a tough one.

I should start off with the caveat that, given the small size and the squishy-bodied nature of the extremely ancient organisms involved, the fossil record for this sort of thing is relatively scanty. That said, scientists think that the first multicellular life-forms, the metazoans, evolved somewhere between 770 and 850 million years before present (BP). By 600 million years BP, a new line of metazoans had evolved, sporting identical right

---

* Although most of the more than twenty thousand members of the phylum Platyhelminthes regurgitate undigested food, some species have an anus, or even several, situated on their backs. The problem for other species is the fact that tapeworms (cestodes), and especially the flounder-shaped flukes (digeneans), are internal parasites that cause serious diseases, like schistosomiasis, in humans and their livestock—nowadays primarily in Africa.

and left sides, rather than radial (a.k.a. circular) shapes. Their embryos also added a third layer to what had previously been a double-layered, embryonic body plan. The more ancient setup consisted of an outer ectoderm, which developed into structures like skin, nervous tissue, the mouth, and the anus, and a more deeply situated endoderm, fated to become the inner lining of the digestive and respiratory systems. The newly evolved third layer, known as the mesoderm, was found between the other layers and served as a source of new building blocks for larger and more complex organisms. Eventually, it would give rise to muscle, connective tissues like cartilage and fat, structures like bones, and a not-inconsequential assemblage of tissues that would become known as the heart.

The next-highest level of organization in a multicellular body is the tissue. Each type of tissue is made up of different cell types, as well as substances collectively known as the extracellular matrix, found outside and between those cells. Together, the cells and the matrix work in a tissue to carry out a specific function or functions, such as supporting the body against the pull of gravity or helping move liquids from place to place. There are but four tissue types: connective tissue (like blood, bone, and cartilage), epithelial tissue (which covers body surfaces and lines cavities like hollow organs and blood vessels), nervous tissue (neurons and their support cells, the glia), and muscle tissue. Within muscle tissue, there are three subtypes: smooth muscle (involuntarily controlled), skeletal muscle (voluntarily controlled), and cardiac muscle, which is, happily, also involuntary, thus freeing us from the bother of remembering to keep our heart beating.

The next organizational level of the body is the organ. Your organs each carry out at least one particular function, and often many more. Each organ is composed of at least two different types of tissue, and some of the larger organs, including the heart, can be composed of all four types. Though the heart, kidney, and liver are more readily recognized as organs, blood vessels actually fall into the category as well, since veins and arteries are composed of epithelial, connective, and muscle tissue, and carry out the function of transporting and distributing blood.

At the top of this hierarchy of body organization are organ systems like the circulatory and digestive systems. These are made up of multiple organs involved in some general function or functions. In the case of our circulatory system, the organs consist primarily of the heart, arteries, capillaries, and veins involved in transporting blood throughout the body.

Like other organs, most blood vessels are made up of layers of cells. Muscle cells, more often referred to as muscle fibers or myocytes, form an internal layer bounded on either side by connective and epithelial tissue. When those muscle fibers contract, the liquid within the vessel is compressed and moves—picture your fingers squeezing the center of an elongated water balloon. Scientists believe that this is how water, and eventually blood, began to be transported from place to place within organisms that were growing increasingly larger over evolutionary time.

How did this process unfold? One hypothesis is that approximately half a billion years ago, some of the cells derived from the newly evolved mesoderm of some unknown organism developed the ability to shorten their lengths—that is, to contract. For this to happen, at some point the contractile proteins within a cell would have had to line up next to each other. Once provided with an energy source, these proteins (like the actin and myosin found in human muscles, including the heart) would have slid past each other in opposite directions. If millions of these molecules did this simultaneously, it would have contracted the cells they were found in, and so, too, the whole structure around those cells. Then when the contractile proteins slid back to their previous positions, the cells would have relaxed and returned to their precontracted length.

Half a billion years ago, though, the first contractile cells would have been far simpler than our own muscle cells (a.k.a. myocytes or

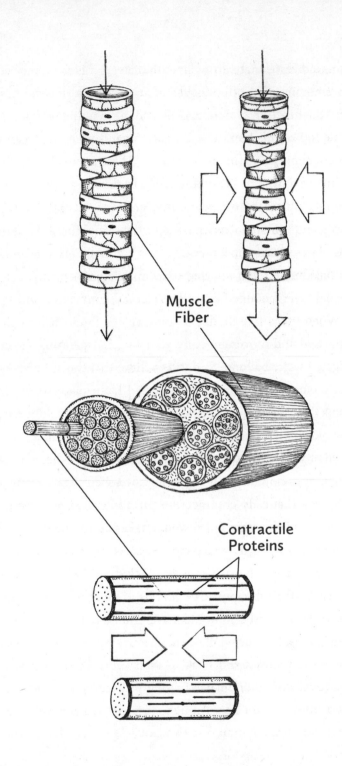

Muscle
Fiber

Contractile
Proteins

myofibers). Additionally, they couldn't have evolved in blood vessels, since neither blood nor the vessels that transport it existed back then— though water certainly did, and it was primarily within water that material could move into and out of an organism. Even now, contractile proteins are found inside normal body cells, where they form a vital part of the cell's internal transportation system. Scientists believe that in some ancient creatures, cells containing ancient contractile proteins may have accumulated into tubes, thus forming primitive circulatory systems. These contractile tubes would have enabled the movement of water and the material contained in that water—and, *much* later, blood—from place to place within increasingly larger organisms. With innovations like contractile circulatory systems in place, the new blobs on the block would have branched off relatively quickly* into myriad forms, like segmented worms, mollusks, and, after a while, even chordates—a subset of which, the vertebrates, makes up the vast majority of this book's readership.

Along the way, those critters equipped with adaptations like these would have outcompeted many of the organisms that lacked such systems, driving them to extinction. Though not all of them. Corals, jellyfish, and comb jellies had already split off from the rest of the invertebrates before the evolution of the muscle-producing mesoderm. Though they never inherited muscle tissue from their ancestors, members of the phylum Cnidaria developed their own evolutionary advantages, like toxins and stinging cells to ward off predators. They also evolved contractile epithelial cells that act like muscle cells. Because of this, they were able to survive and thrive.

Though they were certainly revolutionary, circulatory systems did not evolve in a vacuum. Blood vessels are all well and good, but a significant reason for the success of organisms possessing circulatory systems was

---

* Well . . . quickly as in over the course of a hundred million years or so.

that they had coevolved other organ systems, notably the respiratory system. Evolving and functioning in tandem, these two systems solved the problem of moving large amounts of gases into and out of the body—and as a result, they enabled organisms like chordates to cope with the energy costs associated with increasingly complex behaviors and processes.

Most respiratory systems consist primarily of gas exchange mechanisms, like gills or lungs. Their main function is to facilitate the uptake of oxygen, which is essential for the life-sustaining chemical reactions that occur within the body. These reactions are known as metabolic processes, and they're collectively referred to as an organism's metabolism. One of the most important of these processes is the release of usable energy from the food we eat. As the process of digestion reaches completion, the nutrients in our food are broken down into smaller molecules, like carbohydrates, fats, and proteins. Through a process known as cellular respiration, the sugar glucose (a carbohydrate) can be converted into adenosine triphosphate (ATP), the energy currency of the cell. Muscle fibers and other cells have the ability to break the chemical bonds holding ATP together, and that energy can then be used to fuel things like repair, growth, and muscle contraction. Crucially, the chemical reactions involved in this molecular breakdown and energy release require oxygen. Enter stage right: gills and lungs.

There's more, though. Besides releasing energy, cellular respiration also releases carbon dioxide (CO_2) as a by-product, and this stuff is toxic to many organisms. As a result, the body constantly needs to get rid of CO_2 before it accumulates to harmful levels. Most circulatory systems, therefore, play the dual role of carrying oxygen from the gills or lungs to the cells of the body while simultaneously carrying the waste biproducts of metabolism back to the gills or lungs, where they can be removed from the body. (As a point of emphasis, although many people think that we breathe faster during exercise because of an increased requirement for oxygen, it is as much the need to eliminate excess carbon dioxide that gets us huffing and puffing.)

As respiratory systems were evolving, so, too, were circulatory systems, enabling the movement of a fluid called blood* around the body. The earliest evidence for this dual system dates to approximately 520 million years ago and an arthropod called *Fuxianhuia protensa*, first discovered at the Chengjiang fossil site in southwest China.

Passing through a series of vessels known as arteries, capillaries, and veins, blood back then, as it does now, likely functioned to deliver nutrients, gases, and waste to and from each cell in an organism's body. As importantly, this arrangement allowed these deliveries and pickups to be made *far* from the outer surface of the organism. While diffusion is still the name of the game when it comes to moving these products into and out of body cells, nutrients, gases, and waste now travel through the blood vessels to get there, instead of having to seep their way in and out, cell layer by cell layer, to and from the external environment.

Now jump ahead half a billion years from *Fuxianhuia protensa*, and envision five hundred million tiny saclike alveoli (roughly 0.2 millimeters in diameter) at the end of the bronchial tubes, deep inside your lungs. Each alveolus is surrounded by a meshwork of capillaries, tiny blood vessels with diameters approximately one-tenth that of a human hair. These are the microscopic sites of gas exchange between the respiratory and circulatory systems. Both the alveoli and capillaries have extremely thin walls, one cell layer thick, which enables the rapid exchange of gases. But although they might be tiny, when taken together the alveoli cover a surface of roughly one hundred square meters, allowing the large amounts of air we breathe in to be processed. As we inhale, oxygen diffuses out of the alveoli and into the alveolar capillaries, where it is carried by increasingly larger blood vessels back to the heart (the left atrium this time) and, when

---

* The invertebrate version of blood is known as hemolymph. When discussing invertebrates, the two terms are used interchangeably, as will happen throughout this book.

Fuxianhuia protensa

 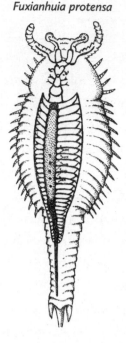 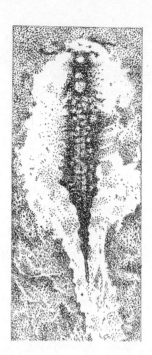

the left ventricle contracts, out to the body. CO_2 moves in the opposite direction, out of the alveolar capillaries and into the alveoli, to be exhaled back out into the environment.

Okay, demonstration time. Ready? Take a breath . . . then breathe out.

That was it. Now read the above paragraph again because that's exactly what happened during this exercise.

This interplay between the circulatory and respiratory systems is one of the many ways in which the organ systems of the body do *not* function like separate chapters in a textbook, which regrettably is how many of us first learned about them. Because this mindset is detrimental to a real understanding of how biological systems work, it is something that I continually warned my Human Anatomy and Physiology students about. I told them that organ systems interact: they cooperate, they depend on each other, and they are basically useless by themselves.

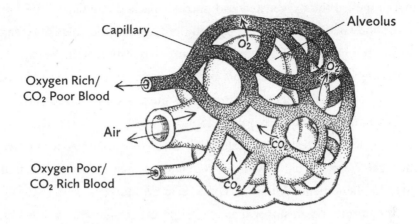

Unfortunately, this loss of synergism does sometimes occur. It is characteristic of diseases like emphysema, where a failure in one system sets off a chain reaction in the others. Emphysema is a degenerative and incurable respiratory disease, characterized by the systematic breakdown of alveoli in the lungs. This results in a reduction in their number, coupled with a loss of their function—which is to serve as the tiny middlemen between the atmosphere we breathe and the circulatory system that moves oxygen and carbon dioxide around the body. The causes of emphysema can range from having a rare inherited deficiency of a lung-protecting protein, to inhaling occupational dusts and chemicals, to the primary reason—smoking cigarettes. Regardless of the cause, the end result is that along with the respiratory system, a key function of the circulatory system is compromised, since the blood returning from the emphysema-stricken lungs is unable to deliver enough oxygen to the tissues and organs of the body for them to function normally.

EVENTUALLY, AS ORGANISMS grew more diverse and more complex, so, too, did their circulatory systems. One evolving feature was the pump that propelled the oxygenated and nutrient-packed circulatory fluid out to the body before returning it, oxygen- and nutrient-depleted and ready for another go-round. Of course, the pump in question is the heart.

As we will now see, the heart is not a single structure shared across the entire animal kingdom. Circulatory pumps evolved separately in different animal groups. They often look and work very differently, and as such, some of the resulting organs don't check off enough boxes to merit the label "heart." What they do share relates to their function, and that's due to a phenomenon known as convergent evolution.

Sometimes, organisms will appear to have a similar adaptation—like the tapered (a.k.a. fusiform) body shapes of sharks and dolphins. These animals are not closely related, since dolphins are mammals and sharks are fish. The key here is that the adaptation was *not* passed down from a single common ancestor of those creatures but, instead, evolved twice or, as it often happens, on multiple occasions (tuna also have roughly the same shape, as do torpedoes) in very different groups of organisms. The explanation for this phenomenon is that fusiform bodies are perfect for generating speed, and this makes them precisely the right shape for fast-swimming predators from very different branches of the evolutionary tree.

Blood feeding in the animal kingdom is another example of convergence, with animals as different as leeches, bedbugs, and vampire bats sharing a suite of similar vampiric adaptations that include stealth, small size, sharp "teeth," and salivary anticoagulants.*

Like fusiform bodies in aquatic predators or vampiric stealth, circulatory systems appear to have evolved on numerous occasions in

---

* Probably the most famous examples of convergence are the wings of insects, pterosaurs, birds, and bats. Each of these airfoils evolved separately yet perform a similar function, allowing their owners to overcome the force of gravity and fly. Gills are also convergent; these gas-exchange organs have apparently evolved multiple times in both invertebrates and vertebrates.

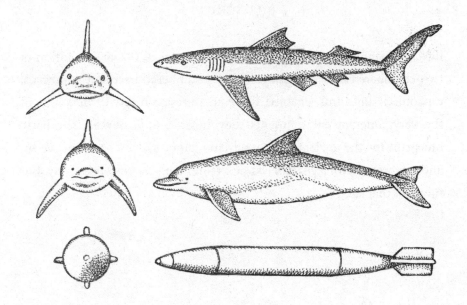

different invertebrate groups. Circulatory pumps and their associated vessels perform what is essentially the same job, and because of this they exhibit similarities even when their owners are not closely related. Their multiple evolutionary origins can also explain why the invertebrate circulatory systems we'll be examining next show such a high degree of variation in form. There are single hearts, multiple hearts, and no hearts, as well as circulatory systems that are either open or closed—the latter a distinction we'll be exploring soon.

Conversely, evolutionary origins also explain why there is less variation in vertebrate organ systems. Most scientists think that all vertebrate circulatory systems can be traced back to a single common ancestor, likely a type of jawless fish that lived about five hundred million years ago.* As a result, some of that ancient vertebrate's adaptations can be seen in all living vertebrates—though many other aspects of these structures

---

* Interestingly, there are specific regulatory genes (small sections of the genetic blueprint) that are shared by both insects *and* vertebrates. This points to the *possibility* of an ancient shared ancestry for *all* circulatory systems.

have changed over evolutionary time. The changes, like the evolution of two-chambered hearts in fish and four-chambered hearts in mammals, crocodiles, and birds, enabled these creatures to meet the demands of the very different environments they inhabit. Still, though, the basic blueprint for the ancient vertebrate circulatory system—arteries, veins, and the presence of a chambered heart—remains in existence today. But more on that later.

I'm different. I have a different constitution,
I have a different brain, I have a different heart.
—CHARLIE SHEEN

I bleed Dodger Blue!
—TOMMY LASORDA

[3]

Blue Blood and Bad Sushi

ABOUT A HUNDRED feet from its older but perfectly serviceable-looking twin, the new boat launch cut a fan-shaped swath of granite stones and concrete through Monument Beach.

"Did the locals put up a fight before they built this thing?"

The question came from my longtime friend invertebrate biologist Leslie Nesbitt Sittlow, and it was directed at Dan Gibson, a fit seventysomething neurobiologist at the Woods Hole Oceanographic Institution in nearby Falmouth, Massachusetts. Leslie and I had met Gibson about five minutes earlier, after hightailing it down from Great Bay, New Hampshire, another coastal environment we had barnstormed during a book research-related New England field trip.

Gibson was currently searching for something in the sand. "I live a couple of miles from here," he replied, "and by the time I heard anything about a new boat launch, they'd already built it."

Getting back to the business at foot, Gibson gestured toward a small half-moon-shaped depression in the sand. Using the open end of a plastic pitcher, he began carefully lathing away thin layers until he'd reached five or so inches below the surface. Then he shot us a grin and reached into the hole. After probing with an index finger for a few moments, the scientist scooped out a cluster of tiny bluish-gray spheres.

The eggs belonged to *Limulus polyphemus*, one of four extant species of horseshoe crabs. From the Yucatán Peninsula to Maine, these claw-bottomed domes are a familiar sight during their annual late spring/ early summer pilgrimages from deeper water into coastal shallows. The females follow the tides in, hunkering down to lay their eggs in hollows they scratch out of the sandy substrate. Gibson explained that horseshoe crabs are quite particular about where they deposit their eggs, since nests needed to be water-covered at high tide but dry and warmed by the sun at low tide. We also knew from our previous day's observations in Great Bay that horseshoe crab males are 20 to 30 percent smaller than the females that they swarm like a convention of rude helmets. Each of the males had jostled for position, attempting to mount a female and latch a pair of club-like appendages onto her shell. Thus situated, he would be in prime position to deposit his milky-looking sperm onto the walnut-sized egg masses on the female's underside. Eventually, two to five clusters, totaling up to four thousand eggs, would be laid during a single high tide, after which everyone would head back into deeper water, presumably to wait for the next tide-initiated love-in. Gibson told us that by the time the season ended, a female horseshoe crab would have deposited somewhere around eighty thousand eggs.

Although the annual mating swarms draw crowds of the curious to beaches all along the Atlantic coast, Leslie and I were actually there to investigate the horseshoe crab's cardiovascular system, particularly its

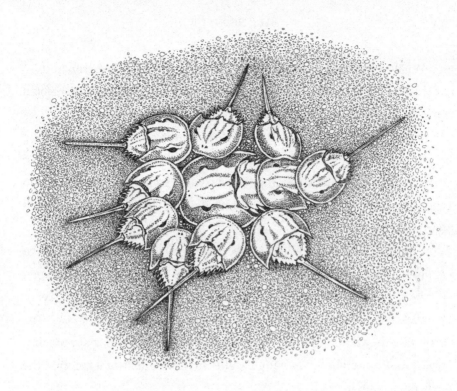

heart and the unique qualities of its blood. And despite the crab-orgy detour, our trip had a more serious tone—namely a major threat to the survival of these ancient creatures, related to the very same aspects of horseshoe crab biology that had drawn us to coastal Massachusetts.

After showing off his find, Dan Gibson carefully placed the egg cluster back into the hole he had dug. Then, with tiny spherical search images neurologically inserted, Leslie and I were handed our own pitchers and instructed to locate additional nests. After scanning the broad concrete ramp that extended out into the shallows for what looked to be a hundred feet or more, we quickly decided to move off in search of a sandier locale. I could see that the longest stretch of Monument Beach was adjacent to a large parking lot. It was just before noon, and the lot currently held a dozen or so cars, inside of which were folks who had stopped by for a bite of lunch or a smoke with an ocean view.

What Leslie and I *didn't* see on the beach were horseshoe crab nests—at least not very many of them, and none where we had been instructed to look, on the beach adjacent to the old boat launch.

When we met up with Gibson a few minutes later, he was looking frustrated. He explained that not only had the launch builders covered fifty yards of prime spawning beach with softball-sized rocks and concrete, but that the previously prime nesting spot had become much harder to reach.

"The border along the old ramp was a calm place for the crabs to approach and lay their eggs, while the rest of this beach is more open to choppy waves. The crabs coming in from deeper water usually swim parallel to the shore until they can find the perfect spot," he told us. "The only way they'll find that old stretch of beach now is if they approach it head-on. If the crabs are moving parallel to the shore, they'll run into the new boat ramp instead."

Horseshoe crabs, though, are famously resilient. With their 445-million-year fossil record—dating back to roughly 200 million years before the first dinosaur—they are the sole survivors of a once-diverse class of arthropods that also includes trilobites, arguably the most famous of ancient invertebrates. You would be hard-pressed to come up with a group of animals that has been around as long as the horseshoe crabs, and because of this, they are commonly referred to as "living fossils."

The pessimistic predictions of horseshoe crab researchers like Gibson are, then, especially troubling. Unfortunately, in addition to habitat destruction, a number of other factors, including one related to the horseshoe crab's own unique cardiovascular system, are threatening to end their spectacular longevity record.

Horseshoe crab eggs, and the miniature larvae that emerge from them around two weeks after fertilization, are an important food source for fish and for migratory birds like the endangered red knot (*Calidris canutus*), a chunky-looking member of the sandpiper family. As a result, the vast majority of horseshoe crab eggs and larvae never survive the approximately ten years it would take for them to develop into sexually active adults. In fact, horseshoe crab expert John Tanacredi told me that he believes that something like one in three million eggs produces a larva that survives to adulthood.

When Europeans came to the New World, they found Native Americans using horseshoe crab parts for food and fertilizer, and for tools like hoes and fishing spear tips. As colonies of settlers sprang up along the East Coast, they harvested horseshoe crabs in numbers that seem almost unbelievable today. In 1856, for example, more than one million horseshoe crabs were collected from a single mile-long stretch of New Jersey beach. This type of population-straining harvest continued well into the twentieth century, with workers stacking the crabs into chest-high walls extending across vast stretches of waterfront, awaiting transportation to fertilizer factories.

That industry, which was centered along the Delaware Bay and coastal New Jersey, finally collapsed in the 1960s due to declining crab populations and the increasing popularity of alternative forms of fertilizer. Unfortunately, the mass collection of horseshoe crabs did not end. Sometime around 1860, American eel fishermen had discovered that chopped-up horseshoe crabs were terrific bait for their eel traps—especially jumbo-sized females laden with eggs. And horseshoe crab harvesting was still rampant in the mid-twentieth century, when some commercial fishermen began to turn to oversized snail relatives called whelks for an alternative source of income. The problem was that whelks also enjoy hacked-up horseshoe crabs, and so crab populations were threatened anew as whelk fishermen sought bait for their pots.

Today, many eel and whelk fishermen still consider horseshoe crabs to be their bait of choice, and the bait industry continues to reduce horseshoe crab populations by around seven hundred thousand individuals per year. But while the American horseshoe crab fishery is fully regulated (at least in theory), there exists a growing problem with poachers and the inability of officials to control the number of animals harvested.

In Asia, the three remaining horseshoe crab species* are under even more serious threat of extinction, and the reasons extend past the eel pot and onto the dinner plate. In places like Thailand and Malaysia, horseshoe crab eggs are considered an aphrodisiac, and so there are restaurants where their roe is the main item on the menu.

But consuming the egg masses, which are usually boiled or grilled, does come with a few drawbacks. For one, people die from eating horseshoe crab roe. Their deaths are somewhat less than gentle, and are almost certainly related to an important aspect of our own circulatory system.

Tetrodotoxin is a deadly nerve-blocking agent that is at least an order of magnitude (i.e., ten times) more lethal than black widow spider venom. Although its infamy stems from its presence (when poorly prepared) in arguably the most dangerous of exotic cuisines, fugu or puffer-fish flesh, several outbreaks of tetrodotoxin poisoning have been traced back to the consumption of horseshoe crab roe. Tetrodotoxin is extremely dangerous because after digestion it accumulates in tissues like muscles and nerves. Although the exact mode of tetrodotoxin's entry into the nervous system is still unknown, its lethality is due, at least in part, to its ability to bypass the protective blockade known as the blood-brain barrier (BBB).

The BBB is regulated in part by a class of starburst-shaped cells called astrocytes. Astrocytes are one of several different types of

---

* *Tachypleus gigas*, *Tachypleus tridentatus*, and *Carcinoscorpius rotundicauda*

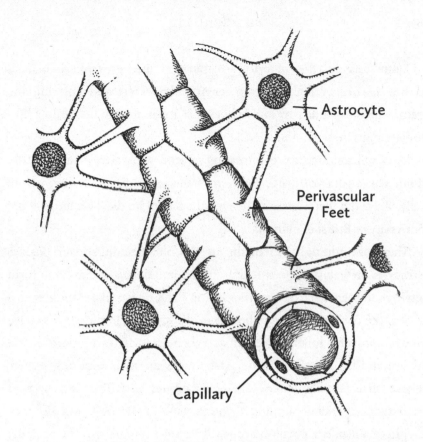

Astrocyte

Perivascular
Feet

Capillary

glial cells (or neuroglia) that assist, support, protect, and repair the
superstars of the nervous system—the neurons. Among several other
duties, astrocytes can be found latched on to capillaries in the brain.
As in the rest of the body, these vessels supply oxygen and nutrients to
tissues while carrying away waste products and carbon dioxide. In the
brain, however, the astrocytes restrict that back-and-forth movement,
allowing the passage of only *some* substances (like oxygen, glucose, and
alcohol) out of the tiny vessels. As for how that works, the astrocytes
have footlike structures, appropriately called perivascular feet, which
act as a barrier, covering the capillary walls. Mostly that's a good thing,
since they prevent harmful materials like bacteria and certain chemicals
from exiting the circulatory system and doing damage to the delicate
neural tissues of the brain.

Unfortunately, the blood-brain barrier also prevents beneficial substances like antibiotics from leaving the blood and entering the brain, which explains why *any* infection of the brain can turn into a life-threatening situation.

"A significant hindrance to treating neurodegenerative diseases at the moment is that most drugs can't cross the blood-brain barrier," wrote Kelly McNagny, a professor in the Department of Medical Genetics at the University of British Columbia.

There are additional elements of the blood-brain barrier besides astrocytes. Notably, these include "tight junctions," seams that form between adjacent cells in the inner lining of blood vessels. If these seams loosen, the consequences can be devastating. For example, studies have shown a probable link between a bacterium associated with periodontal disease and the development of Alzheimer's disease. Some researchers believe that *Porphyromonas gingivalis* evades the BBB and invades brain tissue, possibly slipping through gaps in the tight junctions or hitchhiking inside white blood cells whose roles require that they exit the circulatory system. Experiments on mice have shown that once inside the brain, *P. gingivalis* bacteria release toxic substances called gingipains, which disrupt the functioning of essential proteins, damaging neurons and worsening the effects of Alzheimer's. The infection also causes the accumulation of two distinctive proteins, amyloid and tau, which have historically been considered signs of Alzheimer's—although there is now a growing suspicion that these sticky masses, or plaques, are actually a defense mechanism against *P. gingivalis* and not themselves a cause of Alzheimer's disease. This ongoing research is a potential game changer, since Alzheimer's is the sixth leading cause of death in the United States, killing more people than breast cancer and prostate cancer combined.*

---

* In 2018, the latest year for which data from the CDC were available as of this writing, 122,019 deaths in the United States were attributed to Alzheimer's disease. With COVID-19 deaths far surpassing that number in 2020, Alzheimer's will likely drop a notch, to number seven.

Among the substances that *can* be delivered across the blood-brain barrier is tetrodotoxin, and people who consume horseshoe crab eggs need to know that its presence in roe is unpredictable. It is believed that the crabs ingest certain bacteria that produce the neurotoxin by consuming contaminated shellfish or decayed matter. Symptoms of tetrodotoxin poisoning generally start with a slight numbness of the lips and tongue, not a unique sensation when consuming spicy Thai food. A pins-and-needles numbness of the face might be the first clue to diners that something has gone terribly awry. The real buzzkill follows quickly, in the form of headache, diarrhea, stomach pain, and vomiting. As tetrodotoxin spreads throughout the body, walking becomes difficult as the chemical begins to block the nerve impulses that lead to contraction of voluntary muscles like those of the limbs. Tetrodotoxin can also interrupt the spread of electrical signals through the myocardium, the heart's thick layer of cardiac muscle. As we'll see later, this is the electrical system responsible for the heart's coordinated contraction and relaxation—the heartbeat itself.

Eventually, around 7 percent of those who fall victim to tetrodotoxin poisoning die, reportedly conscious and quite likely completely aware that consuming a week-old California roll, or even a chopstick, would have been a better choice than the horseshoe crab roe or fugu that became their last meal.*

But beyond facing the possibility of being eaten, ground up as fertilizer, or chopped up as bait, horseshoe crabs also face a unique threat to their survival.

THE AMERICAN HORSESHOE crab, *Limulus polyphemus*, and its three Indo-Pacific cousins are not really crabs at all. Like true crabs, however,

---

* Because victims can remain conscious through the paralyzing effects of tetrodotoxin, an ethnobotanist named Wade Davis suggested in 1983 that voodoo practitioners had used it to turn people into zombies to toil as slaves on Haitian plantations. His claim was subsequently hammered into the ground like a tent peg by some science types, who happened to know a thing or two about tetrodotoxin and its *actual* effects.

they *are* arthropods, members of a massively diverse phylum of animals, which includes insects, spiders, and crustaceans, that share the presence of jointed exoskeletons. Crucially for the horseshoe crab, they also share *open* circulatory systems. These are significantly different from the *closed* circulatory systems found in blue whales, humans, and the roughly fifty thousand other species of mammals, fish, amphibians, reptiles, and birds. As we'll soon see, some invertebrates, like earthworms, octopuses, and squid, also have closed circulatory systems, though they are very different from those found in creatures burdened with a backbone.

In closed circulatory systems, blood leaves the heart via large arteries, which branch into a series of increasingly smaller arteries and even-smaller arterioles. The arterioles enter and thread through organs and muscle tissue before they, too, split into even-smaller vessels called capillaries. These tiny tubes make up approximately 80 percent of the total length of the circulatory system, and it is within dense networks known as capillary beds that the mutual exchange of substances between the blood and the body takes place. As discussed earlier, oxygen from the lungs or gills and nutrients absorbed from the digestive system pass through the thin capillary walls and into the surrounding tissues. Meanwhile, metabolic waste products, like carbon dioxide and ammonia, diffuse into the blood, to be carried back toward the heart, first by tiny venules and then by increasingly large veins.

In gilled vertebrates like fish, some salamanders, and all amphibian larvae, the now-deoxygenated blood gets pumped through the gills, where the carbon dioxide diffuses into the surrounding water and a new batch of oxygen diffuses back in. As you may have noticed as a non–water breather, a pretty major tweak to this gas exchange system occurred somewhere down the line to allow for the exchange of oxygen and carbon dioxide with the air instead of water. The nature of this tweak? Lungs. There will be more on this particular story later.

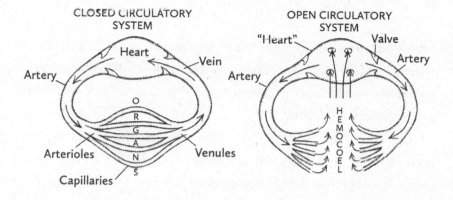

CLOSED CIRCULATORY SYSTEM

Heart · Vein · Artery · Artery · ORGANS · Arterioles · Venules · Capillaries

OPEN CIRCULATORY SYSTEM

"Heart" · Valve · Artery · Artery · HEMOCOEL

Whether oxygenated by gills or lungs, however, one thing closed circulatory systems have in common is that blood is always confined to a closed loop. Not so for most invertebrates, including the horseshoe crab. In their open circulatory systems, fluid (called hemolymph, rather than blood) also leaves the heart through arteries.* But instead of flowing into capillaries, the hemolymph spills out of the vessels and into body cavities called hemocoels, where it bathes the organs, tissues, and cells it comes into contact with. There, the hemolymph drops off nutrients through diffusion while simultaneously picking up waste products. Many of these open circulatory systems also exchange oxygen and carbon dioxide, although as we'll see in the next chapter, insects are a significant exception to this rule.

Although gills are forever linked in our minds to fish, they are the respiratory organs found in many invertebrates, including horseshoe crabs. This is another example of convergent evolution: though vertebrates and invertebrates evolved separately, both use diffusion to draw oxygen into the familiar-looking arrangement of gill membranes that often resembles the overlaying pages of a book. In non-insect arthropods, oxygenated

---

* When describing open circulatory systems, the term "artery" is used more for convenience than for scientific accuracy. Circulatory system purists require that card-carrying arteries have an inner lining of epithelial tissue called the endothelium, which is not present in open circulatory system vessels. For our purposes, the term "artery" is purely functional and describes a vessel that carries a circulatory fluid *away* from the heart (while veins carry blood *toward* the heart).

hemolymph leaves the gills and heads back to the heart via the circulatory system. And in horseshoe crabs, by this point the hemolymph has undergone an additional transformation. It has turned from milky white to a beautiful powder blue.

The "blue blood" of horseshoe crabs and invertebrates like cephalopods, clams, lobsters, scorpions, and tarantulas is due to the presence of a copper-based protein called hemocyanin. Carried in the hemolymph in a dissolved form, the hemocyanin latches on to oxygen whenever it comes into contact with it. When copper oxidizes, it turns blue—and so, too, hemolymph turns blue as it leaves the gills, having undergone the same chemical reaction that gives the copper-plated surface of the Statue of Liberty its famous blue-green tint.

With the exception of the blue bloods mentioned above, the oxygen-carrying molecule in pretty much every other creature with a circulatory system is hemoglobin. Here, though, the oxygen binds to an atom of iron, rather than copper. And unlike hemocyanin, hemoglobin is not free-floating in the blood. Rather, it's carried by a specialized type of cell called an erythrocyte, which spends its roughly four-month life span toting hemoglobin around the circulatory system.* Because the erythrocytes contain iron instead of copper, when they oxidize, they don't give off blue light. Instead, they emit red light. If these cells sound familiar, it's because they're also called red blood cells. And if the oxygen-related color change rings a bell, that's because it's the very same oxidation reaction that occurs when an iron fence is exposed to atmospheric oxygen and turns rusty red.

So why, you might ask, don't humans and other vertebrates have blue blood? The answer likely relates to body size and oxygen-carrying efficiency. Larger bodies require more oxygen, and hemoglobin is better

---

* Hemoglobin is also found in some non–red blood cells, like the previously mentioned astrocytes in the brain.

equipped to provide it: each hemoglobin molecule can carry four oxygen molecules, while hemocyanin can carry only one. So over time, organisms with blood containing hemoglobin were able to evolve into larger-bodied creatures than those utilizing hemocyanin.

We interrupt this chapter for a hemoglobin-related public service announcement: A *serious* problem for humans is that hemoglobin is much more strongly attracted to molecules of carbon monoxide (CO) than it is to oxygen (O_2). This makes even small amounts of this odorless, colorless gas, which is released by things like car engines, gas appliances (like heaters), and woodstoves, *especially* dangerous. In fact, the potential presence of carbon monoxide is so dangerous that if you don't have a carbon monoxide detector in your house or apartment already, or if you know of a loved one who doesn't have one, then take a break from reading this book and go purchase one.

I'll wait . . .

OKAY, WHERE WAS I?

In closed circulatory systems like ours, blood returning from the body enters the heart directly, by way of large veins like the superior and inferior venae cavae. This occurs during the portion of the cardiac cycle known as diastole, when the ventricles relax after contracting and forcing their contents out of the heart during the phase called systole. Since horseshoe crabs have open circulatory systems and don't have veins, oxygenated blood leaving their gills must enter the heart differently, first flowing into a reservoir surrounding the heart called the pericardial cavity.*

Once blood fills the pericardial cavity, how does it get into the horseshoe crab heart? To start, the heart itself is suspended within the

---

* Readers should note that this is *not* how the pericardial cavity in a closed circulatory system (like that mentioned earlier) works. In fact, any blood found within it would be a deadly serious problem.

pericardial cavity by a series of elastic bands called alary ligaments. These stretchy bands run along the length of the heart, and they anchor the outer walls of the heart to the inside of the crab's exoskeleton, or shell. As the heart contracts (during systole), the alary ligaments are stretched like rubber bands, causing them to store elastic energy. Once the heart has contracted and ejected its contents, it relaxes (diastole) and the ligaments' elastic energy tugs the heart back open to its precontraction volume.

Simultaneously, as the volume increases, pairs of valve-like apertures in the heart, called ostia (singular ostium), reopen. The blood that has gathered in the pericardial cavity flows through the ostia, refilling the empty heart—moving from the higher pressure of the pericardial cavity to the lower pressure of the newly emptied organ. Then the process of filling and emptying the pericardium and the heart repeats.

A neat system, to be sure, but as horseshoe crab expert and University of New Hampshire zoology professor Win Watson explained to Leslie and me, horseshoe crab circulation gets an assist from the action of another organ system, and in a manner that turns out to be quite familiar. This discovery started with the observation that horseshoe crabs' so-called book gills fan back and forth in a rhythm that is synced with the movement of blood into the pericardial cavity.

As Watson described the mechanics, I thought back to a paper on galloping horses that I had read during my PhD work at Cornell back in the 1990s. In that study, functional morphologists Dennis Bramble and David Carrier had suggested that during the gallop (a pace during which all four feet are off the ground at the same time), the accompanying backward-and-forward movement of a horse's liver within the abdominal cavity turned that massive organ into a "visceral piston," which assisted in the process of breathing, and thus the efficient exchange of oxygen and carbon dioxide.

Bramble and Carrier hypothesized that as the bulky liver slides backward (see horse figure A), it pulls the dome-shaped diaphragm, to

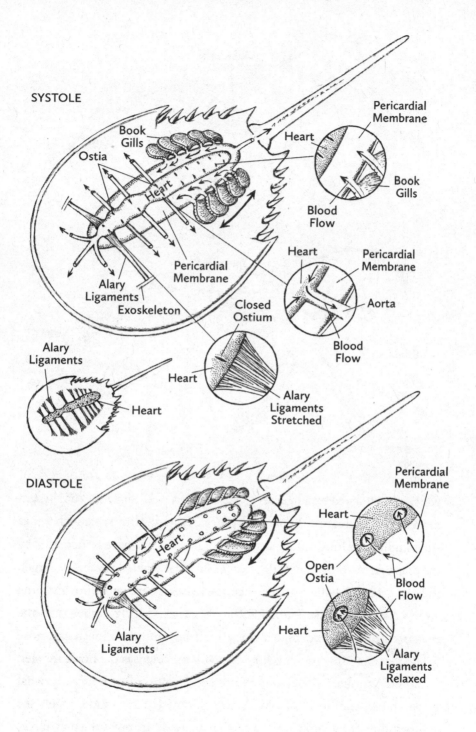

SYSTOLE

Book Gills

Ostia

Heart

Alary Ligaments

Pericardial Membrane

Exoskeleton

Closed Ostium

Heart

Alary Ligaments Stretched

Alary Ligaments

Heart

Pericardial Membrane

Heart

Book Gills

Blood Flow

Heart

Pericardial Membrane

Aorta

Blood Flow

DIASTOLE

Heart

Alary Ligaments

Pericardial Membrane

Heart

Open Ostia

Blood Flow

Heart

Alary Ligaments Relaxed

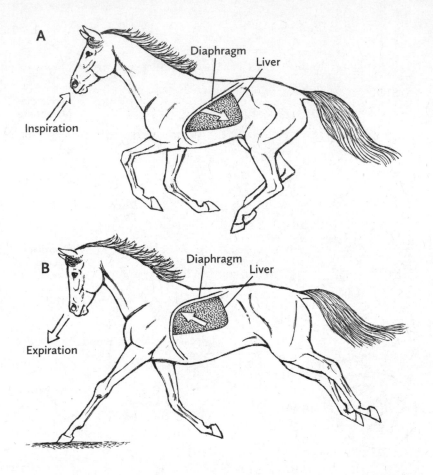

A

Diaphragm Liver

Inspiration

B

Diaphragm Liver

Expiration

which it is attached by a ligament, backward as well. Since the diaphragm makes up the rear wall of the thoracic cavity (the sealed chamber surrounding the lungs and heart), the volume of that space is increased by the diaphragm's movement. Physics tells us that when a space gets larger, the air pressure within that space decreases—here meaning that the atmospheric air pressure outside the horse is suddenly higher than the pressure inside the thoracic cavity. Air rushes into the mouth and nose to equalize the pressure, thus helping to fill the lungs as the horse inhales.

Sound familiar? That volume-pressure relationship is exactly what helps empty our heart of blood during ventricular systole, when the contraction of the ventricles causes an increase in pressure that forces blood out of the heart. During ventricular diastole, the exact opposite happens. Here, as the ventricles relax, the pressure inside them drops, causing the ventricular volume to increase and allowing the chamber to fill with blood coming from the atria.

With that in mind, it should be easy to figure out how the visceral pump works during exhalation. Bramble and Carrier explained that as the horse's forelimbs stride forward (see horse figure B), the liver moves in the same direction, slamming against the diaphragm and causing it to bow forward. This *decreases* the volume of the thoracic cavity and, you got it, *increases* the pressure inside the cavity. This increase compresses the horse's lungs, the way a hand would squeeze water out of a sponge. But instead of water, here the compressed lungs squeeze CO_2-laden air back into the atmosphere.*

So why does this adaptation make sense? As we've already seen, muscle contraction requires energy. According to Bramble and Carrier, the key benefit of the visceral piston is that in a galloping horse, inspiration and expiration take place with less energy cost to the animal.

Similarly, horseshoe crab blood returning to the heart is aided by the book gills, which are already busy waving back and forth as they exchange oxygen and carbon dioxide with their watery environment. Like the backward-and-forward movements of the horse liver, the back-and-forth movements of the horseshoe crab gills drive the blood within them toward the pericardium, thus decreasing the energy that would have been required to move blood into the pericardium by a separate means.

Open circulatory systems have long been regarded as relatively simple and thus somehow inefficient. But as we've just seen in the rather complex workings of the horseshoe crab circulatory system, this is not the case. Instead, this mindset is just one more unfortunate bias against pretty

---

* In humans, the pressure and volume relationships are similar, but changes in thoracic cavity volume occur primarily because of the up-and-down movement of the muscular diaphragm as it contracts and relaxes.

much any organism that doesn't wear blue jeans and carry a cell phone.*

One complex and unique feature of horseshoe crab circulatory systems in particular has to do with immunity. Invertebrates do not have the mammalian equivalent of *acquired* immunity, which is the part of the immune system in which specialized cells called lymphocytes and bits of protein called antibodies recognize and combat foreign invaders like bacteria, fungi, and other pathogens. This immune response is turned off (or "suppressed") once the invaders are gone, leaving behind memory cells, which remain in circulation and can rapidly crank up the immune response should they encounter the same foreign invader again. This is why, for example, you don't get the same strain of flu twice—your already-primed immune response destroys the pathogen before you can get sick again.

Although invertebrate immune systems are different, scientists now understand that they are quite spectacular in their own right. Horseshoe crabs, for instance, have evolved their own version of immune cells. And although they are unlike anything seen in humans, they have undoubtedly saved thousands of human lives.

The story of the Atlantic horseshoe crab's first turn toward medical relevance occurred in 1956. That's when Woods Hole pathobiologist Fred Bang determined that certain types of bacteria caused horseshoe crab blood to clot into stringy masses. He and his colleagues hypothesized that this was an ancient form of immune defense. Eventually, they determined that a type of blood cell called an amoebocyte was responsible for the clot formation.† As their name implies, amoebocytes resemble amoebas, the blobby single-celled protists that make pseudopods so popular and dysentery so unpopular.

---

* Our completely off-base depiction of Neanderthals as brutish, apelike losers, fit to be driven to extinction by modern humans, comes to mind.

† Amoebocytes occur in other invertebrates (like land snails), but while these cells may also be involved in clotting and response to blood-borne toxins in non–horseshoe crabs, there has been relatively little research related to this topic.

Bang, and those who followed up his research, hypothesized that the clotting ability of the amoebocyte evolved in response to the bacteria- and pathogen-rich muck that horseshoe crabs plow through for pretty much their entire lives. Their army of blood-borne amoebocytes can wall off foreign invaders, isolating them in prisons of gelatinous goo before they can spread their infections.

As a result, horseshoe crabs are not only disease-resistant but have an impressive ability to survive *extreme* physical damage. The most lethal-looking wounds are quickly plugged with amoebocyte-generated clots, allowing banged-up individuals to carry on as if they hadn't just lost a fist-sized section of shell to an outboard motor propeller. This unique defense-and-repair system may be at least partially responsible for the horseshoe crabs' record of having been around for nearly half a billion years, a period during which they've survived a total of five planetwide extinction events.

We now know that the amoebocytes do their thing by detecting potentially lethal chemicals called endotoxins. These are associated with gram-negative bacteria, a class of microbes that includes pathogens like *Escherichia coli* (food poisoning), *Salmonella* (typhoid fever and food poisoning), *Neisseria* (meningitis and gonorrhea), *Haemophilus influenzae* (sepsis and meningitis), *Bordetella pertussis* (whooping cough), and *Vibrio cholerae* (cholera).

Oddly, the endotoxins are not themselves responsible for the myriad diseases associated with these bacteria. Nor are they protective products— released, for example, to combat the bacteria's own enemies. Instead, these large molecules form much of the bacterial cell membrane, helping to create a structural boundary between the cell and its external environment. Endotoxins are also known as lipopolysaccharides, since they consist of a fat attached to a carbohydrate. These molecules become problematic for other organisms only after the bacteria have been killed and sliced open, or lysed—something that can happen when the immune system (or an antibiotic) is engaged to fight off a gram-negative bacterial infection. At this point, the bacterial cell contents spill out and the lipopolysaccharide components of the membrane are released into the environment.

Unfortunately, although the disease-causing bacteria may have been conquered, the sick host's problems are not over. The presence of endotoxins in the blood can cause the rapid onset of fever, one of the body's protective responses to a foreign invader. Such fever-inducing substances are called pyrogens, and they can lead to serious problems (like brain damage) if they drive body temperatures too high for too long. Further complications can also arise from the body's dangerously overblown immune response—a condition healthcare professionals have been forced to deal with during the coronavirus pandemic. In the worst cases, exposure to endotoxins can lead to a condition known as endotoxic shock, a cascade of life-threatening symptoms that range from damage to the lining of the heart and the blood vessels, to dangerously low blood pressure.

After our trip to find horseshoe crab eggs at the beach, Leslie and I accompanied Dan Gibson to the Woods Hole lab, where he prepared a microscope slide of fresh horseshoe crab blood. We were soon examining live horseshoe crab amoebocytes.

"They're all full of granules," I said, noting the sand-like particles that packed the cell interiors.

"Those are tiny packets of a protein called coagulogen," Gibson said. As their name may suggest, coagulogens cause coagulation, or clotting. "When the amoebocytes encounter even the slightest amount of endotoxin, they release their packets of coagulogen, which quickly transforms into a gel-like clot."

Because endotoxins can cause such a dangerous response in humans, in the 1940s the pharmaceutical industry began to test its products for the presence of these substances, which can also be released by accident during the drug manufacturing process. One of the first methods developed was the rabbit pyrogen test, which became an industry standard. Here's how it worked: In what definitely sounds like a job for "the new guy," baseline rectal temperatures were taken for lab rabbits involved in the test. Next, the lab technicians injected rabbits with the batch of whatever drug was being tested, often doing so via an easily accessible ear vein. They then recorded rectal temperatures every thirty minutes for the next three hours. If a fever developed, it would signal the potential presence of an endotoxin in that particular batch.

Having discovered that horseshoe crab blood would clot in the presence of endotoxins, in the late 1960s, a colleague of Fred Bang's, hematologist Jack Levin, developed a chemical test, known as an assay, that would come to replace the laborious and controversial rabbit pyrogen test. Essentially, Levin and his colleagues sliced open horseshoe crab amoebocytes to collect the clot-forming component, a substance they named Limulus amoebocyte lysate (LAL). Not only could LAL be used to test for the presence of endotoxins in batches of pharmaceuticals and vaccines, researchers eventually discovered that it also worked on instruments like catheters and syringes, medical devices for which sterilization might kill bacteria but also could accidentally introduce endotoxins into patients receiving medical care.

While this discovery was presumably greeted by relief within the rabbit community, horseshoe crabs and their fans were somewhat less than thrilled, especially when another Woods Hole researcher quickly established a biomedical company that began extracting horseshoe crab blood on an industrial scale. Three more such companies soon sprang up along the Atlantic coast, turning the production of LAL into a multimillion-dollar industry. As a result, today nearly half a million horseshoe crabs are hauled out of the water each year, many during spawning season. Most are transported to industrial-sized lab facilities, not in tanks of cold salt water, but in the back of open pickup trucks. Upon arrival, the crabs encounter teams of mask- and gown-clad workers, who scrub them with disinfectant, bend their hinged shells in half ("the abdominal flexure position"), and strap them to long metal tables, assembly line–style. Large-gauge syringes are then inserted directly into the horseshoe crabs' hearts. The blood, blue-tinted and with the consistency of milk, drips down into glass collecting bottles. And in a move that would make Count Dracula envious, the collection continues until the blood stops flowing, usually when around 30 percent of it has been drained.*

In theory at least, the horseshoe crabs are supposed to survive their ordeal, and once bled, by law they must be returned to the approximate area where they were collected. But according to Plymouth State University neurobiologist, Chris Chabot, an estimated 20 to 30 percent of the crabs die during the roughly seventy-two hours from collection to bleeding to return.

"It's significant that the gill-breathing crabs are held out of the water for the entire time," Chabot told Leslie and me. We were visiting the scientist and his colleague, zoologist Win Watson, at the University of New Hampshire's Jackson Estuarine Laboratory.

Also of potential significance, Chabot explained, is the fact that no one knows whether previously bled specimens suffer any short- or long-term effects after being returned to the water—or even whether they survive. (The Atlantic States Marine Fisheries Commission [ASMFC] has been formally managing horseshoe crab populations since 1998, but various policies have hampered its ability to access mortality rate numbers in horseshoe crabs harvested for biomedical companies.) With this in mind, Chabot and his research team have been trying to determine the effect that the harvesting process has on horseshoe crabs once they are returned

---

* It's likely that the needle interrupts the circulation of blood back to the heart—no surprise—so only the blood in the heart and the blood that gravity drags back into it can drain out.

to the water. To do this, he and his students collected a small number of specimens and subjected them to conditions mimicking those the crabs face during encounters with the biomedical industry.

Chabot and his students observed listlessness and disorientation in their subjects, which they hypothesized was due in part to the fact that after bleeding, the crab's body can't deliver as much oxygen as it requires. "It takes weeks to replenish the amoebocytes and the hemocyanin they've lost," he told us.

Chabot also explained that with many of their protective amoebocytes being lysed in a test tube somewhere, things like wound repair and a return to environments infested with gram-negative bacteria made for a pretty grim outlook for those horseshoe crabs headed home after a long day on the assembly line.

Watson confirmed that the combination of three days spent out of the water, at high temperatures, coupled with significant blood loss, can make for a lethal combination for horseshoe crabs. What's more, he added, since crabs are usually collected during the mating season, and often *before* mating occurs, any death rate would have the potential to affect the size of future generations—especially since the larger female crabs are preferentially selected during collection. And given that the crabs have slow maturation times, the extent of the problems that are brewing may not become apparent to researchers, or anyone else, for a decade. According to the ASMFC, the New York and New England regions are already starting to see a decrease in the abundance of horseshoe crabs.

Watson and Chabot both suggested that some fairly simple steps could be undertaken to improve mortality numbers, thus helping sustain horseshoe crab populations without hurting the LAL industry. The first step would be to delay the harvesting of horseshoe crabs until after the mating season. Their second suggestion was to transport specimens to and from biotech labs in cool water tanks rather than stacking them up, dry and hot, on boat decks and in the backs of trucks. This, the horseshoe crab mavens explained, would not only prevent heat stress but also keep the thin, membranous "pages" of their book gills from drying out.

From talking to Watson and Chabot, it is clear to me that they fully appreciate the importance of LAL to the medical community and to the patients whose lives it saves. These researchers are simply trying to improve the odds for a species that has been coping with threats to its existence long before humans showed up and added pollution, habitat destruction, and overharvesting to the horseshoe crab shit list.

Although the steps Watson and Chabot suggested would go a long way toward improving horseshoe crab mortality, there is another harvesting-related risk. This one stems from the fact that each horseshoe crab heartbeat is initiated and controlled by a small mass of neurons called a ganglion, located just above the heart. Its job is to stimulate each section of the heart to contract in the right order in response to minute electrical pulses.

These *neurogenic* hearts are found in crustaceans like shrimp as well as segmented worms like earthworms and leeches. They differ significantly from the *myogenic* hearts seen in humans and other vertebrates, which beat without being stimulated by external structures like ganglia or nerves. Instead, the stimulus for myogenic contraction originates in small regions of specialized muscle tissue called cardiac pacemakers, located within the heart itself.

The absence of these pacemakers in neurogenic hearts may at least partially explain why Aztec art never depicts priests as holding the still-beating hearts of newly sacrificed lobsters or horseshoe crabs. That's because their neurogenic hearts would have stopped beating the moment they were severed from the ganglia controlling them.

Meanwhile, thanks to pacemaker cells, human hearts have the ability to generate a continuous sequence of electrical signals. These begin at a location in the right atrium called the sinoatrial (SA) node, and speed through the heart along highly specific routes called conduction pathways. Moving like ripples of water after the splash of a pebble, the signals travel from the right atrium to the left atrium, both situated within the uppermost "base" of the heart. As the ripple begins to move downward toward the ventricles, another patch of pacemaker cells, called the atrioventricular (AV) node, slows down the signal, the slight lag time allowing the ventricles to fill with blood. The electrical signal from the AV node continues down toward the pointy apex of the heart. As it does, the muscles making up each ventricle are stimulated to contract in turn.

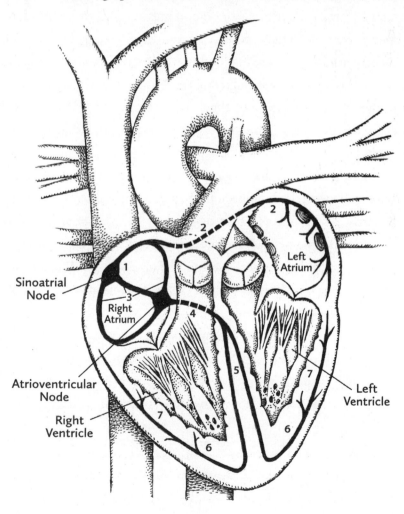

But while our myogenic heart *initiates* its own beat, a pair of nerves control the rate and the strength of contraction. These are the vagus nerve, which slows down the heartbeat, and the cardiac accelerator nerve, which . . . well, you know. They work as part of the autonomic nervous system (ANS), which goes about its considerable duties without your consent or voluntary input.

There are two divisions of the ANS. One, the sympathetic division, prepares you to deal with real or imagined threats with a host of responses, including increased heart rate and blood pressure. This is often referred to as the "fight-or-flight response." As your heart rate speeds up, your ANS also causes an increase in blood flow to your brain and leg muscles. This occurs as blood vessels supplying those areas receive a signal to start vasodilation (i.e., widening of their inner diameters). Simultaneously, blood is diverted *away* from the digestive tract and kidneys through vasoconstriction of the tiny blood vessels that normally supply them.* The reasoning here is that digesting Cheerios and producing urine becomes somewhat less important when you are suddenly confronted by a grizzly bear or the prospect of speaking in front of an audience.† Instead, the extra blood heads to the leg muscles through their wide-open capillaries—preparing you for a sprint. The blood flow is also increased to the brain, presumably enabling you to figure out what to do if running away doesn't work.

The second division of the autonomic nervous system is the parasympathetic division, which takes over during normal (a.k.a. grizzly-bear- and public-speaking-free) conditions. This is the "rest-and-repose" alternative of the ANS. It slows down the heart rate, directing blood

---

* The reduction or increase in flow to blood vessels is made possible because of the respective contraction or relaxation of smooth muscle fibers embedded in and encircling their walls.

† Remember, the ANS treats perceived threats like real ones—which is also why you respond the way you do during a good scary movie.

flow to the organs slighted by the fight-or-flight response, like those that handle digestion and urine production.

Interestingly, if the nerves that control the ANS are damaged, or if their impulses are blocked (attention fugu fans), the heart does not stop beating—which would be quickly fatal. Instead the SA node takes over regulation of the heart rate, setting the pace internally at around 104 beats per minute.

The problem for a horseshoe crab getting the hypodermic Dracula treatment is that its heart has no such ability to pace itself. Its heartbeat is solely governed by the ganglion situated above it.

Watson explained that the ganglion activates motor neurons, which communicate with the heart muscle by releasing a neurotransmitter called glutamate. This chemical messenger fits like a key into neurotransmitter-specific locks found on the surface of the heart. These locks are known as receptors, and the resulting lock-and-key arrangement directs the cells making up that muscle to contract.*

"The problem is," Watson said, "that if you stick a needle into a horseshoe crab to drain its blood and you hit the cardiac ganglion by mistake, you'll likely kill the animal."

"So, workers bleeding specimens in these biomedical facilities have to take the location of the cardiac ganglion into consideration when they insert their needles, right?"

Watson shook his head. "Bill, I doubt any of them even know about it."

IN AN ATTEMPT to be as fair as possible, I contacted several of the major biomedical facilities that process horseshoe crab blood. After listing my professional credentials, I explained in an email that I was fully aware of the important role of the LAL biomedical industry and that I was

---

* Removal of the neurotransmitter (which is taken back up into the motor neurons) leads to muscle relaxation.

interested in presenting *both* the conservation and the industry sides of the story. The initial response was deafening silence.

Eventually, a former student who "knew a guy who knew a guy" was able to put me in touch directly with someone at a biomed facility. I sent another inquiry, making sure to mention my student's name. Soon after, I received a nice letter explaining that they were sorry but that company rules did not permit on-site interviews, and that for proprietary reasons nobody was permitted to see the rooms where the horseshoe crabs were being bled. The letter writer *did* assure me in no uncertain terms that the horseshoe crabs were doing well—*so* well in fact that I expected a few of them might be sending me their own follow-up notes, telling me not to worry because everything was cool.

I was also able to dig up another company's prepared impact statement, written to address the "misleading suggestions" that (1) the American horseshoe crab population was in danger and (2) that the production of LAL was a primary cause of horseshoe crab mortality. Running counter to a number of peer-reviewed scientific papers, the company's findings were that *Limulus* populations were not only stable but they were actually *increasing*. This claim likely draws on figures from the Delaware Bay, where conservation efforts *have* led to an increase in horseshoe crab populations. But the impact statement appears to ignore population declines reported elsewhere along the Atlantic coast. It also concluded that the biomedical industry had only a minor impact on horseshoe crab mortality, with an accompanying bar graph used to point the finger at the *real* culprits—the whelk and eel fisheries. There is no argument that these particular industries remain serious problems to horseshoe crabs, though their designation as the *primary* problem is troubling.

There are, however, some promising developments, as I learned from biologist John Tanacredi, the director of the Center for Environmental Research and Coastal Oceans Monitoring at Molloy College. At the site of an old oyster hatchery on Long Island's South Shore, Tanacredi and his team maintain the only captive breeding colony of *Limulus* in the United States. In addition to this small-scale but locally popular effort, he and his colleagues are working hard to protect *Limulus polyphemus* by having it designated as a World Heritage Species by the United Nations. But even if they're successful (and the odds seem rather long), Tanacredi believes there will be local extinctions of horseshoe crabs, and possibly worse, if (1) harvesting by the bait and biomedical industries is not curtailed or at least better regulated by states, (2) their consumption as "exotic food" persists, and (3) critical habitats, especially breeding sites, continue to be destroyed through development or pollution.

Perhaps, though, the best answer to the dilemma facing horseshoe crabs originated with the work of Singaporean biologist Jeak Ling Ding in the 1980s. Ding sought to insert the horseshoe crab gene responsible for LAL's powerful response to endotoxin into the DNA of a microorganism. Similar recombinant DNA technology had already allowed drug companies to produce human insulin in large vats of yeast. Eventually, Ding and her research team were able to identify the gene behind the production of "factor C," the substance in horseshoe crab blood responsible for clot formation. They used a virus to inject factor C into cultures of insect gut cells (a popular cell type for this sort of work), which became tiny factories, cranking out the clot-generating LAL. Ding's patent for a recombinant factor C test kit was approved in 2003, but the pharmaceutical industry paid little interest. At the time, the test kit had only one supplier, and that company was still waiting for FDA approval. Because of this, the biomedical industry was apparently reluctant to switch from horseshoe crab-derived LAL, a product it had been successfully using for decades.

Recently, though, a second company has begun to produce recombinant factor C. Though the majority of biomedical companies producing

LAL have not yet adopted the new test kits, one has started offering it for sale in addition to its crab-based kits. And, in great news for horseshoe crab lovers everywhere, the pharmaceutical giant Eli Lilly has begun using recombinant factor C to quality-test its new drugs. One can only hope that this is just the beginning of what will become a full-blown transition to noninvasive technology to detect endotoxins, and that someday soon, stringing up horseshoe crabs to drain their blood will go the way of the rabbit pyrogen test.*

---

* In a sad setback, as drug companies scrambled to find ways to prevent or treat the coronavirus in 2020, the use of horseshoe crab blood derivatives to detect endotoxins in sterile labs skyrocketed, and kits using the new noninvasive technology were relegated to the back burner.

As an insect increases in size, demand for oxygen will increase in proportion to length cubed, but rate of supply will only increase as length squared . . . The upshot of all this is that Mothra is going to have to add a lot of tracheal tubes to maintain a sufficient oxygen supply.
—MICHAEL C. LABARBERA,
THE BIOLOGY OF B-MOVIE MONSTERS

[4]

Insects, Sump Pumps, Giraffes, and Mothra

HAVING LEARNED ABOUT the collegial relationship between the circulatory and respiratory systems, it might come as something of a shock that in many invertebrates, most notably the vast majority of insects, the circulatory system does *not* carry oxygen or carbon dioxide. Instead, oxygen-rich air enters the body through tiny holes called spiracles, then flows through a series of smaller and smaller tubes (called tracheae and tracheoles) until it eventually reaches the body tissues. The air takes a reverse trip as it exits, this time minus much of its oxygen and having picked up CO_2, both through the process of diffusion.

This tracheal system explains why many species of insects are able to exhibit active (and sometimes hyperactive) lifestyles without the connections between the circulatory and respiratory systems seen in other groups of animals. Interestingly, that connection may once have existed in insects, since a few species like stoneflies have the oxygen-carrying pigment hemocyanin in their hemolymph. This suggests that some ancient (a.k.a. basal)* insects may have retained an ancestral blood-based mechanism for gas exchange that was lost later in their evolution as spiracles took over the job. Additional evidence for this hypothesis comes from a study in which the copper-based pigment hemocyanin was found in the embryonic hemolymph of a grasshopper, but not in later developmental stages.

As it happens, though, insect circulatory systems are also unusual for one additional, very unexpected reason: insects lack hearts.

How can a circulatory system possibly function without a heart? Well, like the horseshoe crab and many creatures with open circulatory systems, each insect possesses a dorsal vessel that runs along the midline of its entire body.† Here, though, the blood vessel *itself* comes equipped with ostia, the intake valves we recently saw in the horseshoe crab heart. The dorsal vessel, therefore, acts somewhat like a heart, in that nutrient-rich hemolymph enters through the ostia and is expelled by contraction of the vessel's muscular walls. Once the hemolymph leaves the dorsal vessel, it enters chamber-like hemocoels throughout the body, bringing it into contact with the head and major organs. The hemolymph is then routed toward the rear of the body, delivering nutrients to the back-end organs and waste to the excretory system. Picking up another batch of nutrients from the digestive system, the movement of the body and an assortment of auxiliary "hearts" found in the wings, antennae, and legs return the hemolymph to the dorsal vessel, which it reenters as the ostia open between contractions.

---

* The term "basal" implies a group near the base of the evolutionary tree of whatever group (i.e., taxon) is being discussed. Basal groups may be extinct, like jumbo-sized dragonflies at the base of the dragonfly tree, or living (i.e., extant), like the soon-to-be discussed bristletails, which are basal on the insect tree.

† Before we go on, in order to clarify the term "dorsal," please lie down on the floor and imagine that you're an insect, or an earthworm, or any four-limbs-on-the-floor species. The side of your body in contact with the floor is called the ventral surface, while the side facing the ceiling (since hopefully you're not doing this outside where passersby can see you) is the dorsal surface.

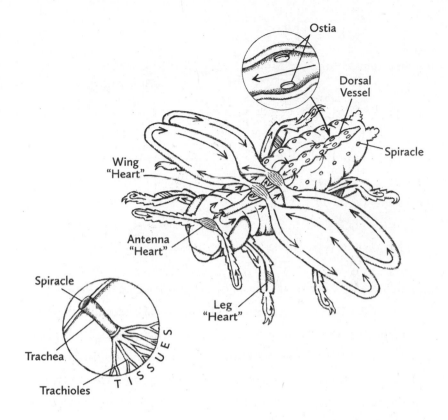

In another example of the way organ systems serve multiple purposes, as the dorsal vessel contracts, the pressure that develops within it helps maintain body shape and contributes to locomotion, reproductive behavior, molting (shedding of the exoskeleton), and hatching. This open system also plays the more traditional role of the circulatory system by supplying backup energy to the insect. It does so by carrying chemical energy from storage depots, called fat bodies, to the organs, where it can help meet the insect's metabolic needs during energy-sapping processes like flight.*

With nearly a million species known, insects display some weird variations on the highly generalized circulatory system described above.

---

* As seen across the animal kingdom, when stored fat is needed, it is broken down into energy-rich molecules of fatty acid, which are transported by the circulatory system to the regions where they're required. There, cells break the chemical bonds holding the molecules together, using the former energy for a variety of purposes.

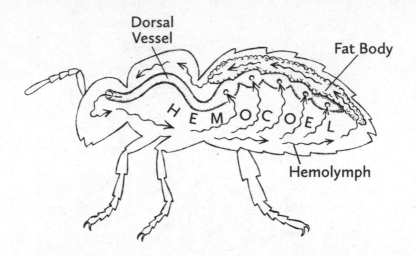

One such example occurs in a basal insect group called bristletails (order Diplura), which have specialized valves in the dorsal vessel that allow blood flow to alternate direction. As we touched on when discussing prolapsed valves in the human heart, backflow is usually a no-no. In bristletails, though, bidirectional flow allows hemolymph to more effectively reach both the head and tail regions. Most insects' dorsal vessels struggle to pump hemolymph to distant dead-end structures like legs, wings, and antennae, but only bristletails evolved this particular solution. More commonly, evolution provided what might appear to be a jerry-rigged response, in the form of auxiliary hearts, those previously mentioned dead-end structures. Without all of the gear typically associated with true hearts, these tiny muscle-driven pumps help drive hemolymph into hollow, elongated appendages, like wings, legs, and antennae, that would otherwise receive inadequate flow. Note: For those insect-oriented grad students looking for a research project, much about the mechanism behind these pulsatile organs remains unknown.

ONCE HEMOLYMPH IS on the move inside an insect's open circulatory system, what keeps it from backing up? As hinted at by the bristletail tale

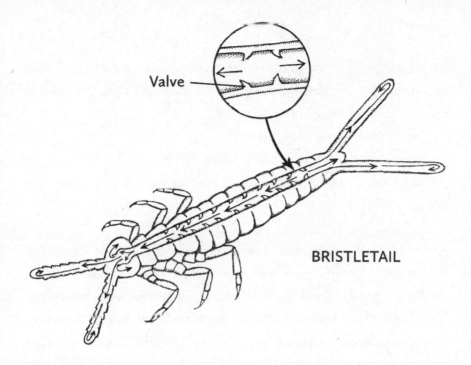

Valve

BRISTLETAIL

above, the mechanics of backflow prevention are pretty much the same as those found in animals with closed circulatory systems. It is also the same system found in many flood-prone basements.

In each case, the starting point is a pump, be it a contractile dorsal vessel, a heart, or the electric motor in a basement sump pump. As in a biological system, the sump pump converts energy (in this case, electrical energy from an outlet or a battery) into mechanical energy (in this case, through the movement within the motor, like a spinning fan). This mechanical energy can be used to do work, like overcoming the gravity keeping water in the sump pit, a hole dug into the basement floor where water can accumulate for various reasons—none of which falls under the category of "fun." If the pump is powerful enough, the water is driven through the pump, up a hose, and out into your neighbor's yard. When the electrical power is cut off or when the water gets too far from the pump, gravity tries to pull the water back toward the sump pit again. If it's a decent pump, though, the water *doesn't* flow back down into your basement. That's because the pump has valves, which allow the water to move in only one direction—out.

So do blood vessels work this way?

Essentially, the answer is yes, although you should try to forget the parts about your neighbor's yard and the hole in your basement floor.

AS PREVIOUSLY MENTIONED, when compared to the hearts of vertebrates, the circulatory pumps found in invertebrates are *highly* variable in appearance and function. Blood might be pushed out into the body by peristaltic/pulsatile blood vessels (earthworms), tube-shaped hearts (horseshoe crabs), sac-shaped hearts (acorn worms), or even multichambered hearts (snails). Some invertebrates, like squid and their cephalopod pals, even have closed circulatory systems and *multiple* hearts, which vary in both anatomy and function. Though it would be nearly impossible to cover all of these systems, several of them stand out as interesting examples.

Technically, earthworms and their relatives (a.k.a. annelids, or segmented worms) do not have hearts, but rather have a series of five paired contractile vessels called aortic arches, pseudohearts, or circumesophageal vessels (so named because they wrap around the esophagus). As in insects, earthworm circulatory systems and respiratory systems do not overlap—that is, their hemolymph doesn't carry oxygen or carbon dioxide. But instead of a tracheal system for the passage of air, segmented worms exchange gases directly through their thin, moist skin, through a process known as cutaneous respiration. Note: Because earthworms breathe through their skin, they can drown in rain-soaked soil. This explains why they risk being out and about on rainy nights, much to the delight of early birds and fishermen.

In the mostly slimy-skinned animals that practice cutaneous

respiration, oxygen in the air diffuses through the skin's outermost layer, the epidermis, and into an extensive network of capillaries located in the next layer down, the dermis.* From there, the now-oxygenated blood moves into a larger, dorsal vessel that spans the length of the worm. The rhythmic contractions of the dorsal vessel propel the blood forward into the paired aortic arches. Lined up in parallel, these arches encircle the front section of the body and they contract in a synchronized, wavelike manner known as peristalsis. This is the same rippling-tube process that pushes food down your esophagus, sloshes it around in your stomach, and squeezes it through your small intestine.

In the earthworm, peristaltic contractions drive the oxygenated blood downward and into a ventral blood vessel. From there the blood branches off into capillaries and is distributed to the body and organs. The deoxygenated blood eventually returns to the dorsal vessel through small vessels downstream of the capillaries, allowing it to circulate through the body of the worm in an unbroken loop—and making this a classic example of a closed circulatory system in an invertebrate.

Currently, there is strong support for the hypothesis that vertebrate hearts evolved from peristaltic blood vessels similar to aortic arches, although no one believes that vertebrate hearts evolved from the system currently seen in earthworms.†

WHILE CEPHALOPODS LIKE squid and octopuses may not have five pairs of aortic arches, they do also operate in multitudes, possessing

---

* The dermis (or dermal layer) differs from the superficially located epidermis by being richly vascularized and metabolically active. In many organisms, the epidermis functions primarily as a physical barrier against the external environment, and its uppermost cells are dead by the time they reach functional maturity. It should come as no shock that the epidermis is *extremely* thin in earthworms, and other animals that use cutaneous respiration, like frogs.

† By the way, if you're still lying on the ground since participating in that dorsal/ventral surface demonstration, please get up now.

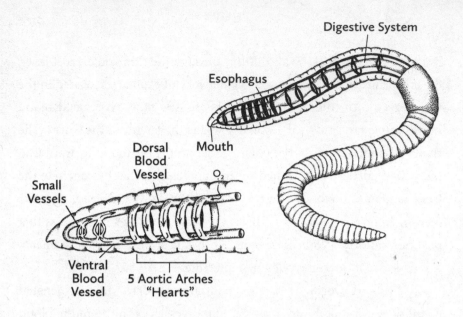

Digestive System

Esophagus

Mouth

Dorsal Blood Vessel

O_2

Small Vessels

Ventral Blood Vessel

5 Aortic Arches "Hearts"

a trio of hearts. The first two hearts, a pair of branchial hearts, receive deoxygenated blood returning from the body. Their contraction propels this blood to the gills, where it picks up oxygen extracted from the surrounding water. Leaving the gills, the oxygenated blood is routed to the third heart, a single systemic heart, which pumps it throughout the body. This highly efficient closed circulatory system likely came about as an evolutionary response to cephalopods developing their characteristically active lifestyles. Equipped with intelligence, jet propulsion, and superb predatory skills, these creatures require relatively larger amounts of oxygen than similarly sized organisms of the couch-potato variety.

At this point, a word of warning might help stomp out a common error that many nonscientists run into when looking across and within the animal kingdom. When observing the significant variation seen in the circulatory systems of insects, earthworms, and squid, it is easy to classify one as "better" than the other—and all of them as "inferior" to those systems possessed by humans. It's something that many scientists

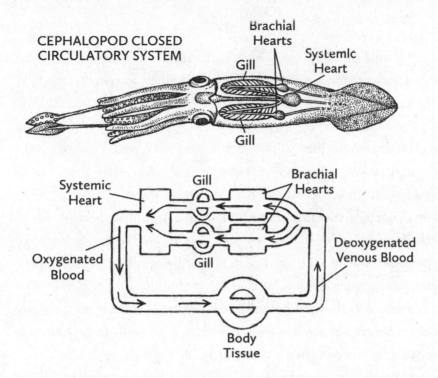

CEPHALOPOD CLOSED
CIRCULATORY SYSTEM

Brachial
Hearts

Systemic
Heart

Gill

Gill

Systemic
Heart

Gill

Brachial
Hearts

Oxygenated
Blood

Gill

Deoxygenated
Venous Blood

Body
Tissue

did until the mid-twentieth century, and because of this, older scientific literature has plenty of purple prose—as man "triumphed" and humans "reached the pinnacle" of whatever topic was being discussed. But rather than considering nonhuman circulatory systems as second-rate or somehow faulty, we should think of them as *functionally equivalent*, each having evolved over hundreds of millions of years to satisfy the nutrient, waste, and gas exchange requirements of its owners and the environmental conditions in which those organisms lived or still live.

What is more, none of these organ systems are perfect. Most are just modified versions of previously existing structures, sometimes with different parts co-opted for new roles. Far more often than not, evolution does not invent; it tinkers with what is already there, tweaking some structures and fashioning new purposes for others. With this in mind, there should be no bragging rights associated with the fact that some circulatory systems are quite complex while others are relatively simple. The key here is that all of them work.

There *are*, however, limits to what an open circulatory system can do. The reason for this is that every anatomical system must deal with the basic laws of physics and with the constraints these laws impose upon them. In other words, in evolution all things are *not* possible. For example, something shaped like a cow cannot fly because of the constraints placed on fliers by the laws of aerodynamics. In open circulatory systems, the constraints also turn out to be quite significant, particularly when it comes to size. It's due to the laws of physics that there are no eagle-sized houseflies or golf cart–sized horseshoe crabs. Large animals are simply composed of too many cells to be efficiently supplied by an open circulatory system.

As usual, this is due largely to diffusion. Closed circulatory systems have extensive interweaving capillary networks, providing a tremendous amount of surface area for the exchange of gases, nutrients, and waste between the blood and the tissues of the body. Open circulatory systems have no such thing. As we've already seen, their exchanges occur across chamber-like hemocoels. Unfortunately for any horseshoe crabs aspiring to mammoth size, the hemocoel walls don't have enough surface area to supply layer upon layer of tissues made up of millions and millions of cells.

Gravity is another constraint on organisms with open circulatory systems, and it explains why there are no creatures of this type equivalent to the giraffe. The reason is that the pumps found in open circulatory systems never evolved to be strong enough to force blood upward against the very significant force of gravity encountered by animals as tall as giraffes—or even as tall as humans.

Giraffes (*Giraffa camelopardalis*) are the tallest living mammals (up to eighteen feet for males), and in order to force blood up to their treetop-level heads, their hearts generate the highest blood pressure seen in any mammal. Normally, it's around 280/180 mm Hg, which is more than twice the blood pressure of a human, which is normally around 110/80 mm Hg. There will be more on the circulatory systems of these amazing creatures shortly, but for now let's make a stop to clear up an important, though potentially confusing, issue.

Some readers may be wondering what those blood pressure numbers I just mentioned actually mean. The first number reflects the force applied to blood vessels during ventricular contraction, while the heart is pumping blood out to the body. This is called systolic pressure. The second number represents the force applied to those same vessels while the heart is relaxed and the ventricles are filling with blood. This is the diastolic pressure. Similar to other pressure measurements, like barometric pressure, these numbers can be envisioned as the height in millimeters that a column of mercury in an open-ended U-shaped glass tube will rise against gravity when a force is applied to one end of that tube. In the case of barometric pressure, that force is generated by the atmosphere; with blood pressure, it is generated by the heart as it contracts and relaxes.

The life-threatening effects of hypertension (or high blood pressure, i.e., generally 120/80 mm Hg and above) in humans are well-known, and recent studies have shown that both systolic and diastolic pressures are important predictors of heart attack, stroke, and other bad cardiovascular juju.* But more on that later.

---

* A recent study also shows a clear link between high blood pressure and the risk of dementia. But while low blood pressure might seem desirable, hypotension (<90/60 mm Hg) can also lead to problems, like confusion, light-headedness, and fainting. Extreme hypotension can result in shock and even death.

On the opposite side of the blood pressure coin from giraffes is the ocean-dwelling family known as hagfish (*Myxine* spp.). Affectionately known as "slime eels" or "snot snakes" (though they're not eels *or* snakes), hagfish are often found atop "The Most Disgusting Animals in the World" lists. This probably has nothing to do with the fact that they have the lowest aortic blood pressures of any vertebrate—between 5.8 and 9.8 mm Hg— and more to do with their feeding habits (they feed by burrowing into large dead animals) and their ability to quickly fill a five-gallon bucket with slime if molested. The hagfish can be thought of as the fishy equivalent of an "anti-shrew." That's because, unlike the constantly on-the-go shrew, the hagfish has extremely low metabolic energy requirements and a lifestyle that would make the laziest person you know seem like an Olympic gymnast who has just consumed a pot of coffee.

Given the hagfish's grim feeding habits, it comes as a bit of a surprise that this charming creature is considered an aphrodisiac in South Korea, where fishermen catch it using a technique that might be described as "somewhat less delicate than fly-fishing." To land some hagfish, you'll need to follow these angling instructions: Tie a rope to a dead cow and sink it hundreds of feet down to a mud-bottomed seafloor.* Tie the free end of the rope to a buoy. Then go home. Come back in a week or so. Haul up the corpse, then split Elsie open, and collect your low-pressure prize. Hopefully, that will consist of dozens of hagfish and pounds of their slime, a protein-based goo composed of strands that are stronger than nylon and thinner than a human hair.

Unlike giraffes and humans, hagfish and most aquatic creatures are *relatively* unaffected by gravity. Mostly, this is because the water surrounding a hagfish, or any fish for that matter, is extremely dense.

---

* A hole-punctured barrel baited with rotten fish can be substituted, if your bait shop is out of cows.

Because of this the water exerts an upward force on the creature that acts in opposition to gravity, a phenomenon known as buoyancy. Because air is *less* dense than water, the benefits of buoyancy are minimal for terrestrial creatures, and so they must continually deal with the downward force of gravity. In fact, gravity explains why problems commonly arise with the return of venous blood from extremities like the legs and feet—even in humans with our normally powerful hearts. This is because blood pressure in capillary beds is much lower than anywhere else in the body, usually around 20 mm Hg or less. Physics tells us that an increase in surface area leads to a decrease in pressure, and the capillary beds have a much larger total surface area than the arteries and arterioles leading into them. What's more, without this pressure drop, the arterial blood would blow out the ultrathin walls of the capillaries it enters. The problem is that once blood *leaves* the capillary bed, it remains under low pressure—and if the capillary bed in question is in your toe, this makes it hard for the blood to fight gravity as it flows back toward the heart.

As a result, humans have evolved an additional assist to boost venous blood flow back out of the legs. It comes from the contraction of calf muscles, like the gastrocnemius and the soleus. The bellies of these muscles (that is, the thickened middle portion of the muscle) surround some of the large veins that carry blood back to the heart from the feet. When these muscles contract—for example, when you point your foot downward—they compress the veins and the blood flowing within them. This increases the pressure within these vessels (imagine squeezing an elongated water balloon again), which drives the blood upward and back toward the heart. Known as the "skeletal muscle pump," this machinery is constantly at work, since separate bundles of muscle fibers within leg muscles alternate their contractions on a regular basis and without your permission.

As you might expect, the long legs of a giraffe present plenty of circulatory system–related problems. We will return to that in a minute, though, because it's their necks, which reach lengths of up to six feet, where serious challenges related to venous return to the heart arise and are overcome. As giraffes lower their heads to drink, it's easy to imagine that there would be a danger of blood pooling in the vessels of the head and brain. Luckily, this is prevented by a series of approximately seven valves in each of the two jugular veins, which carry deoxygenated blood from the head toward the heart. Just as valves prevent water leaving your basement from flowing back into the sump pump, blood leaving the lowered giraffe head cannot flow back into that head. And to provide some extra lift for the gravity-defying blood, giraffe jugular veins have more muscle in their walls than is typical of those of most mammals, and their contraction helps move the venous blood upward when the head is down.

On the arterial side, the problems faced by the world's tallest mammal are quite different. When giraffes lower their heads, you would think that the already-high-pressure blood would get even more of a boost from gravity, and flow Colorado River rapids–style into the head. Instead, that blood, which is carried by the carotid arteries, enters a dense network of vessels at the upper portion of the neck. Known as the rete mirabile (Latin for "wonderful net"), this system increases the vessel surface area, leading to a decrease in blood pressure. If this sounds familiar, that's because it's similar to the way blood pressure drops in the capillary beds. In this case, the rete mirabile prevents the sudden increase in pressure that would occur when the giraffe lowers its head to drink—a position that can bring it a dozen or so feet below the heart. Once the giraffe raises its head, the vessels of the rete constrict, sending the blood within them off to the brain.

As I mentioned, our lanky friends face yet another problem—when it comes to their extralong legs. Due mostly to gravity, the arteries running through giraffe legs can be exposed to pressures as high as 350 mm Hg. Such an immense pressure would typically leave the limbs vulnerable to edema, an abnormal accumulation of fluid, commonly referred to as "water retention." This fluid buildup occurs when plasma, the liquid portion of the blood, is forced through thin capillary walls and into the surrounding tissue. Evolution has solved this problem for the giraffe, though, in the form of thick, tight-fitting skin on their legs. This arrangement works on the same principle as compression stockings worn by humans. Both prevent edema by decreasing the flow of blood into limb vessels.*

An array of similar pressure-related adaptations can be found in other long-necked creatures, like okapis, camels, and ostriches—many serving as additional examples of convergent evolution. Clearly, there are challenges that come with being tall, and evolution has tweaked a number of previously standard anatomical features to meet these challenges.

Sticking to the topic of supersized creatures for a bit longer, we exist in a world where constraints imposed by the laws of physics dictate that many of the movie monsters of my childhood could never actually exist. Mothra, the zeppelin-sized lepidopteran, immediately comes to mind. Although the open circulatory systems found in insects work very well for the small and lightweight, we've seen that they simply do not cut it for the extralarge. But once again, there *are* exceptions.

The most spectacular outliers may be the approximately 120 species of king crabs (family Lithodidae), which can reach body weights of eighteen pounds and with a leg span of almost six feet. Another outsized aquatic outlier is the giant clam, *Tridacna gigas*, which can reach more than four feet in width and tip the scales at around 550 pounds. Their ability to

---

* Foot-related footnote: If an adult suddenly goes up in shoe size, it could indicate swelling caused by edema. This should be checked out by a physician since it *may* indicate a heart problem. For example, an increase in blood pressure might be forcing blood plasma out of capillaries and into the surrounding tissues, causing them to swell.

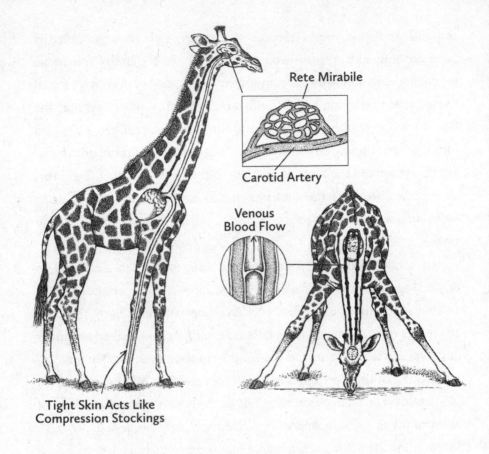

Rete Mirabile

Carotid Artery

Venous Blood Flow

Tight Skin Acts Like Compression Stockings

attain such size relates to their stationary (a.k.a. sessile) existence, with relatively low energy expenditure and subsequently low energy demands.*

King crabs, however, have a more active lifestyle. The key factor enabling them to grow to outsized outliers is that they live in a marine environment, where gravity is far less of a limiting factor than it is in air. In the ocean, a giant crab experiences the pull of gravity on its body, but

---

* Contrary to popular belief, giant clams are not dangerous to humans, since their shells close too slowly to entrap an arm or leg. And even if they were trying to snag you, they are the only bivalves unable to completely close their shells.

because of buoyancy, the total force acting to pull the crab downward has been reduced. This means that it requires less effort/energy for the crab to stand and move around in its watery environment, and its energy and nutrient requirements can be met with an open circulatory system. But because there's far less of a buoyancy effect in air, if you were to place that king crab on a beach, it would not be strong enough to support its body against the pull of gravity.

So, yes, there are some exceptions to the size rule, but such phenomena are to be expected in an animal kingdom whose diversity can surprise even the experts.*

From bristletails with two-way blood vessels to giant squid with a trio of hearts, the amazing degree of diversity exhibited by invertebrates is clearly reflected in their hearts and circulatory systems. Though the fossil record for soft-tissue structures has been less helpful than it has been for researchers studying things like shells and bones, it is abundantly clear that circulatory systems evolved on multiple occasions and in a variety of different animal groups. So while we can examine relationships between animal groups, and the workings of organs like those that make up their circulatory systems, what remains difficult to identify are the origins of circulatory structures like the hemocoel, auxiliary hearts, or, for that matter, the hemolymph that fills them.

As THE CHAPTERS that follow turn back to vertebrates, the path of evolution will become easier to trace. That's because there are a far more manageable number of circulatory system models to study, and a relatively

---

* My favorite example of animal diversity is the existence of over 350,000 species of beetles—a fact I learned as a child, and which got me wondering about Noah's specimen-collecting skills. This was followed by questions like who got to clean up the mess left by roughly three thousand rodents (a male and female from each of the approximately fifteen hundred species)?

clear fossil record for the transitions taking place between fish, amphibians, and terrestrial vertebrates like reptiles, birds, and mammals. There
are, after all, only around sixty-five thousand species of vertebrates alive
today, while there are roughly five and a half times as many species of
beetles. Of course, vertebrate circulatory systems still vary, with many
differences having emerged in the transition from aquatic to terrestrial
lifestyles. And, again, many of these adaptations will help to illustrate the
constraints and trade-offs that vertebrates face—creatures, whose habitats range from inky-black ocean depths to hunting grounds a thousand
feet above the earth's surface.

It is a great thing to have a big brain,
a fertile imagination, grand ideas, but the man with these,
bereft of a good backbone, is sure to serve no useful end.
—George Matthew Adams

[5]

On the Vertebrate Beat

Growing up on the South Shore of Long Island, I spent a considerable portion of my childhood and teen years fishing off neighborhood docks and beaches. Far less exciting than the blue claw crabs, blowfish, and flounder that populated my childhood were the omnipresent rubbery blobs that clung to the piers and docks near my home. They're called sea squirts, for their propensity to spurt streams of water from their funnel-like siphons, usually when they are detached from their moorings for examination, which I have been responsible for on occasion. I had no idea however, that from an evolutionary perspective, these potato-shaped orbs were more closely related to my friends and me than they were to the blue claw crabs we hunted with nets and spotlights.

"Protochordate" is an informal term used to describe several invertebrate groups, thought to be the closest relatives of the backboned vertebrates. The protochordates include a few marine species, such as the tadpole-like lancelets (a.k.a. cephalochordates) and the miniature jelly barrels known as sea squirts (a.k.a. urochordates or tunicates). While it's difficult to connect the physical appearances of adult sea squirts to blue whales or humans, scientists believe that their larvae offer a decent approximation of what the ancestor of the first vertebrate looked like. For the purposes of our tale, we'll now learn how evolution has driven the transition of the vertebrate heart from a simple tube in our distant protochordate cousins into a diversity of forms, including the four-chambered heart seen in blue whales and the pushy bipeds responsible for their near extinction.

Looking nothing like their adult forms, larval sea squirts have tails that propel their elongated bodies, giving them a sort of headless-tadpole look that has never quite caught on with real tadpoles. Eventually, the swimmers, which were initially thought to be a completely separate species from the adults, latch on to a dock or some similar substrate. (See sea squirt figures of larval and adult stages.) Their tails are resorbed by their increasingly gelatinous-looking bodies, and they sprout a pair of in- and out-current siphons, living out the remainder of their lives filtering plankton and detritus.

While its transformation from mobile to sessile is interesting, perhaps the most fascinating thing about the sea squirt is a hypothesis about its heart. Scientists like Annette Hellbach, from the Max Planck Institute of Biochemistry, believe that the tubular heart of the sea squirt is a precursor of the vertebrate heart, particularly due to the electrical conduction system that allows both types of hearts to generate their own rhythmic beats. Hellbach and her colleagues discovered that the sea squirt heart "beats from one end to the other, stops for a short while and then starts to beat in the other direction."[*] The researchers also identified cells along the tubular heart that respond to heart rate–decreasing chemicals, very much like the cells found in the previously discussed cardiac pacemakers found in humans and other backboned bodies.

---

[*] This appears to be an alternate method to reverse the direction of blood flow than that seen in the bristletail's unique two-way valves. These adaptations promoting bidirectional blood flow would be another example of convergent evolution.

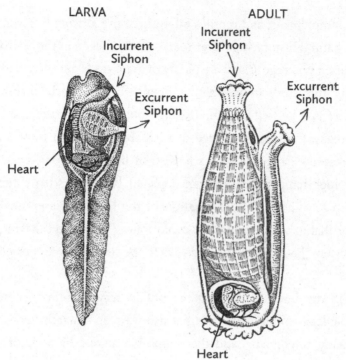

SEA SQUIRT

LARVA

ADULT

Incurrent
Siphon

Incurrent
Siphon

Excurrent
Siphon

Excurrent
Siphon

Heart

Heart

We interrupt this story for brief evolution-themed advisory: Readers should not fall into the trap of believing that the term "precursor" (as used above) implies that this species of sea squirt itself evolved into what might be considered the first backboned swimmer. Rather, the thought is that an ancient *ancestor* of modern sea squirts *may* have evolved enough adaptations (e.g., dropping the adult blob stage and evolving a support rod running down its dorsal side) that on the off chance some humans were to find its fossil half a billion years later, they might decide that the creature shouldn't be classified as a protochordate anymore—but instead that it was a chordate,* quite possibly an ancient fish

---

* Chordates are named for the presence (at some point in their lives) of a rodlike structure called the notochord, running down the dorsal length of their bodies. In vertebrates, the largest group of chordates by far, the notochord has largely been replaced by the vertebral column, with the only remaining traces of it found within the cartilaginous intervertebral disks.

Fish, amphibians, and reptiles offer a glimpse at what is perhaps the greatest natural history story ever told, one that began in the sea roughly half a billion years ago. There are plenty of great reads out there dedicated to telling this story in fascinating detail—I particularly recommend *Your Inner Fish* by Neil Shubin. Through fits and starts, cataclysms, and lucky breaks, the branching tree of vertebrate evolution took some of these creatures from aquatic to terrestrial lifestyles, from dwelling in shallow high-temperature, oxygen-depleted backwaters, to a tentative but ultimately successful colonization of the land, sea, and air. But in order for that to occur, significant evolutionary tweaks to the typical fish organ systems had to evolve—*especially* to the circulatory and respiratory systems.

Along the lines of yet another public service announcement, it would be dead wrong to assume that fish were on a march to becoming semiaquatic amphibians, and that amphibians were likewise driven to evolve the landlubber characteristics of reptiles and mammals. Similar reasoning has led some uninformed types to ask why modern chimps don't evolve into humans. In short, evolution does not work like that.

Instead, researchers believe that the way the transition from water to land occurred was that a relatively small group of fish species (known today as the elpistostegids) already had simple lungs with which they could exchange gases between their circulatory systems and atmospheric air. These lungs had themselves evolved from antigravity buoyancy bags, called swim bladders, found in basically all fish except for sharks and their dorsolaterally flattened pals, the rays and skates. Initially, lungs had enabled these ancient species to colonize swampy, low-oxygen aquatic environments. Stumpy lobe-fins helped them paddle around these shallow-water environments. They supplemented the oxygen from their gills by swallowing gulps of air, which filled their swim bladders—which also just happened to be covered by a dense network of capillaries. Diffusion did the rest.

Then, before you could say "semiterrestrial vertebrate" (bringing us to around 375 million years ago), critters like the crocodile-headed *Tiktaalik* were making short trips onto land. They then used their previously existing stumpy fins for a brand-new purpose—walking. Similar to the way ancient horses survived and thrived because they evolved the ability to eat stuff that nobody else ate, *Tiktaalik* and its descendants would have found plenty of terrestrial snacks, with absolutely no competition from other vertebrates—since all of them were still living in the water. As with the grass-munching horses, this ability to exploit a resource (in this case, *many* resources) that nobody else could exploit became a formula for evolutionary success. Predictably, it also led to an explosion of species diversity, as some vertebrates eventually evolved from semiterrestrial amphibians into more landlocked reptiles and, later, some of the reptiles evolved into what we now classify as mammals. Most fish, though, remained fish. And with the exception of a relatively few species of walking catfish, mudskippers, and their ilk, the majority of fish species never ventured out of the pool. But what an interesting pool it turned out to be. Since those ancient times, fish have evolved into nearly every aquatic environment, from mud puddles to the deepest marine trenches.

Fish also possess the closest thing vertebrates have to an invertebrate heart, thus allowing scientists to get a glimpse of what the earliest vertebrate hearts may have looked like. Most significantly, fish hearts possess only one atrium and one ventricle. As a result, their circulatory systems are not arranged in two separate loops like other vertebrate systems, but in one continuous circuit.

Although fish hearts have only a single atrium and ventricle, they have two additional compartments through which blood exits and

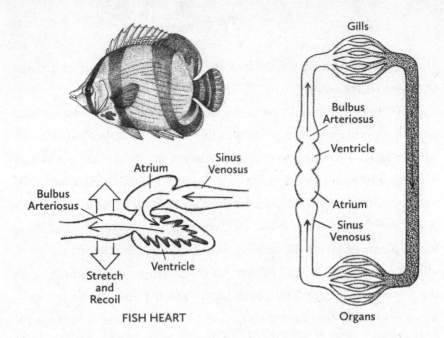

Gills

Bulbus
Arteriosus

Ventricle

Atrium

Sinus
Venosus

Sinus
Venosus

Atrium

Bulbus
Arteriosus

Ventricle

Stretch
and
Recoil

FISH HEART

Organs

enters. These four chambers are arranged in a roughly straight line. The venous blood headed back into the heart from the body first enters the sinus venosus, a large collecting chamber that passes blood along into the thin-walled atrium. The atrium contracts and sends the blood on to the thick-walled ventricle, which it exits through the bulbus arteriosus, an often-pear-shaped structure composed primarily of smooth muscle and stretchy fibers consisting of the proteins elastin and collagen. When the ventricle contracts, the bulbus arteriosus fills with blood, its expandable walls s-t-r-e-t-c-h-i-n-g to accommodate the volume. Once filled, the walls of the bulbus recoil, pumping blood out of the heart and toward the gills at a *constant rate and pressure*, even when the heart is relaxing. This function is vital because of the delicate nature of the feathery, thin-walled gills. Without the bulbus, the contraction of the ventricle would lead to sudden increases in blood pressure, and the gills could be damaged.

The advantages of elastic (a.k.a. potential) energy were significant enough that they endured throughout the evolution of the mammalian heart, and several of our largest arteries are referred to as "elastic arteries" for this reason. Like the bulbus arteriosus, their walls are rich in elastin, the same springy fibers found in places like the skin.* In mammals, an example of an elastic artery is the aorta, which stretches as it fills with blood from the contracted left ventricle. Then, as its walls recoil, the energy stored within is applied to the blood exiting the left ventricle.

Older humans (and other mammals) are often faced with a condition called arteriosclerosis, during which the large elastic arteries can stiffen and lose their elasticity. Commonly known as "hardening of the arteries," it occurs for several reasons, including fibrosis—a pathological (i.e., disease-related) response to injury in which fibrous, nonelastic tissue replaces elastic or contractile tissue in the vessels. Also imparting a negative effect on the vessels is the process of calcification—the buildup of calcium in body tissues, in this case in the form of inflexible deposits within the vessel walls. Without the help of elastic vessels, the heart must work harder to deliver blood to the body, a condition that often leads to serious health problems.

THE DEMANDS OF a transition to life on land, even a partial transition, eventually led to the evolution of a three-chambered heart (two atria and one ventricle) in amphibians. And even though there is some mixing of oxygenated and deoxygenated blood in the single ventricle, this tri-chambered design also carried over into the majority of reptiles.

In most amphibians, deoxygenated blood travels from the body to the right atrium. Oxygen from the gills or lungs returns to the left atrium along with blood that has been oxygenated through the process of cutaneous respiration. Because amphibians have thin, moist skin and an abundance of blood vessels located just below it, oxygen is able to diffuse out of the air through the skin and into the body. This

---

* So-called muscular arteries contain less elastin and more smooth muscle fibers, in a layer called the tunica media.

cutaneous respiration, plus a series of valves and flaps within the heart (that maintain a *partial* separation of the oxygenated and deoxygenated blood), more than makes up for the mixing of blood that occurs in the single ventricle. So efficient is cutaneous respiration for small vertebrates dwelling in damp environments that it is the sole method of gas exchange in the largest family of salamanders, the plethodontids, which are both lungless and gill-less.

While amphibians like frogs, toads, and salamanders have at least some aquatic stages of their lives (and some spend pretty much all of their lives in water), reptiles do not have this requirement—although some of them, like sea turtles, have reevolved aquatic lifestyles. However, by land or by sea, they all have a significant difference in the overall circulatory system function: in reptiles, lungs completely took over the role of gills. This was possible because, unlike amphibians, reptiles never go through an aquatic larval stage (like the tadpoles seen in frogs and toads). This important difference was first identified in the early nineteenth century, and, as a result, frogs, toads, and salamanders were placed into their own phylogenetic class, Amphibia, as distinct from the animals of the class Reptilia, with which they had previously been lumped.

From an evolutionary perspective, the reptilian emphasis on terrestriality wasn't a bad thing, since it meant that these critters were no longer as dependent on finding an appropriate body of water in which to mate or lay their eggs, or to have their tadpoles develop—especially since they didn't have tadpoles. This enabled them to move into habitats located farther from bodies of water, providing opportunities to exploit new types of food while decreasing the risk of running into predators. Because of this, however, reptiles lost their ancestors' moist skin, since it became vitally important to conserve the water contained in their bodies and not allow it to evaporate. As a result, their skin is most often dry, sometimes scaly, and rarely a suitable place for cutaneous respiration to occur.

As mentioned earlier, one characteristic that both amphibians and most reptiles *do* share is the presence of a three-chambered heart, where oxygenated and deoxygenated blood mix. But while this feature provides a clue to their close evolutionary ties, there are in fact a number of smaller variations between the hearts of the various "herps."* The key difference, however, between (noncrocodilian) reptile hearts and amphibian hearts is that in the reptiles, the single ventricle is at least partially divided by a wall-like septum.

As with the previous description of insect circulatory systems, please be aware that what follows is a generalization. Okay, stay with me now and hold on *tightly* to the next figure. In lizards, as the atria contract, two streams of blood (deoxygenated blood from the right atrium and oxygenated blood from the left atrium) enter the left side of the ventricle. Remember, it has a partial septum. Within the left side of the ventricle, the deoxygenated blood sticks to the right and the oxygenated blood to the left (and, yes, here's where some mixing of O_2-rich and O_2-poor blood takes place). When the ventricle contracts, the deoxygenated blood is directed through an opening in the septum and into the *right* side of the ventricle, and from there into the pulmonary artery (which is, thankfully, sitting close by) and on to the lungs. Simultaneously, as the ventricle contracts, the oxygenated blood in that chamber gets pumped out through a pair of aortae and on to the body.† *Whew!*

In one order of reptiles, the crocodilians (alligators, crocodiles, and slender-snouted gharials), there is complete separation of the pulmonary and systemic circuits. The same is true of birds—actually, a closely related class of vertebrates. Crocodilians and birds are the only known survivors

---

* Despite their separation by scientist types, reptiles and amphibians are still referred to as herps by much of the nonscience community. It is an abbreviated version of "herpetiles," from the Greek *herpetón*, meaning "creeping animal."

† There are slight differences in the heart and blood flow pattern in turtles and snakes.

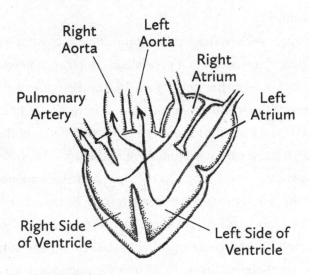

Right Aorta

Left Aorta

Right Atrium

Pulmonary Artery

Left Atrium

Right Side of Ventricle

Left Side of Ventricle

LIZARD HEART SCHEMATIC

of the Archosauria, the group that most famously includes dinosaurs. Their four-chambered hearts are similar, though not identical, to those found in mammals.

With four chambers separated by backflow-preventing valves, and with a septum separating the left and right sides, the hearts of crocodiles, birds, and mammals form not just a pump but a pair of pumps and a pair of circuits. In the pulmonary circuit, oxygen-depleted blood returns from

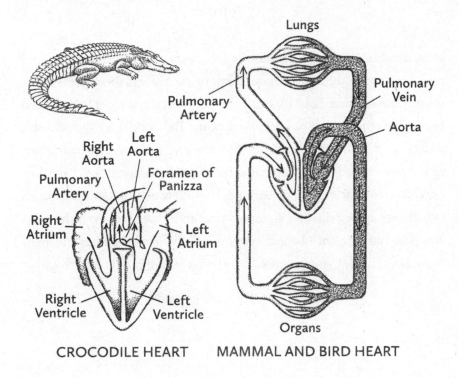

Labels in figure:

Lungs

Pulmonary Artery

Pulmonary Vein

Aorta

Right Aorta

Left Aorta

Pulmonary Artery

Foramen of Panizza

Right Atrium

Left Atrium

Right Ventricle

Left Ventricle

Organs

CROCODILE HEART   MAMMAL AND BIRD HEART

the body to the right atrium, passes into the right ventricle, and is sent off to the lungs. In the systemic circuit, the oxygen-rich blood returns from the lungs to the left atrium and passes into the thick-walled left ventricle, which contracts, propelling it out to the body. As a result, the mixing of oxygen-rich and oxygen-poor blood does not occur, and the oxygenated blood is not diluted by its deoxygenated counterpart.

In the end, though, whether vertebrate hearts have two, three, or four chambers, and whether some degree of mixing occurs between oxygen-rich and oxygen-poor blood or it doesn't, these systems work very well indeed for the creatures that possess them.

ONE THING THAT all animals share is that in order to survive, they must be well adapted to their habitats—the environments in which they live. Often, though, conditions within those environments can change— sometimes abruptly, sometimes significantly, and very often both. For some creatures, extreme conditions are the norm: arid deserts, humid rain forests, the thin air of mountaintops, or the crushing pressure of deep oceans. Other habitats experience significant seasonal or sudden changes in temperature or water availability. The circulatory system plays a key role in the ability of animals (including humans) to cope with environmental extremes. Conveniently, many of the adaptations that allow organisms to deal with these extremes also allow us to better understand how hearts and circulatory systems work, and to see that even the most complex and efficient of these systems will fail if pushed beyond their limit. In an already compromised heart, that failure can be catastrophic.

Cold hands, warm heart.
—TRADITIONAL PROVERB

Call me Jack-the-Bear,
for I am in a state of hibernation.
—RALPH ELLISON, *INVISIBLE MAN*

[6]

Out in the Cold

I'VE SPENT MUCH of my thirty-year career studying bats. Though most people have, thankfully, moved beyond the stereotypical portrayal of bats as bloodsucking flying mice (only three of fourteen hundred species feed on blood, and all species are more closely related to humans than they are to rodents), to many, bats remain shrouded in mystery. One of the things most folks *do* know about these mostly nocturnal mammals is that many of them hibernate. As we'll soon see, hibernation is primarily a circulatory strategy: the energy-burning system that transports oxygen and nutrients must slow down to accommodate the often-long periods when the conditions are frigid and there is no food to be found.

By coincidence, though, as I began to research the topic of cold adaptation in bats, I got sidetracked. Long Island and New York City were being slapped by what local meteorologists were calling "a dangerous deep freeze." Part of what earns any old freeze the label of "dangerous" are concurrent reports of people dying from heart attacks and, less frequently, hypothermia—a condition in humans in which core body temperature falls below 95 degrees Fahrenheit.

It is not difficult to understand why someone with a heart condition might fall victim to the exertion of shoveling snow, especially the heavy, wet stuff so common in the northeastern United States, where a typical driveway can hold several tons of snow after a storm. Much of the uptick in heart attacks is thought to occur because the motion of shoveling snow, especially the lifting involved, causes the heart to beat faster and more strongly. As with any exercise, this increases blood pressure, leading to the potential for damage to what is all too often an already faulty pump. What is less obvious is the way that the cold escalates the situation.

When exposed to low temperatures, as one would experience wading into a snowdrift with a shovel, the human body attempts to conserve heat in core organs, like the brain, heart, lungs, and liver. It does this by reducing blood flow to capillary beds in peripherally located structures, like the arms, the legs, and the tip of your nose, directing it instead toward the previously mentioned essential body parts. This occurs through a process called localized vasoconstriction—that is, a closing off of blood vessels in specific regions of the body. This vessel shutdown occurs as tiny muscular valves, called precapillary sphincters, receive a message from the brain to close. As the bands of muscle constrict, the blood upstream from the valves bypasses the capillary beds, like cars on a highway driving past a temporarily closed exit. This allows the blood to bypass the capillaries (via vessels called metarterioles) and flow through the tissues without supplying them.

In the warmth of your home, something similar takes place after a meal, but in this case, blood is diverted *to* a different set of capillary beds, namely those located within the walls of your digestive tract. Here the precapillary sphincters do not get the signal to close. As a result, blood enters into the digestive system capillary beds, picking up nutrients

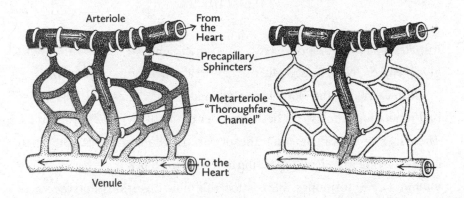

Arteriole   From the Heart

Precapillary Sphincters

Metarteriole "Thoroughfare Channel"

To the Heart

Venule

absorbed through the inner lining of the stomach and intestines (hello, diffusion!) and returning the nutrient-rich blood back to the heart for distribution to the body.

To be clear, though, the route the venous blood takes as it returns from the digestive tract isn't *quite* that direct. Instead, it takes a detour to the liver, via a vessel called the hepatic portal vein. Inside the liver, cells called hepatocytes remove the sugar from the blood and assemble it, building block–style, into a starch-like molecule called glycogen, which can be easily stored nearby. Only then does the still-nutrient-laden blood exit the liver and make its way, via the inferior vena cava, to the right atrium of the heart. This process is the reason that we don't overdose on sugar after eating a half dozen Twinkies.

As for the fate of the glycogen stored in the liver, it can be quickly broken down by the very same hepatocytes and returned to the blood as glucose. Generally, this happens when alarm-like chemoreceptors determine that "blood sugar levels" are too low, something that commonly occurs between meals. These structures are embedded in walls of the carotid arteries (which supply blood to the head) and the aorta. As their name implies, chemoreceptors are stimulated by changes in the concentration of chemicals (like glucose, oxygen, or carbon dioxide) flowing past them in the blood, and when the glucose level drops too far, they relay that information to the brain via nerve impulses. The brain then initiates a suitable response (e.g., "Too much glucose . . . Store some as glycogen," or "Not enough glucose . . . Break down some glycogen").

Also carried in the blood is a waxy lipid called cholesterol. Despite its bad reputation, cholesterol has a number of vitally important functions. These include making up a significant part of cell membranes, helping to conduct nerve impulses, and acting as a building block for substances like vitamin D, sex hormones, bile (which aids in fat digestion), and the stress hormone cortisol.

Cholesterol is carried in the blood attached to lipid-transporting proteins. These come in two forms: high-density lipoproteins (HDLs) and low-density lipoproteins (LDLs). Normally, these lipoproteins can pass into and out of blood vessel walls, but LDLs sometimes get stuck, causing a buildup of fatty deposits inside those walls. These buildups are known as atherosclerotic plaques, and the disease they cause is atherosclerosis. Atherosclerosis is dangerous because the accumulation of plaque can decrease blood flow. An easy way to demonstrate this concept is to turn on your garden hose and then step on it. As you do, resistance increases within the hose, thus reducing the flow of water. What will happen if you keep your foot on the hose for several minutes? If the pressure increase upstream from your foot is great enough, the hose could split—highly dangerous in a blood vessel, since that rupture can lead to uncontrolled bleeding and even death.

So, what does all of this digestion-related information have to do with cold-weather-related heart problems?

Coronary arteries are the blood vessels that supply the heart muscle— the myocardium. Because active digestion routes blood toward the organs near your waistline, it reduces coronary blood flow. For people whose coronary arteries are already narrowed by disease, and whose hearts might be receiving barely enough blood to function under *non*stress conditions, the added stressor of physical exertion in the cold can lead to a significant decrease in the amount of blood being delivered to the heart muscle, and therefore a dangerous drop in the amount of oxygen and nutrients it receives.*

What's more, our blood cholesterol levels are often at their worst—and most dangerous—in the winter. Cardiologist Parag Joshi and his team at Johns Hopkins looked at cholesterol levels in 2.8 million Americans between 2006 and 2013. They determined that the tendency to eat more and exercise less during cold months led to a 3.5 percent increase in LDL cholesterol (so-called "bad cholesterol") levels in men, and a 1.7 percent increase in women. Additionally, the reduced levels of sunlight during the winter months can lead to a decrease in vitamin D, which is thought to lower blood LDL cholesterol levels.

The take-home message should now be clear: *avoid overexertion in the cold*, especially if you've just eaten a big meal, since more blood is being diverted to your digestive tract, or if you've recently started eating more high-cholesterol foods like processed meat, fried food, fast food, or desserts. Best, too, for smokers to be careful, since nicotine causes resistance-increasing vasoconstriction. Oh, and no pre-snow-shoveling cocktails either, since alcohol is also a potent vasoconstrictor. Instead, when you decide to clear away what the big winter storm left in front of your house, stay warm, take frequent breaks, use a smaller shovel, and push the snow instead of lifting it. Better yet, hire a kid to clear your driveway and sidewalk.

Still determined to shovel that driveway yourself? Well, okay. But before heading out into the cold, a few more words of wintry warning. In addition to heart-related issues associated with stressful exercise in frigid temperatures,

---

* According to statistics, things are worst, cardiac-wise, after breakfast. A 2011 study looking at eight hundred heart attack patients in Madrid, Spain, showed that more heart attacks occur in the morning (between 6 a.m. and noon) than at any other time. Also significant is the fact that, on average, these attacks resulted in 20 percent more tissue damage to the heart.

problems can occur if the cold conditions overwhelm the body's attempts at internal temperature maintenance. In humans, for example, once the core body temperature falls below 95°F, the body begins to lose more heat than it can replace through mechanisms like shivering and decreasing blood flow to the digestive tract and extremities. As hypothermia occurs, organ systems, like the circulatory and nervous systems, begin to shut down, and dangerous effects start to set in. These include decreases in coordination and cognition, as well as slowdowns in reaction time. The old saying "The cold makes you dumb" is actually quite apt, since judgment also becomes impaired. As the body cools down, an energy-conserving desire to stop moving sets in, and the dulled mind may be unaware of the inherent danger of falling asleep. The final stage is characterized by slow or absent pulse and respiration. Death can follow. The take-home message once again is: hire the kid (or the guy with the plow).

MOST ORGANISMS CAN'T simply throw down the snow shovel and head for their warm homes, and many have evolved unique mechanisms to deal with exposure to cold temperatures and the stresses that accompany them. For warm-blooded species, including those that *do* wield snow shovels, the body compensates for low external temperatures by working to hold internal temperatures relatively steady. Normally, our body temperature is maintained at around 98.6°F.* Primarily, this occurs as an indirect result of metabolic processes like digestion and muscle contraction, since each produces heat as a byproduct of its chemical reactions. Something very similar occurs when you turn on your car engine. Gasoline contains chemical bond energy. When mixed with air and ignited in a small space (one of your car's cylinders), a controlled explosion results and the chemical bond energy is converted into the mechanical energy that spins your

---

* The normal range is between 97°F and 100°F.

tires. Since no energy conversion is 100 percent efficient, some energy is lost during the process, here in the form of heat. You can demonstrate this yourself by asking someone you don't like to place a hand on your car engine a few minutes after starting it. What they feel is the energy that has been lost during the conversion from chemical to mechanical energy. You can explain this to them once they stop screaming at you.

In the body, as heat is released, mostly during muscle contraction, it radiates out of the tissues where the reactions take place and into the adjacent thin-walled capillaries, warming them and the blood within. The warm blood flows back to the heart and is circulated throughout the body. As that happens, the heat leaves the blood and moves into the cooler surrounding tissues.

But what keeps the temperature of the human body constant? Why don't our bodies cool down when we step outside on a cold morning? The reasons relate to a section of the brain called the hypothalamus.

The hypothalamus is the command center for the autonomic nervous system, the portion of your nervous system that regulates most bodily functions without your conscious input. Those include the maintenance of the body's internal environment, including body temperature.

Upon receiving nerve impulses from temperature receptors in the skin, the hypothalamus acts as a sort of thermostat to keep the body temperature stable. Detecting frigid temperatures, the hypothalamus initiates the previously described shunting away of blood from the peripheral structures like the fingers and toes. It also reduces blood flow to the skin, where superficial blood vessels allow heat to be quickly lost to the environment. Additionally, the hypothalamus sets off a series of involuntary heat-releasing muscle contractions, better known as shivering.

Interestingly, some temperature receptors in the skin "learn" to ignore inconsequential stimuli, which explains why stepping into a hot shower

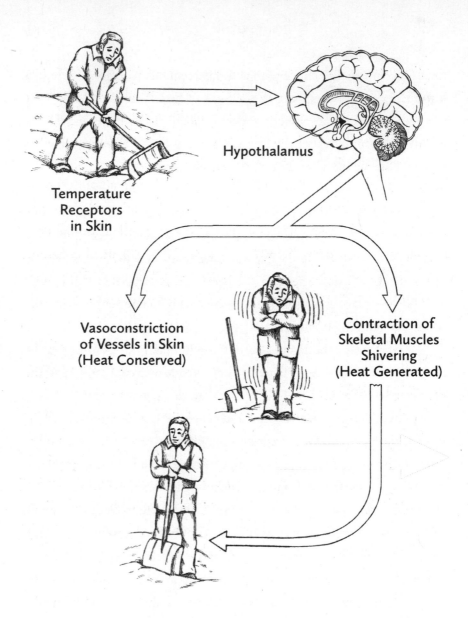

Hypothalamus

Temperature Receptors in Skin

Vasoconstriction of Vessels in Skin (Heat Conserved)

Contraction of Skeletal Muscles Shivering (Heat Generated)

might be painful at first but then becomes comfortable. The phenomenon is known as thermal adaptation. Something similar occurs on a tactile level when you put on socks. Initially, the brain receives signals from touch and pressure receptors in the skin of the feet and ankles, and you feel your socks being pulled on. Very soon, though, the nervous system begins to ignore these unimportant tactile stimuli, allowing you to concentrate on more important things, like making sure that you've put on socks that match. Sensory adaptation can also be related to smells or sounds. Fortunately, this process has its limits. For instance, the nervous system doesn't adapt to stimuli that can be harmful, like putting on a sock with a burr inside or increasing blood pressure.

The ability to internally maintain stable body temperature is known as endothermy, and those that exhibit it, like mammals and birds, are endotherms. This ability differentiates them from ectotherms, like fish, amphibians, and most reptiles. These so-called cold-blooded creatures require externally supplied energy (usually from the rays of the sun) to keep their bodies at a temperature at which tissues and organs can function properly.

Regardless of *how* consistent body temperature is maintained, it's important that it *be* maintained. The myriad chemical reactions (that is, metabolic processes) taking place in the body can occur only when things like temperature are held within a very specific range. So how do ectotherms deal with the cold, specifically temperatures that would normally freeze their bodies and the liquids like blood found within them?

As described previously, hemoglobin is an iron-containing molecule whose primary function is to pick up oxygen in the lungs or gills, then transport it and drop it off in the tissues. And, yes, a byproduct of this oxygen/hemoglobin interaction is the distinctive red color of vertebrate blood. But for those of you who might be wondering if there's an exception to the red-blooded vertebrate rule, the answer is also yes, and that blood belongs to the Antarctic icefish (family Channichthyidae). Known to nineteenth-century whalers long before researchers snagged one in 1928, icefish are the only known vertebrates that lack hemoglobin as adults. Because of this, their blood is nearly clear.

I first heard about these unique creatures during my three-semester tenure as an undergraduate marine science major at Southampton

College on Long Island. Howard Reisman, my ichthyology professor there, taught me that icefish blood not only lacked hemoglobin, but it also contained a unique array of antifreeze proteins that allowed the fish to survive in temperatures that would normally freeze a body solid. Similar to the antifreeze in a car radiator, these substances function by chemically lowering the temperature at which freezing occurs. In the icefish, the antifreeze proteins restrict the growth of ice crystals in tissues, *including* blood, and in hollow structures like the heart and blood vessels. This opens some exciting avenues to medical researchers, who are experimenting with using antifreeze proteins to prevent damage to tissues and organs that are stored on ice before their use in transplants and related procedures.

Interestingly, this characteristic led a European food company to patent a strain of yeast that had been genetically modified to produce the very same antifreeze proteins found in icefish blood. In an amusing twist on the original function, the company currently uses the stuff to prevent crystals from forming in ice cream. More specifically, the edible antifreeze spares frozen dessert munchers from having to deal with the crunchy ice that can form when ice cream's tiny crystals melt and then refreeze into larger, less palatable crystals. The antifreeze proteins work by latching on to the surface of the smaller crystals, thereby preventing them from clustering together into jumbo chunks.

Admittedly, my primary interest in icefish blood had nothing to do with improving mouthfeel for ice cream lovers. Instead, I wanted to know how icefish were able to evolve this weird bit of biology and still obtain enough oxygen to survive. According to University of Alaska Fairbanks icefish expert Kristin O'Brien, the explanation involves their habitat and a related quirk of physics, as well as their anatomy and behavior.

Icefish inhabit the deep waters of the Southern Ocean, also known as the Antarctic Ocean, since it encircles that particular continent. There are relatively few fish species living there and even fewer predators (mostly seals and penguins). Because of this, icefish face little or no competition for the krill, small fish, and crabs they feed upon. They are also ambush predators, which means that their movement consists of short and infrequent bursts of speed. Without much in the way of extended physical activity, their bodies require less oxygen.

The cold water itself offers the hemoglobin-free icefish an additional benefit: it holds more oxygen than warm water. This is because molecules in cold water move more slowly than in warm water. When the molecules move faster, it is easier for oxygen to break free from the H_2O molecule and escape. As a result, cold water ends up hanging on to more oxygen, which is useful for the organisms that require it.

Research suggests that the very first hemoglobin-free icefish ancestors were the result of a mistake—a genetic mutation that occurred sometime around five million years ago. Fortunately, because of their oxygen-rich environment, this mutation didn't immediately doom the ancient fish to extinction. According to O'Brien, what it *did* do is force an extensive remodeling of the icefish cardiovascular system. This evolutionary tweakage resulted in the fish having four times the blood volume and three times the blood vessel diameter of a similarly sized red-blooded fish, as well as a heart more than five times larger than one might expect. This means that although icefish blood pressure and heart rates are low, the *volume* of blood that leaves the heart with each contraction is high. Additionally, once the blood reaches muscles and organs, extremely dense capillary beds help improve the efficiency of gas exchange. Finally, in one innovative evolutionary twist, icefish have no scales covering their bodies, and so oxygen uptake occurs not just through the gills but also directly through the skin.

So, yes, originally, perhaps icefish ancestors were lucky to have lived where they did. Now, though, they have successfully compensated for the species' lack of hemoglobin—the vital oxygen carrier found in the blood of pretty much every other vertebrate in existence.

WHILE ICEFISH MAY avoid the risk of freezing entirely through the production of antifreeze proteins, other species survive by *allowing* themselves to freeze. When temperatures plummet, the hearts of frogs like the North American wood frog (*Rana sylvatica*) can stop beating for several weeks at a time. This occurs because they're frozen solid, as are other vital organs, like the liver. Then as spring approaches and air temperatures climb, the frogs thaw out, as do their hearts, soon to resume their prerefrigerated pulsations.

I spoke to Miami University of Ohio biologist Jon Costanzo, an expert on the phenomenon. He began by telling me that while there is a great deal of public interest in the broader topic of freeze tolerance, only a few researchers are currently working on it. According to Costanzo, the topic peaked in popularity in the 1990s, centering around the cryopreservation of human organs and tissues, but since then research has more or less dead-ended.

I flashed back to a rumor I had heard as a child: Walt Disney's body had been frozen after his death in 1966. Supposedly, it remained in a state of cryogenic preservation in a top-secret facility located beneath Disneyland's Pirates of the Caribbean attraction. I remember being disappointed to learn that, according to his family members, Uncle Walt had in fact been cremated two days after passing away from lung cancer.

But why does freeze tolerance work in wood frogs but not in woodsmen? I put the question to the frozen frog maven, who told me that most creatures' tissues are too damaged by the formation of ice crystals to defrost intact. "Imagine jagged ice crystals forming between the tissues, as well as among and inside cells," Costanzo said. "They tear everything up." So while the buildup of extracellular ice can be problematic, ice forming within the cells themselves is generally fatal.

In addition to causing structural damage from crystallization, freezing leads to excessive cell shrinkage, through loss of liquid water; messes with membranes and other cellular components; depletes energy-containing molecules; and prevents the elimination of cellular waste products, which can accumulate to toxic levels.

So how does the wood frog survive the fatal facts of frost?

"As the wood frog is chilling, it cools down to a temperature below its freezing point," Costanzo told me. "These are woodland frogs, so they've already situated themselves beneath the leaf litter on the forest floor. Of course, there are ice crystals all around, because it's cold out, and these crystals eventually permeate through the frog's moist skin."

Costanzo reminded me that the freezing of liquid water is an exothermic reaction, which means that it releases heat. As a result, the frog's body temperature actually spikes during the early hours of freezing. Its heart rate climbs rapidly as well, nearly doubling, as its heart pumps cryoprotectants out to its body. These are substances that either stop tissues from freezing (like the antifreeze in icefish blood) or prevent damage to cells when they do freeze.

One of these cryoprotectants is actually a very common substance: glucose, the high-energy sugar released into circulation by the liver. As the frog's body freezes, its liver begins breaking down starchy stores of glycogen into glucose. It does so at a tremendous rate, pumping more than eighty times the usual amount of the sugar into circulation. This rush of glucose prevents ice from forming inside cells by allowing water to leave them via a version of our old friend diffusion—this time called osmosis. The water moves from a higher concentration in the cells to a lower concentration in their sugar-saturated surroundings. This prevents the cells from swelling and rupturing when they freeze. We'll learn more about this movement of water in just a bit.

Zoologist Ken Storey has theorized that the glucose release is an exaggeration of the body's fight-or-flight response, in which stress causes the brain to signal the liver to dump glucose into the circulatory system. There, the high-energy glucose molecules can be used as an emergency energy source for things like "fighting" or "flighting." Apparently, a similar alarm goes off when wood frogs begin their transition to frog-sicles.

Recent research by Costanzo and his colleagues focused on another cryoprotectant, nitrogen, and their data suggested that some gut bacteria remain active in their frozen hosts. The bacteria release an enzyme that frees up nitrogen from any urine the frogs have in their system. As with glucose, nitrogen is thought to protect against freeze/thaw damage, possibly because it freezes at such a very low temperature (-346°F or -210°C).

According to Costanzo, all of this previously described action occurs during the first phase of freezing. After several hours, the initial spike in body temperature subsides and the frog starts to cool down again. Eventually, its heart stops beating and the blood freezes within the vessels. The wood frog remains in this condition for most of the duration of the freezing episode, which can last anywhere from half a day during a cold snap to several months in the parts of the frog's home range that extends into the Arctic circle.

"During this time," Costanzo said, "there's no breathing and no heart function."

I asked him if anyone had determined whether electrical activity in the frozen frog brain ceases as well. A flatlining electroencephalogram (EEG) is used to indicate when someone can be declared clinically dead, and I was curious whether the waking frogs were technically zombies. Costanzo responded with a laugh. He told me that he had heard that the frog's brain flatlines when frozen, but he couldn't point to a reference that might support or refute that particular phenomenon.*

I also discovered that it isn't just circulating cryoprotectants that safeguards the wood frog from the effects of freezing. Another factor appears to be a massive redistribution of water within the amphibian's body.

Normally, water can move in and out of cells through osmosis. Since bodily fluids (like blood and intracellular fluid) are mostly water, it's vital that the levels of stuff dissolved in those fluids remain stable. Otherwise,

---

* Grad student research project, anyone?

the cells would dehydrate or swell as water passed in and out on its way to equalize the concentration.

In the frozen frogs, osmosis is prompted by an increase in the concentration of glucose throughout their bodies, which means there's more water inside the cells than outside. Once the water leaves the cell, though, it also seems to move outside of the organ itself, and as a result the organs dehydrate during freezing. "The liver and the heart, for example, will lose more than half of their normal water content," Costanzo told me.

"And where does all that water end up?" I asked him.

"In the coelom," the frozen frog maven answered, "the body cavity where the guts are stuffed."

I tried to envision it. "So how much water are we talking about here?"

"If you were to freeze a wood frog and then dissect it, it would look like a snow cone in there."

I remember thinking to myself that this was clearly a scientist who had seen such a frog-flavored frozen treat, and a moment later he confirmed my suspicion. With an enthusiasm that can be *fully* appreciated only by someone who has, for example, eaten a human placenta in the name of science, I listened as Costanzo described how his team had carefully scooped out and weighed the frosty frog ice to determine what percentage of the total body weight it represented.

It's worth noting that surviving dehydration isn't entirely atypical for frogs. Terrestrial frogs and toads often show a tolerance to drying out. It's possible that evolution has simply enhanced that system to improve freeze tolerance. Ultimately, Costanzo and his collaborators concluded that this process is a way for the frog to freeze a lot of its body water without risking as much organ damage, since most of the ice would accumulate outside of these vital areas. When spring rolls around and the frog thaws out, the resulting water returns to rehydrate the cells and the excess glucose returns to the liver to be processed back into glycogen and stored.

Interestingly, it's not clear to Costanzo, or to any other researchers, what exactly stimulates the heart to start beating again upon thawing. There's no particular temperature or particular time during the thawing episode at which the heartbeat kicks in. But whatever its stimulus, the reanimation process is something that did *not* occur in lab studies on a related terrestrial genus, the northern leopard frog (*Lithobates pipiens*).

"There were a couple of heartbeats after thawing," Costanzo said. "And it looked like [the heart rate] was going to pick up and take off. But then it just crapped out." He told me that there wasn't a clear explanation for the different reactions in the two genera but that it likely relates to an absence of the protective adaptations that evolved in *Rana sylvatica*.

I asked Costanzo if defrosted wood frogs appeared to suffer any ill effects. He explained that while there is no evidence that freezing affects the frogs' longevity, it *does* affect mating performance, since newly thawed frogs show little interest in the opposite sex.

In lab experiments performed on male wood frogs placed in plastic testing arenas, Costanzo and his team showed that test subjects had little interest in sex for up to twenty-four hours after thawing. And even after that, they competed badly in a lab setting against control frogs that had never been frozen. One hypothesis is that the bodies of the thawed frogs are too busy getting rid of copious amounts of glucose to deal with the rigors of mating. Another significant negative factor could be that while some of the glucose used in the freezing process is derived from glycogen stockpiled in the liver, additional glycogen comes from the breakdown of the frog's own body. This form of autocannibalism can reduce the mass of hop-generating leg muscles by up to 40 percent during hibernation.* As

---

* A similar effect can be seen during starvation, as the body literally feeds on itself, breaking down structural proteins in skeletal muscle and elsewhere and converting the products of that breakdown into glucose. This produces the characteristic wasted appearance of starvation victims.

a result, newly thawed male frogs may be physically incapable of chasing down females—until, that is, their limb muscles are restored to their former size and function.

THE TEMPORARY DENT in the thawed-out wood frogs' sex life is an example of the kind of trade-offs associated with *all* adaptations. For example, closed circulatory systems may transport more gases, waste, and nutrients, but they're more expensive (energy-wise) to maintain. Additionally, their complexity makes them more susceptible to breakdown. This is the hallmark of evolutionary biology: organisms that gain an advantage will almost certainly pay a price. Arguably, the most famous example of a trade-off relates to sickle cell disease, the most common of inherited blood disorders.

As many of us learned in high school biology class, genes are tiny sections of our genetic blueprint that control the development of one or more traits (like hair color or blood type). They come in pairs, with each gene located on one of two similar (i.e., homologous) chromosomes. Humans have twenty-three pairs of homologous chromosomes, and in total these contain somewhere between twenty thousand and twenty-five thousand genes. You may also remember that we get one chromosome in each pair— and therefore one gene in each pair—from our mother and the other from our father. This is something that occurs when one of Dad's sperm cells fuses with one of Mom's eggs to produce a single cell that multiplies, develops, and differentiates into us.

As it turns out, the genes that lead to sickle cell are a problem only if someone has two copies of the gene for the disease. And in some ethnic groups, it's incredibly common to possess *one* copy. Approximately 8 percent of African Americans carry the sickle cell trait, meaning that they have received the mutant hemoglobin gene from either their mother or their father—but *not* both. These "carriers" typically live normal lives, with no sickle cell–related health woes.

When a person is born with *two* of the mutant hemoglobin genes, though, serious problems result due to the production of an abnormal form of hemoglobin called hemoglobin S.* Unlike normal hemoglobin, hemoglobin S forms long fibers that cause the red blood cells carrying them to twist into rigid crescent moon (or sickle) shapes. These sickle cells are able to carry less oxygen than cells containing normal hemoglobin, and as a result, less oxygen gets delivered to tissues.

The *most* significant problem is that the deformed sickle cells are not malleable enough to slip into tiny capillaries like normal red blood cells do. Instead, they end up obstructing those vessels. This logjam blocks the blood supply to certain areas of the body, like the extremities. It also stimulates pain receptors, the body's way of alerting us that something is wrong. Eventually, this blockage leads to life-threatening damage to organs, like the kidneys.†

Frequently, my anatomy and physiology students asked me why natural selection hasn't eliminated the mutant hemoglobin gene over time. Their rationale is that prior to modern medicine, carriers of the mutant hemoglobin gene were much less likely to survive to adulthood, and therefore less likely to pass on their defective genes to subsequent generations. So why then, they wondered, didn't that mutant gene disappear over time? As it turns out, the answer relates to the most severe of evolutionary trade-offs.

---

* If both parents carry the defective gene, their offspring have a 25 percent chance of picking up two defective genes themselves and developing sickle cell disease.

† Sickle cell disease can manifest itself in various ways, including tissue damage and vascular obstruction. Sickle cell anemia, in which tissues do not get enough oxygen, is one such manifestation. So the two terms are not synonymous.

The first clue is that sickle cell disease is most common in people whose ancestors came from Africa, the Arabian Peninsula, the Mediterranean, and South and Central America. These are regions where the incidence of malaria is high—and as it turns out, sickle cell carriers are more resistant to malaria than people who do not carry the mutant gene. Therefore, in places where the mosquito-transmitted killer is often the number one threat to human life, having a single copy of the mutant hemoglobin gene actually *increases* reproductive fitness, since more carriers avoid dying from malaria and pass their genes on. For this reason, the mutation remains in the gene pool. Unfortunately, the cost of this particular benefit is lethally steep for those who carry *two* copies of the mutant gene and who develop sickle cell disease.

Such is the nature of trade-offs.

As we've seen, low ambient temperature (a.k.a. the air temperature of the immediate surroundings) can present a significant hurdle to survival, and this has led to some particularly notable evolutionary trade-offs. Some species seasonally migrate enormous distances to warmer climes, while others have evolved thick coats as insulation from the cold. Still others, like icefish and wood frogs, have evolved extreme responses to frigid conditions. Far more frequently, though, animals undergo winter changes that fall into the realm of torpor and hibernation, both of which present major challenges to vertebrate hearts and circulatory systems.

Torpor, a phenomenon something akin to "hibernation lite," occurs when there is a controlled and pronounced reduction in metabolic rate— the rate of energy expenditure by the body. Bouts of torpor generally last for less than a day, while hibernation can be thought of as multiday bouts of torpor, interrupted by periodic arousals.

For a long time, scientists thought that torpor was a mammalian adaptation, but recent studies point to something quite different. I spoke to Miranda Dunbar, an associate professor of biology at Southern Connecticut State University, whose research focuses on bat hibernation. She explained that many experts now see mammalian torpor as a vestige of an earlier evolutionary feature.

While mammals are endotherms, capable of generating and maintaining their own internal body temperatures, this was a relatively late-evolving feature in the vertebrates. Until roughly a quarter billion years ago, most if not all backboned creatures were ectothermic, a condition in which the regulation of body temperature depends on external sources. Like today's fish, amphibians, and reptiles, those ectotherms presumably coped with low ambient temperatures by situating their bodies to maximize the amount of heat they can gain from their environment. Anyone who has seen a turtle sunning itself on a rock has seen an ectotherm in action. The same thing can be envisioned for a chilly chameleon or a cool cobra.

As some reptiles began to evolve into what we now call mammals, newly acquired adaptations like torpor and heterothermy (the ability to adjust body temperature to match ambient temperature) stuck around. While endotherms are less affected by ambient temperature than their ectothermic kin, they need a lot of energy to power their metabolic processes and keep their body temperatures stable. When winter comes and air temperatures drop, the energy requirement only increases—at the same time that food becomes scarce.

"So," Dunbar told me, "during the winter, with little to no energy available, the animals enter into a torpid state that can be prolonged seasonally into hibernation."

I wondered how hibernation might have originated in bats, given that their ancestors apparently evolved in the tropics.

"It does seem counterintuitive," Dunbar told me, "to think that bats would be using hibernation and torpor in the tropics, but that turns out not to be the case."

She explained that in very hot climates, small mammals like bats evolved the ability to turn off the chemical reactions that would normally help them maintain stable body temperatures. Because the environment already provides a source of heat, they don't need to expend energy from nutrients to stay warm. This slowdown of metabolic processes is actually similar to what happens when bats go into torpor in colder climes. Presumably, as they moved out of the tropics, evolution simply tweaked these abilities to deal with a different temperature extreme.

Although bats are famous for hibernating in thermally stable places like caves and mines, many of them can make do almost anywhere. "We've seen them hibernating under loose tree bark and in tree hollows," Dunbar told me. "In man-made structures and even on the ground, underneath leaf litter. But probably the most bizarre example are bats in Japan that hibernate in the snow."

In a recent paper on the Ussurian tube-nosed bat (*Murina ussuriensis*), researchers noted that snow-hibernating bats were typically found only after the winter thaw had begun. In twenty-one out of twenty-two instances, they were discovered alone in small cone-shaped depressions in the melting snow. All of them were curled up in spherical or semispherical postures, which are optimal for body heat conservation. If authors Hirofumi Hirakawa and Yu Nagasaka are correct in their conclusions, this is only the second recorded instance of a mammal hibernating in the snow. Polar bears (*Ursus maritimus*) are the other, although whether they're actually hibernating is up for debate.

Male polar bears are active year-round, but although females come close to hibernation as they nestle into winter dens with their cubs, their body temperatures never make the extreme drop normally seen during this physiological state. The tweak appears to be an adaptation that allows the females to nurse their cubs. So even though adult female bears eat nothing during this period of up to eight months, and their metabolic rates drop (e.g., heart rate can decrease from forty to eight beats per minute), true hibernation requires a drop in body temperature for an extended period. It is likely, then, that tube-nosed bats can be considered the only card-carrying snow hibernators.*

Temperature aside, this dip in metabolic rate is an important feature of hibernation, as it allows bats and other hibernators to use less oxygen and nutrients over the course of the winter. Just as the bear's heart rate drops precipitously, a bat's heart rate can go from five hundred to seven hundred beats per minute down to as low as twenty. During this time, just as in cold humans, blood is diverted from the limbs into the core of the body, to supply and warm the most vital organs. A significant difference, though, is that the hearts of hibernators have evolved to function at low temperatures and low oxygen levels that would cause fibrillation—a disastrously rapid, irregular, and unsynchronized contraction of the heart's muscle fibers—in nonhibernators like humans.

Since there's no food to be found, the nutrients that hibernators do use come from an accumulation of a substance known as brown fat, which in bats is stored in small deposits between the shoulder blades. Unlike most fat, brown fat can be broken down through a chemical reaction that produces heat *directly*, without using up energy on intermediate steps. Brown fat is also found in newborn humans. Infants are especially vulnerable to the cold, because it takes a while for a newborn's temperature regulatory mechanisms (like the ability to shiver) to get up and running.

---

* A little-known fact about hibernators is that they periodically awaken—an energetically costly behavior, but necessary in some species to get rid of metabolic waste.

What's more, because babies are small, their body surface area is relatively large—the same surface-to-mass problem that bats and other small animals run into. As a result, babies can lose heat around four times faster than adults. Premature and low-birthweight babies are especially prone to temperature regulation–related problems because they have less brown fat to burn. This is part of why preemies often spend the first weeks of their life inside warm incubators.

The presence of brown fat in babies also explains their chubby phase, which lasts until most of this tissue gets burned up. In adult humans, very little (if any) brown fat remains, with small amounts generally limited to places like the upper back, the neck, and the vertebral column.

In bats and other hibernators, brown fat is rationed over the entire hibernation period, metabolized to release heat whenever the hibernator's core body temperature decreases past a certain set point. Problems can arise when hibernating animals are aroused unexpectedly by disturbances (such as inquisitive humans). Since a small portion of brown fat is burned during every arousal, there is real threat that the animals will starve to death if their fat stores are burned up before improving weather conditions allow them to seek alternative forms of energy (i.e., food).

One interesting side effect of hibernation is that it causes animals to live longer, and to age more slowly. Bats can live for over twenty years in the wild, an unusually long span for such a small mammal, the vast majority of which have extremely short life spans, typified by the pygmy shrew (*Sorex minutus*). These tiny predators weigh in at around one-fifth of an ounce and burn through their hyperactive lives in around eighteen months.

Miranda Dunbar told me that she'd recently read a study published by the Committee on Recently Extinct Organisms at the American Museum of Natural History, which concluded that over the past five hundred years, there have been sixty-one species of mammals with confirmed extinctions.

"Of those, only three were hibernating bats," she said. "So they must be doing *something* right."

The creatures outside looked from pig to man,
and from man to pig, and from pig to man again;
but already it was impossible to say which was which.
—GEORGE ORWELL, *ANIMAL FARM*

[7]

Ode to Baby Fae

WE'VE NOW LEARNED quite a bit about animal hearts and their associated vascular circuitry. Although the anatomical differences between some of these systems appear vast, across the animal kingdom their functions are roughly equivalent—to pump blood, or its invertebrate equivalent, around the body. Our attention will soon turn inward, to our own hearts. But before we do, let's see how similarities of anatomy and function have led to additional, unexpected connections between nonhuman and human hearts.

In the early 1980s, Leonard Bailey (1942–2019), a cardiothoracic surgeon at Loma Linda University Medical Center in California, began experiments in which he transplanted lamb hearts into baby goats. Eventually, his subjects not only survived to adulthood but they were able to reproduce and have their own "kids." This encouraged Bailey and his team, who hoped that they might someday use the same technique for transplants involving a very specific subset of human newborns—namely, those who had been given no chance of survival due to what were at the time inoperable heart defects. In this case, the donor hearts would come from baboons, since they're genetically, developmentally, and physiologically similar to humans. From a cardiovascular perspective, baboon hearts are nearly identical to human hearts, and, crucially, they even have similar A, B, and AB blood types, though type O is rare.

In October 1984, Bailey's colleague, neonatologist Douglas Deming, contacted the mother of one such baby and told her that there was a transplant protocol that might save her child. After traveling from her home in Barstow, California, to Loma Linda and meeting with Bailey, the young woman initially wondered if he was a mad scientist. But after he spent long hours carefully reviewing his previous research with her, the desperate mother gave her permission for the surgery, knowing full well the potential for heartbreak.

The child who would become known to the world as Baby Fae was born on October 14, 1984, with a congenital heart defect known as hypoplastic left heart syndrome (HLHS). Invariably fatal at the time, HLHS is characterized by an underdeveloped left ventricle and faulty or completely closed mitral (a.k.a. bicuspid) and aortic valves. The symptoms include difficulty breathing and feeding, as well as the blue or purple tint to the skin, lips, and nails that is the hallmark of an inadequate oxygen supply.

The surgical team chose the most compatible of six baby baboons that were available to be donors, and on October 26, the surgery was performed at Loma Linda University Medical Center. A team of doctors removed the baboon heart in a separate ICU operating room, placed it in a bath of cold saline, and carried it to the room where Baby Fae was surrounded by a well-coordinated surgical team led by Bailey. Baby Fae's heart, which was in terrible condition, was removed, and the donor heart was placed into her chest. According to immunologist Sandra Nehlsen-Cannarella,

HYPOPLASTIC LEFT HEART SYNDROME

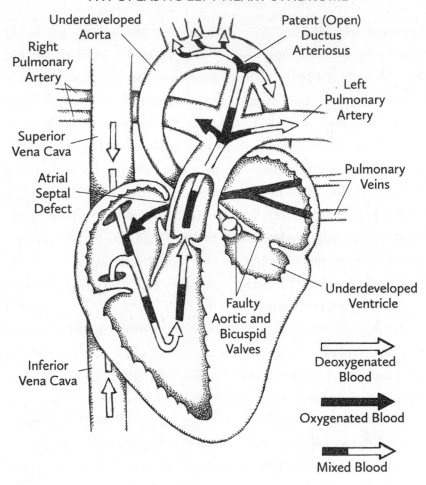

one of the doctors on the surgical team, it fit perfectly. Bailey and his colleagues efficiently sewed the heart into place with no setbacks. Then the big moment came as they rewarmed the baby and let the blood flow through the newly implanted heart. Moments later, it began beating.

"There wasn't a dry eye in the room," Nehlsen-Cannarella told filmmakers for the 2009 documentary *Stephanie's Heart: The Story of Baby Fae*. "We were all just choked up to hear that heartbeat." She added that nothing about the heart looked odd or reminiscent of its previous owner. "It looked like a perfectly normal human heart," she said. "A heart is a heart."

But as Baby Fae recovered and the doctors who cared for her took turns camping out in her room, something else was taking place. The story had become a worldwide media sensation, complete with hundreds of picketers at the hospital and outside Leonard Bailey's home. The protesters questioned the ethics of transplanting a baboon heart into a human, taking all sorts of angles, from pro-animal rights to anti-transplants. What's more, after reporters discovered the identity of Baby Fae's parents, they began hounding them around the clock and reporting on completely irrelevant and borderline slanderous issues related to their backgrounds. Bailey and his team were also singled out, with their sincere attempts to save a dying baby becoming buried under a barrage of claims that they had performed the surgery only for the publicity it was now generating. Thankfully, the media storm also generated sympathetic feelings among some, and Baby Fae's young mother was soon receiving hundreds of letters of support from the public.

Within a few days, Baby Fae was awake, off her ventilator, and eating. Her medical team and her parents were ecstatic. Of course, as was typical for transplant patients, she was also being treated with an immunosuppressant, a relatively new one called cyclosporine, to prevent her body from rejecting her new organ.

Near the end of the second week, postsurgery, problems began to develop as Baby Fae's body began showing signs of what Bailey and his team had first assumed was a "rejection episode." This is a common enough occurrence after organ transplants that the physicians had expected it, and they responded accordingly, stepping up immunosuppression. But when things did not improve, Bailey's team began to suspect that this was actually an autoimmune response—a situation in which the body's immune system actually attacks its own healthy cells and tissues. In this case, what they were now fighting was a body-wide shutdown of Baby Fae's organ systems.

Baby Fae passed away just shy of three weeks after the transplant, on November 15, 1984. At a press conference, Bailey opened by grieving the loss of a precious life and said, "Her unique place in our memories will derive from what she and her parents have done to give rise to a ray of hope for the babies to come."

Although the actual cause of Baby Fae's autoimmune response and subsequent death was initially a mystery, Bailey later disclosed that it was due to mismatched blood type between the human patient and the baboon donor—Baby Fae had type O blood, and the donor baboon had type AB. Bailey called this "a tactical error with catastrophic consequences."

"If Baby Fae had the type AB blood group, she would still be alive today," he told the *Los Angeles Times* in 1985.

He explained that the decision to perform the transplant was based on the mistaken belief that blood type incompatibility wouldn't be much of a problem, and could be circumvented by the use of immunosuppressant drugs. Unfortunately, this turned out to be a tragic error for reasons that will be discussed more fully in an upcoming chapter on blood transfusions.

Despite the tragedy of Baby Fae's death, her case was the beginning of what Sandra Nehlsen-Cannarella referred to as a "revolution in transplantation." Baby Fae's story served to inform the public about the fate of infants with fetal heart abnormalities, and so it also drove home the desperate need for organ donors of all ages. As a result of an increase in neonatal heart donations, Bailey's team soon moved away from what he termed trans-species transplants (and now more commonly known as xenotransplants) and on to neonatal human-to-human transplants. He was able to perform 375 human-to-human heart transplants at Loma Linda University Children's Hospital between 1984 and 2017.

Other researchers carried on xenotransplant research, however. And although primate hearts are indeed similar to human hearts, scientists ultimately determined that they were not a good donation option. The primary reason is that primates (including baboons, chimps, and gorillas) do not produce many offspring, thus limiting any potential supply of their organs.

What researchers *did* decide was that pigs would work. Not only are their hearts very similar to human hearts in size, anatomy, and function, but female pigs can produce large numbers of piglets. While there is a problem with tissue incompatibility, that issue is being addressed by efforts to genetically alter experimental pigs using the genome editor CRISPR. Not only can this engineering technique be used to prevent pig organs from being rejected by the human immune system, it can also strip out the genetic sequences that lead to porcine endogenous retroviruses (PERVs) that could potentially be transmitted to humans.* Recently, researchers have started transplanting these genetically modified pig organs into nonhuman primates, and there is an expectation that preclinical studies would begin in 2021 or soon after.

Today, the prognosis for newborns suffering from hypoplastic left heart syndrome has improved tremendously from 1984. In addition to the options of human-to-human heart transplants and immunologically safe xenotransplants, a series of three remarkable heart surgeries (together known as a staged reconstruction) can be performed.

During the first step in the procedure, which takes place within several days of birth, the right side of the heart, which usually receives deoxygenated blood before sending it on to the lungs, is surgically converted to carry out the role normally performed by the left side of the heart—that is, receiving oxygenated blood from the lungs and pumping it out to the body. Doctors use surgical patches, grafts, and other

---

* Transmission of primate viruses to humans would be a similar concern if baboon hearts had continued to be transplanted.

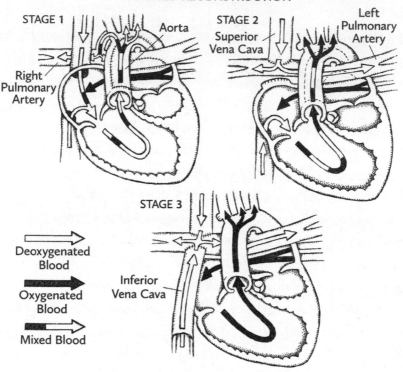

STAGED RECONSTRUCTION

STAGE 1

Aorta

Right Pulmonary Artery

STAGE 2

Superior Vena Cava

Left Pulmonary Artery

STAGE 3

Deoxygenated Blood

Oxygenated Blood

Mixed Blood

Inferior Vena Cava

modifications* to ensure that enough blood gets sent to the lungs that the newborn can survive until the second procedure.

In the second stage of the surgery, which occurs within six months of birth, the superior vena cava is reconfigured to bypass the heart completely, bringing deoxygenated blood directly from the upper body to the lungs. This partially frees up the right side of the heart for its new job.†

Finally, when the child is between one and a half and three years of age, the inferior vena cava is remodeled. As a result of these surgeries,

---

* More specifically, a connection is formed between the left atrium and the right atrium, and the aorta is connected to the right ventricle.

† The superior vena cava (which brings deoxygenated blood from the upper body to the right atrium) is disconnected and attached directly to the pulmonary artery (which normally brings deoxygenated blood from the right ventricle to the lungs).

all deoxygenated blood returning from the body is sent directly to the lungs, and the right ventricle can fully take on the role of the left ventricle, pumping oxygenated blood out the aorta to the body!

As amazing as lifesaving surgical procedures like staged reconstruction are, just as incredible is the prospect of using a ready supply of genetically altered pig hearts and other organs to help eliminate the long and often fatal transplant waiting lists that currently exist.

So how did our knowledge of the heart advance so far?

As readers will soon see, the short answer is: we were slow learners.

PART 2

What We Knew and What We Thought We Knew

A NOTE FROM THE AUTHOR

ANCIENT INTERPRETATIONS OF the anatomy and physiology of human organ systems are as surprising as they are variable. They can also be quite confusing. A primary reason for this is the fragmentary nature of ancient medical texts and papyri. Some of them are tattered pieces of much larger works. Others are compendiums of information written by multiple authors, sometimes working in different centuries and often contradicting each other.

What's more, none of this information can be accessed directly: all of the information available to us comes by way of modern translations. Many of these works describe intricate structures, like veins, arteries, and nerves, and complex conditions, like angina and myocardial infarction, and the job of a translator is subjective, an attempt to accurately interpret words from an ancient language and rephrase them in another. Doubtless, not everything makes it through as originally intended.

There is also the inescapable fact that ancient physicians and scholars got a lot wrong. They were employing rudimentary instruments, when they had them, and working within stringent social and religious boundaries. What's more, many of them were not locked into specializations, as are those of us who work in the sciences today. The ancient scholar-physicians wrote poetry, commented on political and social issues, and were experts in other fields of science, like physics and math. Given these differences, it is far wiser to appreciate the things that the ancient physicians got right rather than judge what they got wrong. We can also take some of these mistakes as a reminder of the dangers inherent in lockstep reasoning. Recycled by scholars, teachers, and physicians, without critique or revision, ancient medical knowledge was taught and practiced by rote, and, as a result, much of it went uncorrected for centuries.

All we know is still infinitely less than
all that remains unknown.
—WILLIAM HARVEY,
EXERCITATIO ANATOMICA DE MOTU CORDIS
ET SANGUINIS IN ANIMALIBUS

[8]

Heart and Soul: The Ancient and Medieval Cardiovascular System

WHEN THE ANCIENT Egyptians prepared human bodies for burial, they removed each of the organs in turn. The heart, known as *ab* (sometimes *ib*) and *haty*, was treated with reverence during embalming procedures, for it was believed to hold a record of all of the deceased's good and bad deeds. It was preserved in a jar, or placed back inside the body so that it could be weighed in the afterlife against a feather of Ma'at, the goddess of truth and justice, who would judge whether its owner had lived virtuously.* The brain, on the other hand, experienced no such funerary weigh-in. Instead, it was unceremoniously yanked out of the nose with a hook and discarded—a clear indication that the ancient Egyptians thought little of its function or importance.

If we place ourselves into the mindset of the ancient Egyptians, the heart as the seat of the soul makes perfect sense. Writing on the topic in

---

* If the heart weighed less than the feather of Ma'at, the deceased went on to live forever in the afterlife. If it weighed more, it was immediately eaten by a monster named Amam ("The Devourer"), waiting at the base of the scale.

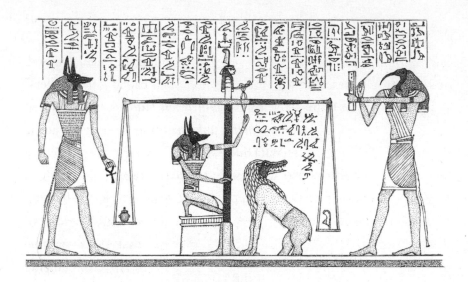

1978, University of Cambridge historian Roger K. French rationalized it this way: Living things were warm. They breathed, and they moved, both innately and in response to external changes. The heart, too, was warm, and it moved. That movement was innate, could be related to breathing, and clearly reacted to external changes—speeding up, for example, when a person was placed in danger. "The heart and its associated vessels were central to the physiology of the living body in Egypt," French wrote. "The pulse was the heart 'speaking' through the vessels; the vessels carried from the heart the secretions and humours necessary to every part; the vessels were responsible for pathological conditions; and they carried the 'breath of life' and the 'breath of death.'"

For those philosophers looking for the seat of the soul or souls, the heart was the answer.

Some modern translations of the Egyptian *Book of the Heart*, originally dating from around 1555 BCE, indicate that Egyptian physicians may in fact have had an impressive grasp of heart-related pathologies, like heart attacks and even arterial aneurysms. In such aneurysms, weakened arterial walls balloon out dangerously. Generally, they occur in medium to large arteries, with the most commonly affected vessels being the thoracic and abdominal aortae, and the iliac, popliteal (behind the knee), femoral, and carotid arteries. Arterial aneurysms are often called "silent killers" because of their asymptomatic nature and due to the fact that ruptured aortic aneurysms (and a closely related condition known as aortic dissection) kill 75 to 80 percent of those stricken. An aortic dissection is a tear in the inner lining of the aorta resulting in a leakage of blood, which then accumulates between the tunics (layers) of the artery. The buildup of pressure leads to a high risk of rupture. It's believed that 90 percent of deaths from ruptured aortic aneurysms and aortic dissections could be prevented through ultrasound screening, which can detect the bulging vessels before they rupture.*

Historian and author John Nunn sounded a word of caution, however, regarding the ancient Egyptians' knowledge of aneurysms and other such specific maladies, explaining in his book *Ancient Egyptian Medicine* that medical papyri "are difficult to interpret in terms of modern concepts of cardiology," due both to differences in conceptual frameworks and to the difficulty of translating hieroglyphs with precision.

But although the ancient knowledge of aneurysms remains conjectural, on more firm ground is the belief by Egyptian physicians that air drawn in through the nose passed through the lungs and into the heart, where it was pumped out to the body in arteries, creating the peripheral pulse. Admittedly, the concept sounds strange, but Nunn pointed out that, if for air we substitute oxygenated blood, then "the whole concept is remarkably close to the truth."

Because Egyptian medical information was held in high esteem by other cultures, their beliefs about the circulatory system were subsequently adopted. Significant interactions between the ancient Greeks and Egyptians

---

* Notable deaths from these aortic conditions include physicist Albert Einstein and actor George C. Scott, who died of abdominal aortic aneurysms. Comedians Lucille Ball and John Ritter died from aortic dissections.

were both direct (for instance, the Greek Ptolemaic dynasty ruled Egypt for 275 years) and indirect (many Egyptian literary works were translated and adapted by the Greeks). Because of this, there are numerous similarities in how these cultures came to view the heart.

Hippocrates (c. 460–c. 377 BCE), who is often called "the Father of Medicine" and who lends his name to the modern Hippocratic oath, was the leader of a medical school on the Greek island of Kos. Celebrated throughout history for his philosophical approach and clinical observations, Hippocrates did much to remove medicine from the realm of magic and superstition. Before Hippocrates, the belief was that all illnesses were a form of punishment by the gods and the only way to prevent or get rid of them was to appease the gods with praise, gifts, sacrifice, and prayers. Hippocrates, though, was heavily influenced by Egyptian medicine, which stressed concepts like cleanliness and a healthy diet. His reliance on the ancient Egyptians may also explain his belief in an air-filled system of arteries. He held, for instance, that the trachea was an artery—which explains its original name, the *arteria aspera*.

Whether or not Hippocrates was in accord with the Egyptian belief that the heart was the seat of consciousness is unclear. His works express seemingly contradictory stances, sometimes identifying the heart, sometimes the brain. One possible explanation for this is the fact that historians have found it difficult to determine which of the many works attributed to Hippocrates were actually written by him and which of them originated with his followers and colleagues.

What we know for sure is that in early ancient Greece, shortly before Hippocrates began his work, natural philosopher and medicinal theorist Alcmaeon of Croton developed a set of truly groundbreaking views on how the body functions. Sometime between 480 and 440 BCE, he hypothesized that the brain was the most important organ. Not only was it the source of intelligence, he said, but it was integral to the functioning of sensory organs like the eyes. This stance may have made Alcmaeon the first craniocentrist—a believer that the workings of the body center around the head and what's in it. For many centuries, though, craniocentrism would remain in the shadows of cardiocentrism.

One influential cardiocentrist was the Greek philosopher Aristotle (384–322 BCE). Although he is known as "the Father of Biology," he certainly did not earn the title for holding accurate knowledge of organs like the heart, brain, and lungs. Far more likely it was because of his pioneering work in the field of taxonomy. Aristotle made detailed observations about hundreds of plants and animals, dissecting many of them and using the characteristics he observed (e.g., blood or no blood) to pioneer a system by which all living things could be classified.

In one such observation, Aristotle watched the action of the heart in a live chick embryo. He noted that it was the first organ to develop, and hypothesized that large animals like humans had a three-chambered heart, with right, left, and center cavities.* According to Aristotle, medium-sized animals had two chambers while small animals had but one.

Aristotle also believed that the heart was the most important organ in the body: the seat of intelligence, emotions, and the soul.† With no knowledge of the nervous system, Aristotle claimed that the heart served as a hub for all incoming sensory information, with signals traveling to it by way of blood vessels from organs like the eyes and ears. As for the brain, Aristotle hypothesized a far less lofty role. He believed that it acted very much like a modern radiator, whose job was to cool the heart.

---

* One explanation for this may be that he did not consider the right atrium to be a separate chamber but merely a widened junction of the vena cava as it connected to the heart.

† The Athenian philosopher Plato (born around 425 BCE) believed that the soul had three distinct parts: The *logos* existed in the head and dealt with reason, while the *thymos* (also spelled *thumos*) resided in the chest and was concerned with anger. The lowest soul, *eros*, was found in the stomach and liver, where it controlled the body's baser emotions and desires.

Half a millennium after Aristotle, Claudius Galenus (129–216 CE), better known as Galen, was born in Pergamon, a city on the Aegean coast. Once part of the ancient Greek world, during Galen's time it was a part of the Roman Empire.* The son of a wealthy architect, he studied to become a physician and philosopher. It would be difficult if not impossible to overstate Galen's impact on the field of medicine, since his teachings— and the teachings of those who followed him blindly—would hold sway for roughly the next fifteen hundred years.

Influenced by Hippocrates, Galen traveled widely as a young man and was exposed to an array of medical practices in places like Alexandria, Egypt—a center for scientific and medical advancement. Like Aristotle, whose principles Galen also followed, he was fully convinced of the existence of the soul and its intimate relationship with the organs, which he would soon be observing firsthand.

While working as a physician at a Roman gladiatorial school in his hometown, Galen became fascinated with internal anatomy. Presented with a parade of gashes, slashes, and traumatic amputations, Galen found that he could curb blood loss by applying astringents, like vinegar, to the wounds. These compounds caused blood vessels to contract, thus reducing the flow of blood escaping from them. Galen also used wine-soaked bandages and spice-laden ointments to facilitate the healing process and curb infection. Though he had no clue what infections were or what caused them, the alcohol in these treatments likely helped curb bacterial growth.

Galen referred to wounds as "windows into the body," but after moving to Rome around 160 CE, he found his view clouded by the city's prohibition on human dissection. It was a taboo that had also existed in ancient Greece, except for a short but illuminating period in the early third century BCE. It was then that two physicians, Herophilus of Chalcedon

---

* Now in Turkey.

and his younger contemporary Erasistratus of Ceos were able to conduct vivisections on condemned criminals. Among the discoveries made by Herophilus were that the heart has valves, whose one-way function was subsequently demonstrated by Erasistratus. The younger scientist also described the heart as a pump, and both men made the anatomical and functional distinction between veins (*phlebes*) and arteries (*artēriai*), though they did not stray from the mistaken belief that arteries were filled with air.*

There can be no argument that a ban on human dissection severely hindered the advancement of anatomy and physiology in ancient Greece and Rome. With few exceptions, the prohibition continued throughout the Western world for roughly eighteen centuries after Herophilus and Erasistratus, until it was finally eliminated in fourteenth-century Italy.

In 1992, Yale University historian Heinrich von Staden addressed the question of why the ancient Greeks considered human dissection to be a taboo, concluding that there were two major factors. The first was a set of formidable cultural traditions related to the contaminating and polluting power attributed to corpses. For example, anyone coming into contact with or even looking at a corpse, including the body of a loved one, had to go through a lengthy process of purification, which ranged from bathing to applying various substances to the body (like blood or clay), to fumigation and confessions. Similar rituals were performed at the deceased's dwelling, hearth, water supply, and place of interment. Anyone, therefore, who performed a human dissection would have been acting far beyond culturally acceptable boundaries, polluted and polluting to the point of criminality.

The second factor contributing to the Greeks' human dissection taboo, von Staden wrote, was a set of negative connotations related to cutting

---

* Their discoveries went far beyond those related to the heart and circulatory system. Herophilus studied the brain, cranial nerves, liver, and womb. He also identified the four layers of the eye, including the initial descriptions of the cornea, choroid (a.k.a. the white portion of the eye), and the retina.

human skin. The Greeks, he said, considered skin to be "a magical symbol of wholeness and oneness." Presumably there were exceptions during times of war, when it was okay to pierce, slice, and dice one's enemies.

Hundreds of years later, and faced with similar Roman taboos, Galen was forced to make inferences about the human circulatory system by carrying out experiments on animals. Working on monkeys like macaques, as well as pigs, sheep, goats, and dogs—often in public—his renown grew. To his credit, Galen, like his predecessors, described the heart as a pump with valves, and he disproved the long-held belief that arteries carried air and not blood. He did so by opening up a dog's artery underwater. Since blood and not air could be seen escaping, it proved conclusively that Egyptian and Greek physicians had been wrong in their contention that arteries were part of the respiratory system.

Galen studied additional organ systems as well. He was able to determine the basic functions of the urinary bladder and kidney, and he tested and distinguished the functions of cranial and spinal nerves, thus providing evidence that the brain, rather than the heart, was the control center for what would become known as sensory pathways and motor pathways—the routes taken by information coming from and headed out to the body, respectively.

Mostly, though, what resulted from Galen's legacy were problems— longstanding and well-documented problems. In hindsight, some of the mess can be chalked up to Galen's inability to obtain human cadavers to dissect. For example, his description of kidneys was based on dogs, and it turns out that the position of canine kidneys, with the right kidney higher than the left, is actually reversed in humans.

Far more serious, however, were the errors rooted in Galen's strong beliefs about the *functioning* of the human body. For starters, Galen believed that venous and arterial blood were separate entities, with different origins. Venous blood, he said, was dark and thick. Produced by the liver from ingested food, it flowed to the right side of the heart, which in turn pumped it out to supply the body with nutrients. Some of the blood, however, passed from the right side of the heart to the left side, through invisible pores in the interventricular wall separating the right and left ventricles. There, Galen claimed, it would mix with pneuma, an air-like spiritual essence obtained from the surrounding atmosphere and delivered to the left side of the heart via the trachea and the lungs. He reasoned that the resulting arterial blood would grow brighter and warmer than the venous stuff, becoming something called "vital spirit," which would be distributed to the body via the arteries. The blood reaching the brain would be imparted with "animal spirit," which would then flow out to the body via nerves, which Galen believed to be hollow. Waste materials, described as "sooty fumes," were eliminated via the trachea during the process of breathing.

Whew!

Although his description of the circulatory system is certainly a laundry list of errors, from an anatomical perspective the most serious mistake Galen made may have been his failure to recognize the true connection between the pulmonary and systemic circuits—in other words, the path by which one could trace the movement of blood from the right side of the heart to the left side *by way of the lungs*. By citing invisible pores as the connection between the sides of the heart, Galen set circulatory anatomy on a centuries-long wrong track.

Unfortunately, Galen also subscribed to Hippocrates's six-hundred-year-old claim that the body contained four substances produced by the liver and spleen, called humors: blood, phlegm, yellow bile (or choler), and black bile.* These corresponded to nature's four elements—air, water, fire, and earth—and each reflected two of the four physical qualities—hot,

---

* Galen was apparently unconcerned about not actually seeing black bile. In fact, no one had. The substance does not exist.

GALEN'S CIRCULATORY SYSTEM

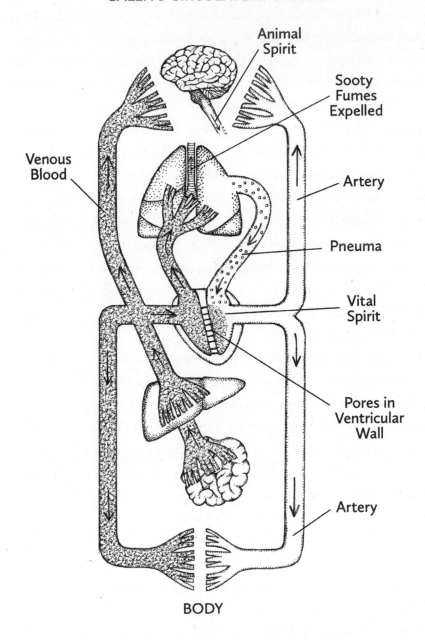

Animal
Spirit

Sooty
Fumes
Expelled

Artery

Venous
Blood

Pneuma

Vital
Spirit

Pores in
Ventricular
Wall

Artery

BODY

cold, wet, and dry. The combinations could be confusing, since they varied from reference to reference. What *was* of prime importance, though, was a need to keep these humors in balance if a person expected to have good physical and mental health, since each humor had a specific effect on the body, analogous to its qualities.

As a result of all this, physicians and so-called barber-surgeons spent centuries prescribing therapeutic purging to counter what they perceived as humoral excesses.* For example, fevers and the flushed cheeks and elevated pulse that often accompany them were thought to result from too much blood. The medical personnel of the day therefore sought to alleviate these conditions by reducing blood volume, and they bled patients early and often. The belief was that calm, cool, and cyanotic (i.e., blue) were preferable to frenetic, feverish, and flushed.

By similar reasoning, Galen believed that the overall composition of the humors resulted in the manifestation of distinct personality traits, depending on the blend. "Sanguine" folks, with blood as their leading humor, were sociable and optimistic, while someone who was "choleric" would be impatient and prone to anger. A "melancholic" person, full of black bile, was prone to sadness, and a "phlegmatic" individual might seem unemotional, calm, and apathetic. One way to gauge the historical significance of these humor-dependent characteristics is the fact that these terms that have reached the present more or less intact, although now they're mostly used as adjectives to describe temporary states of mind rather than to define rigid personality traits.

Although Galen made many errors, the real problem with his work is not the work itself—given the circumstances, many of his mistaken

---

* Barber-surgeons were Middle Age medical practitioners, who, along with cutting hair, performed amputations (since they were already equipped with razors), as well as procedures believed to help keep the four humors in balance, like administering enemas and prescribing emetics (vomit inducers).

theories were understandable. What became a truly devastating development for science was that church leaders in the Middle Ages declared
Galen's word to be divinely inspired and thus infallible, guaranteeing it
a long legacy. Galen's writings were vast in number, with his surviving
works accounting for something like three million words. After the fall
of the Roman Empire, his and other Roman works fell into disfavor, and
so his texts, written in ancient Greek, were not immediately translated
into Latin, which remained the language of scholarship. During the early
Middle Ages, though, they *were* translated into Arabic, principally by
Syrian Christian scholars. Although Galen was not a Christian, he may
have been a monotheist, and the translations that followed, from Arabic
into Latin, likely built upon the Christian leanings of the previous trans-
lators. This quirk of fate made his writings more palatable to the medieval
church, and the consequences were catastrophic.

Due to the church's infatuation with Galen and a handful of other
ancient scientists whose theories were compatible with religious belief,
Galen's mistake-laden views became unchallenged medical doctrine in
Europe and elsewhere for well over a thousand years after his death, in
approximately 216 CE. Until well into the 1500s, and in some instances
beyond that, many physicians seeking truth found it in what they had
read, not what they'd observed. As a result, this church-supported
discouragement of new medical research led to centuries of intellectual
torpor, if not hibernation.

One practice whose unfortunately long life and popularity can
be at least partially attributed to Galen's influence was the previously
mentioned therapeutic bleeding, which lasted until nearly the turn of the
twentieth century. Bloodletting for medicinal purposes began with the
Egyptians, spread to ancient Greece and Rome, and reached its peak in
Europe during the 1800s. Wed to the concept of humorism, physicians
and barber-surgeons wielded specially designed bloodletting instruments

to treat a range of conditions, which included plague, smallpox, and hepatitis. Women were bled to relieve menstruation, and patients were drained before amputations to remove the amount of blood thought to have been circulating in the soon-to-be-former limb. Even drowning victims were bled!

Other patients were deemed to have an insufficiency of blood and were compelled to drink the blood of freshly executed criminals. This practice may have begun in ancient Rome, when people with epilepsy drank the blood of recently slain gladiators. The research of medical historians Ferdinand Peter Moog and Axel Karenberg found that the custom may have stemmed from a general belief on the part of Roman physicians in the supposed curative effects of blood drinking. This claim would have been strengthened by the spontaneous recovery of some epilepsy patients from their seizures, recoveries that had nothing to do with blood drinking but were ascribed to the practice anyway.

Although it seems almost impossible to consider, this sort of thing continued on through the Renaissance and the Industrial Revolution, and well into the nineteenth century. As advances and innovation took hold across Europe and the United States across a range of scientific and nonscientific fields, the same could not be said for many aspects of medicine. And while the use of bloodletting instruments like fleams (picture a pocketknife) and scarificators (boxes containing multiple blades, into which fingers were inserted) was curtailed, they were replaced by a far more ancient bloodletter: the medicinal leech (*Hirudo medicinalis*). Leeches are annelid worms (like earthworms) equipped with sawlike teeth and an array of anticlotting agents in their saliva, and their infamous bloodsucking abilities were used to treat ailments ranging from fever and headaches to mental illness.

The first use of leeches for medical purposes may have begun with Ayurvedic medicine, a whole-body healing system whose roots stretch back three thousand years ago or more in what is now India. The deity Dhanvantari, the Hindu god of Ayurveda, is often depicted (in more recent statues) holding a leech in his hand.

As for how Europeans came to use leeches, one possibility is that the practice may have moved westward along trade routes from the Middle East or Asia. But even if the ancient Egyptians and Greeks swapped medical practices passed down from Indian or Mesopotamian physicians, it is clear that medicinal leech therapy had multiple independent origins, since it was also practiced by the Aztecs and the Mayans. In each case, the idea behind the use of leeches was likely similar to the idea behind humorism: that by balancing the various forms of elemental energy within the body, one could achieve wellness.

Describing what is certainly the most warped use of leeches on record, Pierre de Brantôme, the sixteenth-century French historian, wrote that leeches were inserted into the vaginas of women before their wedding nights. The reasoning for this was that, by doing so the brides could seem like virgins:

> Now the leeches, in sucking, do engender and leave behind little blebs or blisters full of blood. Then when the gallant bridegroom cometh on his marriage night to give assault, he doth burst these same blisters and the blood discharging from them.

According to Brantôme, having hubby maul the mock maidenhead invariably led to an annelid-assisted version of postcoital bliss: "the thing is all bathed in gore, to the great satisfaction of both . . . so the honor of the citadel is saved."

Uh-huh.

Like the similarly common European practice of medicinal cannibalism, the topic of therapeutic bleeding is often swept under the rug, perhaps due to embarrassment. In the United States, for example, few people are aware that in 1799 a team of physicians drained a total of around eighty ounces of blood from former President George Washington in an attempt to treat a throat infection. This was approximately 40 percent of his total blood volume!

The Founding Father was also blistered (a painful technique thought to draw out sickness) and purged at both ends, with enemas and emetics. Reported to have been in horrible pain and weakening rapidly, Washington fell unconscious from what would today be diagnosed as hemorrhagic shock. He died the next day. In looking over the history and the credentials of those doctors who treated President Washington, I realized that these were not the incompetent hacks that one might have initially suspected. They were, in fact, first-rate physicians, which would be expected, given Washington's stature. The problem was the medical community's continued adherence to Galen's deeply flawed directives about balancing the four humors. More than two thousand years after Hippocrates very likely cribbed them from even more ancient Egyptian, Mesopotamian, or Ayurvedic medics, these beliefs remained deeply entrenched within Western medical doctrine.

In the 1800s, the use of leeches received a major boost due to the fact that Napoléon Bonaparte's chief army surgeon, François-Joseph-Victor Broussais, swore by them. Known affectionately as Le Vampire de la Médecine, he reportedly attached thirty leeches to every new patient he saw, no matter what symptoms they presented. Once the malady was established, Broussais was known to treat patients with up to fifty leeches simultaneously, often giving them the appearance of wearing coats of glistening chain mail. Fashion-conscious ladies of the time took notice, and dresses decorated with fake leeches "à la Broussais" became all the rage. Given his popularity, Broussais was responsible for a massive uptick in medicinal leech use in France, which peaked at forty-two million leeches imported in 1833. The demand initiated a decent-sized cottage industry, in which all that was required to generate some coin was an old horse to lead out into a shallow pond, and a basket to collect the leeches once they latched on to their sad equine host.

With the rise of antibiotics, leech therapy faded away in the early twentieth century, only to experience a resurgence beginning in the 1970s. At that time, surgeons were developing microsurgical techniques to deal with the problem of reattaching limbs. Knitting together thick-walled arteries was generally not an issue, so oxygenated blood was able to reach the reattached structures. The problem stemmed from an inability to sew together the thin-walled veins. Instead of returning to the heart, venous blood would pool and clot, and the reattached tissue would inevitably die. Surgeons discovered, however, that if they applied leeches to the area around the reattached structure, the bloodsuckers would set up a sort of auxiliary circulatory system, drawing off the waste- and CO_2-laden venous blood while still allowing arterial blood in to nourish and supply the reattached tissue. Simultaneously, anticoagulants in the leech saliva prevented clot formation. Eventually, the patient's own repair system would produce new veins, and once normal circulation became established, leech therapy, which often employed hundreds of leeches per surgery, could be halted. Things ended less well for the tiny annelid heroes, whose success was usually celebrated with an unceremonious and lethal dunking into a jar of alcohol.*

THANKFULLY, ALTHOUGH THE Western world was held back by centuries of adherence to the teachings of Galen, the rest of the world was able to make its own discoveries.

On the popular game show *Jeopardy*, the question to the answer, "He

---

* Current practitioners of alternative medicine believe that in addition to anticlotting properties, leech saliva contains a range of bioactive substances with therapeutic benefits ranging from anti-inflammatory and anesthetic properties to uses in treating edema and breaking up blood clots.

was the first to correctly track the path of blood to and from the lungs"
would undoubtedly be "Who is William Harvey?" But the truth is that
accurate information about the heart's pulmonary circuit did *not* begin
with the seventeenth-century English physician and had actually been
documented some three hundred years earlier. Given the near fanatical
devotion to the teaching of Galen, some of the explorers who proposed
revised routes for blood's path through the circulatory system did so at
their own peril.

Ibn al-Nafis (c. 1210–1288 CE) was a Syrian-born polymath who studied medicine in Damascus and then ascended to the role of physician in
chief at the Al-Mansouri Hospital in Cairo.* At the age of twenty-nine, he
published his most famous work, *Commentary on Anatomy in Avicenna's
Canon*. Avicenna, the Latinized name for Abu Ali al-Husayn Ibn Sina,
was a Persian scholar in the first century CE, who produced an amazing body of work on a variety of topics. On the medical front, Avicenna
studied Galen's writings and tweaked them for his students, making
corrections based on his own research. He was also deeply influenced
by Aristotle, which explains his belief that the heart, and not the brain,
functioned as the body's control center. Avicenna's most famous work,
Canon of Medicine, is a five-volume medical encyclopedia that integrates
Aristotelian ideas, Persian, Greco-Roman, and Indian medicine, and
Galenic anatomy and physiology. It became the standard medical text in
medieval universities and was translated into Latin, Europe's academic
language of choice, in the twelfth century. Avicenna's *Canon* was still in
use well into the eighteenth century.

In his commentary on Avicenna's work, Ibn al-Nafis addressed a
problem that had been vexing physicians and anatomists for a thousand
years—namely, the invisible perforations in the interventricular wall
that Galen had claimed allowed the movement of blood from the right

---

* His full name was Ala al-Din Abu al-Hassan Ali Ibn Abi-Hazm al-Qarshi al-Dimashqi, but he
was called Ibn al-Nafis.

ventricle to the left ventricle. Ibn al-Nafis, who studied comparative anatomy and may have dissected cadavers, realized that Galen had proposed the unseen pores for one reason: he had not known that large amounts of blood were constantly flowing into the left side of the heart from the lungs. Of the cavity between the left and right ventricles, Ibn al-Nafis wrote:

> There is no passage as that part of the heart is closed and has not apparent openings as [Avicenna] believed and no non-apparent opening fit for the passage of this blood as Galen believed. The pores of the heart are obliterated and its body is thick, and there is no doubt that the blood, when thinned, passes in the *vena arteriosa* [pulmonary artery] to the lung to permeate its substance and mingle with the air . . . and then passes in the *arteria venosa* [pulmonary vein] to reach the left cavity.

With this, Ibn al-Nafis became the first to propose a nonimaginary connection between the right and left sides of the heart. Further research would not confirm his observations until four hundred years later, when Marcello Malpighi used an early microscope to identify the tiny pulmonary capillaries surrounding the lung's microscopic air bags, the alveoli. These capillaries conclusively linked the pulmonary arteries carrying deoxygenated blood to the lungs with the pulmonary veins carrying oxygenated blood back to the heart.*

Although Ibn al-Nafis was almost certainly the first to correctly determine the route of pulmonary circulation, his work unfortunately did not have a lasting influence on Western medicine. It was largely forgotten until 1924, when an Egyptian physician found a copy of his *Commentary* in a Berlin library.

---

* This is the only example in the human body where a vein carries oxygenated blood and an artery carries deoxygenated blood.

Similarly, practitioners of traditional Chinese medicine (TCM) have been using their own determinations about the circulatory system and the heart, which they consider to be "the emperor of all organs," for over two thousand years. In TCM, while the basic functions of the heart are understood in a way that coincides with Western medicine, the heart remains the center of the mind and consciousness. TCM remains largely beyond the scope of this book, though we will see it again.

IF THE CAMBRIDGE-EDUCATED William Harvey (1578–1657) was not quite the cardiac pioneer portrayed in history books, he is certainly the most famous. He may have also been the first Western scientist to proclaim that the human body operates like a machine, with each organ having its own function or functions. Harvey used the scientific method to explain circulation as a natural phenomenon, often bucking political or religious dogma associated with the teaching of the Bible or Galen. Through experiments on the blood vessels of snakes and fish, and the superficial arteries and veins of the human arm, Harvey demonstrated that the circulatory system worked through the laws of physics, and that the movement of blood results from the beating of the heart. This was a controversial revelation in the early seventeenth century, and in some ways helped set the stage for the explosion of medical advances that occurred during the Enlightenment.* Still a man of his time (and a member of the Church of England), though, William Harvey did not dispute the accepted metaphysical role of the heart as the "spiritual member" of the body and the seat of all emotions.

This sort of dichotomy between modern theories and deeply entrenched beliefs helps explain the disconnect that often existed between theory and practice. Additionally, although advances were being made in anatomy and physiology, the same cannot be said for successfully

---

* The intellectual and philosophical movement known as the Enlightenment is usually considered to have lasted from the mid-1600s until the early nineteenth century.

combating disease. The continued use of leeches to "breathe veins" long after humorism fell into disfavor not only serves as an example of the glacial pace at which therapy caught up to theory, it also explains the everything-but-the-kitchen-sink approach to the treatment of conditions whose cures remained out of reach.

Harvey published his classic and incredibly popular work *Exercitatio anatomica de motu cordis et sanguinis in animalibus* (meaning "An Anatomical Study of the Movement of Heart and Blood in Animals") in 1628, but by then he may have been the third or even the fourth person to correct Galen's error. More surprising still is that William Harvey wasn't even the first *European* to accurately map the flow of blood to and from the lungs.

Michael Servetus (c. 1511–1553) was a Spanish physician who came to a similar conclusion as Ibn al-Nafis regarding Galen's invisible pores and the true nature of the link between the pulmonary and systemic circuits. He may, in fact, have cribbed his hypothesis from Ibn al-Nafis, who went uncredited. But whether the ideas were original or not, Servetus wrote the following in his seven-hundred-page book, *The Restoration of Christianity*, published in 1553:

> This communication is made not through the middle wall of the heart, as is commonly believed, but by a very ingenious arrangement the refined blood is urged forward from the right ventricle of the heart over a long course through the lungs; it is treated by the lungs, becomes reddish-yellow and is poured from the pulmonary artery into the pulmonary vein.

Unfortunately for Servetus and his work's legacy, he went far beyond giving a thumbs-down to one of Galen's divinely inspired revelations about the circulatory system. He also filled his magnum opus with blasphemous statements, most scandalously including rejections of both infant baptism and the Holy Trinity. As a result, the Spaniard found himself in the unusual position of infuriating both the powerful Roman Catholic authorities and the newly emergent Protestant ones, all of whom quickly pinned him with the somewhat less than favorable tag of "heretic."

Servetus was arrested on April 4, 1553, but escaped three days later, whereupon he was tried and executed in absentia by the French Inquisition. They burned Servetus and his naughty books in effigy, with blank paper standing in for the actual texts.

Attempting to flee to Italy, he was captured in Geneva, and in a touching show of bipartisanship, the Protestants decided to put him on trial for his life—in person this time. Everyone, it seemed, agreed that Servetus was guilty and deserved to burn. Surprisingly, the preeminent Protestant theologian John Calvin stepped in to plead for mercy, possibly feeling guilty about the fact that Servetus had been arrested while attending a sermon that he was preaching. Regrettably for Servetus, however, Calvin's plea had nothing to do with sparing his life. Instead, Calvin requested that the condemned man be decapitated instead of being burned at the stake. In the end, Calvin was chided in a letter for his undue leniency and Michael Servetus found himself surrounded by flaming copies of his own book—real ones this time. Of his published works, it appears that only three copies have survived, reportedly hidden away for decades to avoid destruction. From a medical perspective, the elimination of *The Restoration of Christianity* from the public eye meant that Servetus's claims about pulmonary circulation were essentially forgotten.

BEGINNING IN THE twelfth century, the Roman Catholic Church had begun to loosen its ban on human dissections as long as they were performed at universities and *not* by clerics. Founded not long after, in 1222, the University of Padua in northern Italy became a preeminent destination for scholars and physicians studying anatomy. By the mid-six-teenth century, it became especially renowned for its anatomy theater and the frequent presence of Belgian anatomist Andreas Vesalius (1514–1564). By this time, the religious, moral, and aesthetic taboos against human dissection that had paralyzed medical research for centuries had lifted, and Vesalius was able to pioneer the field, studying the body in ways that Galen never could. Notably, he produced a large set of exquisitely detailed illustrations of the human body, which he shared with his stu-dents, often in lectures about just how wrong Galen had been. In 1543, Vesalius published the remarkable *De humani corporis fabrica libri septum* ("The Seven Books on the Structure of the Human Body") in which he emphasized that direct observation was key to an understand-ing of human anatomy. In it, Vesalius's skepticism of Galen was often clear, even if it took a revision in his 1555 second edition to change his original quote that blood "soaks plentifully through the [interventricu-lar] septum from the right ventricle into the left" to the following: "I do not see how even the smallest amount of blood could pass from the right ventricle to the left through the septum." He did not, however, put forth his own hypothesis on pulmonary versus systemic circuits. Additionally, Vesalius stressed that there were important differences between animals (upon which Galen based his research) and humans. Though his work contributed to breakthroughs regarding several organ systems, a key con-tribution was his determination that the heart functioned as a pump that circulated blood around the body. If not quite an original assertion, it was incredibly effective coming in the sixteenth century, at a time when the use of mechanical pumps had begun in earnest—generally to move water from place to place.

While Vesalius's research was sanctioned by the University of Padua, thanks to the lifting of the dissection taboo, his often-contradictory stances on centuries of established medical dogma still managed to make him an enemy of the Roman Catholic Church. So, too, did his Bible-opposing (and accurate) observation that men and women shared the same number of ribs. Vesalius died under mysterious circumstances while returning from a trip to Jerusalem. Some have wondered what might have compelled Vesalius to visit the Holy Land, with rumors that he had fled Spain after accidentally performing an autopsy on a living nobleman. Citing a lack of evidence, this story was dismissed by Vesalius biographer Charles O'Malley. He suggested instead that Vesalius may have been using the trip as an excuse to get away from the Spanish court, with the hope of eventually reclaiming his old anatomy chair at the University of Padua. Unfortunately, whatever the reason for his pilgrimage, he died on the island of Zakynthos (in what is today Greece), though no one is quite certain about how that happened. Modern biographers list poor conditions aboard ship, a shipwreck, or a contagious disease as possibilities.

A pupil of Vesalius's named Matteo Realdo Colombo (c. 1516–1559) later became anatomy chair at the University of Padua. In a chapter on the heart and arteries in his 1559 book *De re anatomica* ("On Things Anatomical"), he wrote what stands as a remarkably accurate pre-Harvey description of the pulmonary circulation:

> Between these ventricles there is a septum through which most everyone believes there opens a pathway for the blood from the right ventricle to the left ... But they are in great error, for the blood is carried through the pulmonary artery to the lung and is there attenuated; then it is carried, along with air through the pulmonary vein to the left ventricle of the heart. Hitherto no one has noticed or left in writing, and it especially should be observed by all.

Ultimately, the Persian polymath, the Spanish physician, and the Belgian and Italian anatomists would not be remembered for their

COLOMBO'S CIRCULATORY SYSTEM

HARVEY'S CIRCULATORY SYSTEM

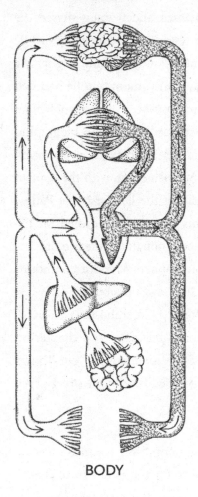

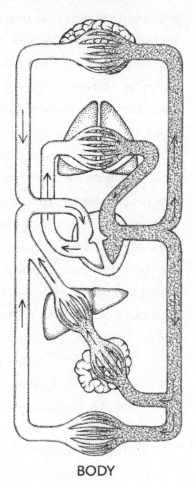

BODY

BODY

significant accomplishments related to the cardiovascular system. Since each of them described, with varying degrees of detail, the correct route of blood from the right side of the heart to the lungs, then back to the left side, I am left feeling that they have been treated unfairly, especially since the works of Ibn al-Nafis, Servetus, Vesalius, and Colombo predated William Harvey's 1628 publication by 389,* 75, 73, and 69 years, respectively.

---

* An approximate date.

Still, though, there is no denying that Harvey's work, if it did not lay the foundations of modern cardiology, certainly helped bring physicians onto the ground floor. His reliance on scientific observations and methodology were blueprints for those who followed him, now equipped with what was essentially a modern perception of how the heart and circulatory system operated and how to study them.

Of course, there were additional floors to build, as researchers were still investigating pulse and blood pressure, and improvising instruments to investigate heart sounds. Still others were studying the exchange of gases between the circulatory and respiratory systems, and a fast-growing list of circulatory system–related defects and diseases. But even before the nature of blood and its path through the body were identified, some seventeenth-century physicians began toying with the idea that instead of simply draining the red stuff in the event of illness, it might make more sense to *add* it.

Practically the whole of the blood was replaced
by beer before life was replaced by death.
—RICHARD LOWER, *TRACTATUS DE CORDE*

One person in 1666 made the recommendation that if a man and wife
did not get along well, each should have a transfusion from the other,
and by thus mixing their blood, they would be made compatible.
— CYRUS C. STURGIS,
"THE HISTORY OF BLOOD TRANSFUSION"

[9]

What Goes In . . .

IN 1614, THE German physician and chemist Andreas Libavius (c. 1540–1616) appears to have been the first person to suggest that a blood transfusion, rather than a bloodletting, might be a way to restore health. Libavius described how this could be accomplished by attaching tubes to blood vessels, but he also emphasized that the inherent difficulties of such a procedure would render any such attempt foolhardy. As it turned out, he wasn't wrong.

To modern readers, the early attempts at transfusions and intravenous (IV) injections will, at best, come off as weird. At worst, they may seem quite horrible. This was, of course, a time when the nature of the circulatory system and the blood traveling through it were unknown, and when much of what *was* known was wrong.

There have been ugly rumors, written and otherwise, that the first blood transfusion took place in 1492, with the recipient being Pope Innocent

VIII. The pontiff is mostly known today for his condemnation of witches and magicians, and with it his 1483 appointment of the infamous Tomás de Torquemada as grand inquisitor of Spain.* Several dubious accounts written in the nineteenth century† contend that in 1492, the pope was on his deathbed, having been in and out of a stupor for years. (Given the man's extreme cruelty, some may consider this to have been a solid career move.) According to these reports, with every effort to revive the pope exhausted, a Jewish physician volunteered to save the religious leader using a new technique. Italian author Pasquale Villari told it like this:

> All the blood of the prostrate old man should pass into the veins of a youth who had to yield up his to the Pope. The difficult experiment was repeated three times, the result being that three boys lost their lives, without the Pope receiving any benefit of it, probably because air had penetrated into their veins.

In 1954, Dutch medical historian Gerrit Lindeboom undertook comprehensive research and turned up no evidence that this blood transfusion had taken place. Of the story's originator, Lindeboom said, "It turns out his vivid imagination framed unhistoric hypotheses." The tale also smells a lot like "blood libel," which would make it one of many centuries-old false allegations asserting that Jews used Christian blood, usually from children, for a laundry list of nefarious purposes.

Given the fifteenth-century belief that drinking human blood had a curative effect, it is far more likely that, if anything, the dying pope was instructed to quaff the blood of these children in a potion—although there is always the possibility that this, too, could be a blood-libelous distortion.

---

* His goal as leader of the Spanish Inquisition was to rid Spain of heretics—particularly Jews and Muslims who had converted to Catholicism in name only—through expulsion, torture, or executions.

† Reviewed by A. Matthew Gottlieb in 1991.

SUCCESSFUL AND *SAFE* blood transfusions would remain out of reach until the turn of the twentieth century. It would take many miserable attempts, though, to stop physicians from trying to introduce a variety of substances, some of them actually blood, into the veins of their patients.

Christopher Wren (1632–1723) was an English mathematician, scholar, and architect, famous for his design of Saint Paul's Cathedral in London. He was also interested in experiments of an anatomical and physiological nature. In a 1656 letter, he wrote:

> The most considerable experiment I have made of late, is this; I injected wine and ale into the mass of blood in a living dog, by a vein, in good quantities, till I made him extremely drunk, but soon after he pissed it out . . . It will be too long to tell you the effect of opium, scammony, and other things which I have tried this way. I am in pursuit of further experiments, which I take to be of great concernment, and what will give great light to the theory and practice of physick.

The reasoning behind using wine for injection comes once again from the ubiquitous Roman physician Claudius Galenus, who believed that it contributed to the formation of blood by the liver. The practice showed up in Christopher Marlowe's play *Tamburlaine the Great*, written around 1587:

> Filling their empty veins with airy wine
> That being concocted turns to crimson Blood.*

The transfusion of alcohol into patients continued into the 1660s, but by then some in the medical community began exploring how they might

---

* According to scholar, poet, and English professor J. S. Cunningham, "concocted" = digested.

transfuse actual blood into humans. In keeping with the long string of feuds between England and France, physicians from the two countries began ignoring each other's transfusion-related work in order to claim priority. Sorting through all of this, two near certainties emerge: In 1665, a British physician and surgeon named Richard Lower (1632–1691) made the first direct blood transfusions, using pairs of dogs and transfusing from the carotid artery of one dog into the jugular vein of the other; in each pair, the recipient dog had been previously bled, nearly to death, and was then rejuvenated by blood transfused from the second dog. And in 1667, the French physician Jean-Baptiste Denys (also spelled Denis) was the first to transfuse blood into human, though the donors of that blood were not human.*

Impressed by Lower's work two years earlier, Denys (c. 1635–1704) built a system constructed out of metal tubing and goose quills and began transfusing the blood of sheep and calves into his patients. One of his first was Antoine Mauroy, who received the bovine option. Mauroy was said to be "a manic-depressive type suffering from a psychosis."† The rationale for the seemingly strange choice of donors appears to be a belief that the "mildness" of the calf blood would cure whatever was causing Mauroy to beat his wife, run around naked, and set houses on fire.

The procedure began with Mauroy being tied to a chair and bled, presumably to rid his body of bad blood while making room for the good stuff. Following this, he received about six ounces of calf blood through a metal tube, which Denys inserted into a vein in his arm. Mauroy complained about some burning in his arm but otherwise exhibited no serious side effects. After nodding off for a bit, the patient awoke and

---

* Cyrus Sturgis, a doctor who presented a paper on the history of blood transfusion at the annual meeting of the Medical Library Association in 1941, claims that donor was a calf, though a sheep is mentioned in other references.

† Denys's first patient appears to have been an unnamed fifteen-year-old, who received a transfusion of sheep blood earlier in 1667.

appeared calm—which many of the onlookers who had watched the procedure much preferred to Mauroy's usual behavior.

Unfortunately, a second transfusion, which was done the following day at the suggestion of Mauroy's wife, was somewhat less successful. This time, upon receiving the transfusion, the patient began sweating profusely and, between bouts of throwing up his recent lunch (reportedly bits of bacon and fat), he complained of severe pain in his lower back, saying also that his arm and armpit were burning. Soon after, he began exhibiting chills, fever, erratic pulse, and severe nose bleeds. Appearing extremely fatigued, he soon fell asleep, only to wake up the next morning looking rather calm (for him) and sleepy. Expressing a desire to urinate, Mauroy reportedly produced "a great glass full of urine, of a colour as black as if it has been mixed with the soot of chimneys."

With the benefit of twenty-first-century hindsight, it is clear that Antoine Mauroy was suffering from his body's multipronged response to incompatible blood. The back pain and black urine were a result of his kidneys dealing with the shock of filtering massive numbers of transfused red blood cells that had been literally torn apart by the immune system in a process known as hemolysis.

In accordance with the medical wisdom of the seventeenth century, Mauroy was then bled, the Galenic equivalent of "Take two aspirin and call me in the morning." Eventually, though, through what was certainly more luck than medical treatment, he began to recover. Of course, Denys took this to mean that his transfusion therapy was a success, and he immediately began treating additional patients.

Meanwhile in England, Richard Lower organized a demonstration of his own for the Royal Society of London. He hired a man named Arthur Coga, who was described by Parliament member and diarist Samuel Pepys as "cracked a little in the head," and paid him twenty shillings "to have some blood of a sheep let into his body." Lower made incisions in the carotid artery of the sheep and an unspecified vein in Coga's arm, then inserted a silver pipe into each vessel, with a length of quills connecting them. According to Lower, nine or ten ounces of sheep blood entered the subject, and soon after, Coga "found himself very well, and hath given in his own Narrative under his own hand, enlarging more on the benefits, he thinks he hath received by it."

Only a few months later, the enthusiasm for blood transfusion was shattered on both sides of the English Channel when the French patient Antoine Mauroy died. According to his wife, the man had resumed his psychotic behavior and thus required another transfusion—but it was later discovered that his conduct had led her to employ a treatment of her own design. This came in the form of a dietary supplement of arsenic, which she added to her husband's meals. Strangely, Madame Mauroy failed to mention this to Denys when the couple approached him requesting a third transfusion. The physician refused to treat the man, noting his patient's less than healthy appearance, but that did not stop Mauroy's wife from suing him and having him arrested for manslaughter when Mauroy dropped dead several days later. Though Denys was found not guilty, the uproar surrounding the case, along with reports of additional patient deaths, all but slammed the door on the practice of human blood transfusions.

In 1668, France banned the procedure, with a proclamation known as the Edict of Châtelet, and England soon followed suit. A pair of transfusion-related deaths in Italy led to a denouncement of the practice by magistrates in Rome. And so all went quiet on the transfusion front, for nearly 150 years.

In 1818, horrified by women bleeding to death after childbirth, English obstetrician James Blundell (1790–1878) began the first successful human-to-human blood transfusions. He did so by using a syringe filled with approximately four ounces of blood drawn from the husband of the patient, and then injecting the blood into a superficial vein in the woman's arm. Reportedly, half of the transfusions he performed had positive results. Unfortunately, given the problems Blundell faced, which included nonsterilized instruments and no knowledge of blood typing, his results often failed, and the well-intentioned practice was soon abandoned.

Although blood transfusions were still frowned upon for most of the nineteenth century, mostly because they frequently had poor outcomes, some rather surprising substances were regularly injected into the blood vessels of both animal and human subjects. Intravenous injection of milk began during a Canadian cholera epidemic in 1854. The physicians who thought this one up were under the mistaken belief that white blood cells were actually red blood cells in the process of transformation. Citing an earlier study, they said they were confident that the "white corpuscles" of milk, which were actually minute globules of oil and fat, would eventually be converted into red blood cells.

In reality, most red blood cells are produced from stem cells in the red bone marrow found in long bones like the femur and the humerus. Approximately two million red blood cells are produced every second, while a similar number of cells at the end of their approximately 120-day life span are recycled by the spleen.

Milk transfusions were performed as late as the 1880s by British surgeon Austin Meldon. According to a short paper he published in the *British Medical Journal* in 1881, Meldon injected milk into twenty patients for illnesses that included tuberculosis, cholera, typhoid fever, and pernicious anemia. He explained that the "very unpleasant symptoms" and even deaths that sometimes followed the procedure could be explained by the milk having gone sour. To remedy this, Meldon recommended that physicians use goat milk, explaining that "it is much more easy to bring that animal in close proximity to the patient, thus avoiding any necessary delay between milking and the injection."

Admittedly, this sounds ridiculous today, but it's easy to see why people might have been willing to accept milk transfusions as a panacea at a time when, for example, a famous and still-in-business pharmaceutical company recommended using heroin to treat children with colds, and cocaine appeared in Sears, Roebuck and Company catalogues. Without clear evidence for what was and wasn't valid medicine, almost anything could be pitched as the next successful cure-all. As for goat milk transfusions, according to Meldon, by following a few common-sense practices, his fellow physicians could prevent "that depression which so frequently follows the operation." One can imagine that these practices included straining out the goat hair before an injection and not letting the donor eat the hospital bedding.

"I look upon it as a much better and safer operation than transfusion of blood," Meldon wrote.

The practice of injecting milk into patients was abandoned around the turn of the century, when so-called normal saline was finally adopted for intravenous use. This is the solution most commonly used in modern IVs, and it consists of nine grams of sodium chloride (NaCl) dissolved in sterile water, creating a 0.9 percent saline solution, which approximates several key properties of blood plasma. Its first usage was during an 1832 cholera pandemic, when British physician Thomas Latta followed through on a hypothesis recently set forth in the *Lancet,* the top medical journal of its day. The author of the article, a newly minted Irish physician named William Brooke O'Shaughnessy reasoned that because cholera victims were dying from dehydration (losing major amounts of body fluid and salt through diarrhea), it would make sense to replenish the body's lost liquid with a solution approximating the salinity of blood. Latta's rehydration therapy was remarkably successful, but it did not gain enough momentum to supplant the standard treatments of the time, namely bleeding, leeches, emetics, and enemas—all of which resulted in *increased* loss of body fluids.

By the early 1880s, a better understanding of human blood chemistry led British physiologist Sydney Ringer to improve upon the earlier recipe for saline, adding potassium to the sodium chloride solution. Lactated Ringer's solution bears the inventor's name, and it is still widely used today.

In 1901, Karl Landsteiner (1868–1943), an Austrian pathologist, revolutionized the ground rules for blood transfusion after discovering the ABO blood group.* In brief, erythrocytes (like other cells) have specific surface proteins called antigens embedded in their cell membranes. These antigens come in two different varieties: A and B. If the surface proteins on a blood donor's red blood cells don't match those of the recipient, the recipient's immune system will attack the donor blood. The result is the previously mentioned hemolysis—literally "blood cell cutting." In addition to putting stresses on the kidney-driven urinary system, incompatible transfusions can lead to a dangerous form of erythrocyte clumping, called agglutination, which can clog small blood vessels and lead to serious medical problems, like strokes and loss of organ function. This, in turn, accounts for the kidney pain experienced by recipients of incompatible transfusions, including the *extreme* repercussions experienced by the seventeenth-century recipients of blood donated by the barnyard set.

Today, problems related to blood clotting and the storage of donor blood have been solved, and we know about the Rh blood group system and the Rhesus (Rh) factor, named for the Rhesus monkeys in which they were first discovered. Most people have the Rh antigen on their red blood cells (making them Rh-positive), while others don't (making them Rh-negative). Problems used to arise if an Rh– mother gave birth to multiple Rh+ children. Having gradually built up Rh+ antibodies during

---

* Adriano Sturli and Alfred von Decastello, who worked in Landsteiner's lab, discovered a fourth blood type, AB blood, a year later. In 1930, Landsteiner won the Nobel Prize in Physiology or Medicine.

her first pregnancy, the mother's immune system would be fully prepared to "attack" the blood of the second Rh+ fetus. Thankfully, modern prenatal screening and treatment prevent such occurrences today.

Additionally, blood is now crossmatched and screened for pathogens and toxic substances before being transfused, thus ensuring that transfusions are as compatible and safe as possible during a myriad of treatments related to surgery, injury, blood disorders, and disease.

Things have come a long way since the grim days of barnyard blood transfusions and the concept of the four humors. But just as centuries of physicians puzzled over the path, function, and replacement of blood, they also struggled to understand and treat the maladies of the heart. The half-century-long decline, and eventual death, of Charles Darwin will serve as a backdrop for this portion of our journey.

You know that running bug people talk about?
Well, I've been well and truly bitten.
—Sadiq Khan

Love bug, leave my heart alone.
—Richard Morris and Sylvia Moy

[10]

The Barber's Bite and the Strangled Heart

Though he was not its originator, Charles Darwin will forever be associated with the phrase "survival of the fittest."* But in 1836, as the twenty-seven-year-old naturalist stepped off the HMS *Beagle* at the end of his five-year voyage, he was not a fit man. For the remaining forty-six years of his life, Darwin would suffer from a laundry list of medical woes, which included heart palpitations, chest pain, dizziness, fatigue, eczema, and muscle weakness. Additionally, he experienced poor vision, tinnitus, sleeplessness, nausea, vomiting, boils, and chronic flatulence.

In 1842, Darwin moved his rapidly growing family from what he referred to as "smoky dirty London" to a quiet country home some

---

* The term "survival of the fittest" was actually coined by philosopher/biologist Herbert Spencer, who used it in his own book, *The Principles of Biology*, published in 1864, after reading *On the Origin of Species*.

fourteen miles southeast of the city.* Not only did Down House provide him with more space (he and his first cousin/wife, Emma, would have a total of ten children), it also signaled the beginning of his near-complete withdrawal from the public eye. Darwin wrote about this period in his autobiography:

> We went a little into society, and received a few friends here; but my health almost invariably suffered from the excitement, violent shivering and vomiting attacks being thus brought on. I have therefore been compelled for many years to give up all dinner-parties; and this has been somewhat of a deprivation to me, as such parties always put me into high spirits. From the same cause I have been able to invite here very few scientific acquaintances.

By avoiding the "excitement" brought on by stress, and even such pleasurable events as a performance of George Frideric Handel's *Messiah*, Darwin was only intermittently successful at lessening the severity and frequency of his symptoms, which were followed by periods of fatigue he unfortunately referred to as being "knocked up." But while the cause of his decades-long illness remains open to debate, many historians have suggested that Charles Darwin was a hypochondriac—someone who not only fears illness but is mistakenly convinced that he or she is unwell. Obsessively preoccupied with his own health, he appears to have tried every available therapy, including some that would now be considered full-on quackery. These treatments included electrical stimulation of the abdomen with shock belts (a.k.a. galvanization) and "Dr. Gully's Water Cure," a form of hydrotherapy in which the patient was heated with a

---

* Darwin's doctors urged him to move to the country to escape London's air, "a destructive malady . . . justly termed *Cachexia Londinensis*, which preys upon the vitals, and stamps its hues upon the countenance of almost every permanent resident in this great city."

lamp until dripping with sweat before undergoing a vigorous rubdown with cold, wet towels.

This particular treatment had been developed by University of Edinburgh medical graduate James Manby Gully. It was based on the popular belief of the day that diseases were caused by faulty blood supply to organs like the stomach and heart. In brief, Gully contended that by applying cold water to the body, the circulatory system would direct diseases *away* from those important organs and *toward* less vital regions, like the skin, where they could be eliminated. The water cure became a favorite of Darwin's, and may in fact have been helpful—although only because the weirdness was coupled with a light exercise regime and a sensible diet. Darwin once wrote of the treatment, "At no time must I take any sugar, butter, spices, tea, bacon, or anything good."

Gully was also an unyielding opponent of the use of drugs to combat disease, resorting instead to medical clairvoyant readings and homeopathy. The latter is a form of alternative medicine created in the 1790s by German physician Samuel Hahnemann and operating on the doctrine of "like cures like" (*similia similibus curantur*). Basically, the idea is that natural substances that can produce symptoms of a particular disease in healthy people can, in tiny amounts, cure people who have that disease. To which I say, go figure.

Darwin was also administered, and in some cases *self*-administered, an array of compounds, which included ammonia, arsenic, bitter ale, bismuth (the active ingredient in Pepto Bismol), calomel (a mercury-containing laxative and horticultural fungicide), codeine (an opioid pain reliever), "Condy's Ozonized Water" (an oxidizing agent used to purify water), hydrocyanic acid (the highly poisonous prussic acid), "oxyde of iron" (a.k.a. rust), laudanum (tincture of opium), mineral acids (minerals unknown), alkaline antacids, and morphine. Shockingly, none of the

treatments did very much to improve Darwin's condition, and there have
even been whispers that some of this crap may have actually worsened
his health.

Ultimately, though, despite decades of chronic anxiety (particularly
a fear of heart disease and impending death), physical ailments, and
personal tragedies, including the deaths of three of his ten children,
Charles Darwin wrote nineteen books, including his landmark studies on
the mechanism of biological evolution. Given the game-changing nature
of *On the Origin of Species*, arguably the most influential academic book
ever written, some have questioned why Darwin spent the last decade
of his life writing about the sex lives of plants, fertilization in orchids,
the movement and habits of climbing plants, and the formation of
vegetable mold through the action of worms. But considering Darwin's
near-obsessive desire to avoid stress that might sicken him further, it
makes sense that these studies, though innovative and scientifically
important, would have been part of a conscious effort to avoid hot-button
topics. It was his immersion in scientific work (and not cold water)
that became Darwin's primary defense against events of a distressing
nature.*

In December 1881, Charles Darwin was seized by a severe attack of
precordial pain, defined as pain originating from nerves in the region of
the chest immediately in front of the heart. His doctors described his heart
condition as "precarious" with "symptoms of myocardial degeneration."

Darwin's letters reflected the realization that he was suffering from
serious heart disease. He wrote to his longtime friend botanist Joseph
Dalton Hooker: "Idleness is downright misery to me . . . I cannot forget

---

* After *Origin*'s publication in 1859, Darwin left the defense of the book to others, most notably
biologist Thomas H. Huxley. Known as "Darwin's bulldog," Huxley vowed to fight "with claws &
beak" for what Darwin referred to as his "damnable heresies."

my discomfort for an hour . . . So I must look forward to Down graveyard as the sweetest place on earth."

Over the next four months, Darwin experienced several such incidents of chest pain, accompanied by nausea and faintness. He was diagnosed with "angina pectoris"—from a Latin term to describe suffocating or strangling pain in the chest. We now know that angina is itself commonly a symptom of coronary artery disease, which occurs when the coronary arteries supplying the heart itself are obstructed by atherosclerotic plaques. Angina can also result when coronary arteries undergo vascular spasm, a sudden and brief constriction resulting from drug or tobacco use, exposure to cold, or even emotional stress.

But whatever the cause, the reduced blood flow leads to oxygen and nutrient starvation in the heart muscle downstream from the vessel blockage—a condition known as ischemia. This, in turn, stimulates pain receptors in the heart, serving up a warning of worse things to come if the plaques or a clot manage to completely block the blood flow to the vessel. The result, then, is a myocardial infarction, better known as a heart attack.

Angina itself can sometimes mimic a heart attack, with the pain perceived as originating from the jaw, neck, back, shoulder, or left arm. Although there are several proposed mechanisms for this "referred pain," most researchers now believe that the neural pathways carrying information from the heart's pain receptors to the brain run quite close to, and may even merge with, similar pathways coming from the other regions, like the jaw or neck. This fools the brain and its owner into thinking that the pain is coming from a location unrelated to the heart.

Incidents of angina usually occur during extremes of exercise or emotion, when the heart rate quickens but the cardiac muscle's suddenly increased requirements for oxygen and nutrients cannot be met. In its stable form, angina subsides several minutes after the physical activity is curtailed or the stress is removed.

Though cardiology was by no means an established field in Darwin's time, by the early 1880s doctors believed that the causes of angina pectoris were rooted in: (1) the heart's status as an organic structure subject to disease and decay, and (2) its connection to emotion and psychological elements—as they were perceived (correctly or incorrectly) at the time. As such, Darwin's doctors covered their bases, prescribing rest and stress-free living.

As for how Darwin actually dealt with his deteriorating pump, among the contents of what we can only imagine must have been a large medicine cabinet were the narcotic morphine, for the pain, and the antispasmodic amyl nitrite.

If amyl nitrite (not to be confused with amyl nitrate, an additive in diesel fuel) sounds familiar, one reason may be that it remains in use today as a treatment for heart disease and angina. The other reason is that it is a commonly used recreational drug. Amyl nitrite is usually inhaled, and it works by vasodilation: that is, expanding the diameter of blood vessels. In doing so, it increases blood flow and reverses vascular spasm. Interestingly, the vasodilatory effect is also accompanied by a brief euphoric state, and so the contents of amyl nitrite capsules (called "poppers") are sometimes snorted recreationally, their psychotropic effects prolonged when combined with drugs like cocaine. Another effect of amyl nitrite is relaxation of involuntary smooth muscle in the anal sphincters—an officially recognized buzzkill for some folks.

Amyl nitrite's status as the primary medication for angina began to change in 1879, when English physician William Murrell published a paper about another compound—one that was already famous, and sometimes infamous, for its spectacular *non*medical effects. Murrell claimed that a drop or two of nitroglycerin in a 1 percent solution was far more efficient than amyl nitrite in the treatment of angina pectoris. Also known as glyceryl trinitrate, nitroglycerin was thought to work much the same way as amyl nitrite, i.e., by dilating coronary blood arteries and increasing flow to the oxygen-starved heart. In fact, its primary mechanism of action is to reduce the volume of blood filling the heart—though this was not determined until well after its initial therapeutic use. With less blood to pump, the heart does not have to work as hard and so requires less oxygen. Though they may not have understood this particular mechanism of action, Murrell and his peers were right about the vessel-widening properties of nitroglycerin, since the body converts it into a potent vasodilator, nitric oxide.

The Italian chemist Ascanio Sobrero first synthesized nitroglycerin around 1846 but it was Sweden's Alfred Nobel who became famous for his work with the compound. Nitroglycerin was also widely used as an explosive, and Nobel began searching for a way to make the compound safer to handle after his younger brother was killed in an explosion at the family's armaments factory. Nobel added stabilizers and absorbents, ultimately producing a substance he called dynamite.

Scientists renamed the medicinal version trinitrin to avoid scaring bomb-phobic pharmacists and their customers. Before his death in 1896, Alfred Nobel, whose considerable fortune was closely tied to his patent for dynamite, called it an "irony of fate" that he had been prescribed trinitrin to treat his own heart condition.

Like amyl nitrite, nitroglycerin is still in use today. Currently administered through transdermal patches and IV infusion, it is most commonly taken sublingually in tablet form. At the first sign of angina, a pill is placed either under the tongue or between the cheek and gum. With an abundance of capillaries located close to a perpetually wet surface, both regions allow for rapid drug diffusion into the circulatory system. This is especially important for compounds that, if taken orally, might be broken down and rendered less effective or even ineffective during passage through the digestive tract, or turned into inactive metabolites by the liver. Another example of a drug usually given sublingually is the antihypertensive nifedipine. Sublingual administration is also often used for patients in hospice care who may be unable to swallow painkillers like morphine, and for those suffering from gastric ulcers or nausea.

On Tuesday, April 18, 1882, Charles Darwin stayed up later than usual, chatting with his thirty-four-year-old daughter, Elizabeth. Just before midnight, he experienced a bout of agonizing pain, alleviated only somewhat when his wife and daughter administered amyl nitrite with some brandy. He spent much of the following day nauseous and in excruciating pain before eventually losing consciousness at 3:25 p.m. Darwin's doctors determined that his final symptoms were "Angina Pectoris Syncope"—unstable angina with a loss of consciousness.* At just before 4 p.m. on April 19, 1882, Charles Darwin died at the age of seventy-three from heart failure.

Given Darwin's fame, dozens of researchers over the course of the past century and a half have attempted to determine the condition or conditions that may have contributed to the great man's ultimate demise. The list of ailments they drew up include agoraphobia (an anxiety disorder); a bacterial infection called brucellosis; chronic arsenic poisoning (arsenicosis); chronic anxiety syndrome; "chronic neurasthenia of a severe grade"†; Crohn's disease (an inflammatory bowel disorder); cyclic vomiting syndrome; depression; extreme hypochondria; gastric ulcers; gout; lactose intolerance; an inner ear

---

* "Unstable angina" is defined as angina that occurs at rest, and becomes more severe and more frequent.

† Neurasthenia was an ill-defined Victorian fad disease of unknown etiology, characterized by physical and mental exhaustion.

disorder called Ménière's disease; panic disorder; mitochondrial encephalomyopathy; lactic acidosis and stroke-like episodes; a maternally inherited neuromuscular disorder; a psychosomatic skin disorder; and repressed homosexuality.

In 1959, during the centennial anniversary of the publication of Darwin's most famous work, Saul Adler, an Israeli scientist who specialized in tropical medicine, concluded that Darwin's health woes were almost certainly *not* psychological in origin. Indeed, he believed they had begun decades earlier and thousands of miles from Down House—on the very journey that made him famous.

IN 1908, THE Brazilian physician Carlos Chagas was invited by officials from the Central Railroad of Brazil to visit a village called Lassance. The rough-and-tumble town, known for its droughts and poor soil, is located along the banks of the São Francisco River in the state of Minas Gerais. It was the terminus of a new railway line, and so it was packed with railroad workers, most living in rugged, unsanitary conditions. Chagas had been summoned because many of the workers were falling ill and dying from what was believed to be malaria.

Though much has been written about malaria, the term "tropical scourge" cannot convey the horrible devastation brought on by this most lethal of all mosquito-transmitted diseases. During the French effort to build the Panama Canal several years earlier, the malaria/yellow-fever tag team killed an estimated twenty-two thousand workers.[*]

Malaria is spread when an uninfected female *Anopheles* mosquito bites an infected person and acquires the protozoan parasite *Plasmodium* from their blood. The mosquito then transmits the parasite to the next person

---

[*] According to the World Health Organization (WHO), malaria killed 405,000 people in 2018, a devastating number, and 67 percent were children under five years of age. Around 93 percent of all malaria-related deaths in 2018 occurred in Africa.

it bites. Once inside the human body, the parasite enters the circulatory system and travels to the liver, where it replicates and can lie dormant for up to a year. Once the protozoan exits the liver and infects red blood cells, it undergoes additional reproduction, with the blood stage of the parasite responsible for symptoms that include high fever, shivering, chills, headache, nausea, vomiting, and body aches.

Chagas, who specialized in the disease, set up a simple lab, but he soon realized that he wasn't dealing with malaria at all. Instead, the malady he was observing seemed far more reminiscent of African sleeping sickness, a deadly disease transmitted by the tsetse fly. The stricken Brazilian workers Chagas saw exhibited an array of acute symptoms, which ranged from fever, headache, pallor, and difficulty breathing to abdominal and muscle pain. And many arrived at his makeshift lab with swollen purple eyelids. Although most patients seemed to recover quickly from the illness, approximately 30 percent of them went on to develop a far more serious set of chronic symptoms. These included severe digestive tract issues, like an enlarged esophagus and colon, neurological problems, like stroke, and cardiac-related issues that included irregular heartbeat, cardiomyopathy (a disease of the heart muscle), and congestive heart failure, a generalized term for a chronic condition in which the myocardium doesn't pump enough blood to meet the requirements of the body.

Chagas determined that the culprit was *Triatoma infestans*, a bloodsucking insect known in Portuguese as *o barbeiro*—the barber—presumably for its habit of nicking victims on the face. Like mosquitoes, *Triatoma* and its relatives don't have chewing mouthparts, and so technically they do not bite. Instead, they employ a pair of hypodermic needle–like stylets to pierce the skin and an underlying blood vessel. Saliva containing an anticlotting agent (a.k.a. an anticoagulant) is then injected into the blood. Finally, the insect snorks up a blood meal as if through a straw.

"Knowing the domiciliary habits of the insect," Chagas wrote, "and its abundance in all the human habitations of the region, we immediately stayed on, interested in finding out the exact biology of *o barbeiro*, and the transmission of some parasite to man."

The physician soon discovered his parasite: a protozoan that spent part of its life cycle in the hindgut of the insect. Initially, Chagas believed that *o barbeiro*'s "bite" transmitted the pathogen to humans, but the route of infection turned out to be something a bit more unsavory. To clear additional space in its digestive tract and to eliminate excess fluid weight filtered from its meal, *Triatoma* defecates while gorging on blood.

The common vampire bat (*Desmodus rotundus*), another obligate blood feeder and messy eater, exhibits a similar weight-reducing strategy. As these bats feed, their high-powered kidneys allow them to quickly eliminate the excess liquid from their blood meals by urinating, even while their victims are being drained. Because vampire bats need to consume a truly enormous amount of blood each night—up to 50 percent of their body weight—carrying any excess weight would make it difficult for the bats to make their unique flight-initiating jumps.

Back at the insect poop pile, the parasite in *Triatoma*'s feces enters its human host when it is rubbed into the puncture wound or a nearby mucous membrane, typically one associated with the eyes and mouth. The characteristic swelling of the victim's eyelids observed by Chagas occurs after *o barbeiro*'s fecal material is accidentally rubbed into an eye. Alternately, animals and people can be infected orally, through consumption of food or drink contaminated by infected *Triatoma* feces. The infection can also be passed from mothers to their babies at birth.

Chagas named the protozoan *Trypanosoma cruzi*, in honor of his mentor Oswaldo Cruz, who made important contributions to discovery of trypanosome-transmitted diseases. Subsequent studies have shown that *T. cruzi* enters the blood through capillaries in the previously mentioned mucous membranes, eventually invading the lining (endothelium) of blood vessels supplying the heart. From there, the invader gains access to cardiac muscle cells.

In approximately 20 percent of those infected by *T. cruzi*, this microbial blitzkrieg leads to irreversible structural and functional damage to the heart and associated vasculature.* Making matters even worse is the fact that unlike other trypanosomes, *T. cruzi* is an intracellular parasite, which means that it enters into normal cells and multiplies there, instead of remaining in the blood, where it could be more easily treated with medications like antibiotics. Recent studies have shown that although *T. cruzi* appears to be absent from the blood of chronic victims, the parasite often persists deep within the heart musculature. There, decades after the initial infection, destruction of the myocardium becomes the major cause of mortality for what we now call Chagas disease, with detection of the parasite possible only postmortem.

Chagas and the researchers who followed him also determined that the protozoan's creepy-crawly insect host belonged to a neotropical family called Reduviidae, which includes more than one hundred species. Some reduviids exist as ambush predators, while others feed on the blood of nesting rodents and sleeping mammals. A few nonreduviid species, including the notorious bedbug, have also become adapted to living with humans, attacking victims as they sleep. But while bedbugs are not yet known to transmit disease, the barber bug, *Triatoma infestans*, and its Central American cousin, *Rhodnius prolixus*, deposit their trypanosome-laden feces onto a multinational list of victims. Housing with thatched roofs and adobe walls is prime habitat for the bugs, and so most of the people they infect are those living in poverty.† Oral transmission through feces-contaminated food and drink is also a serious problem.

In Peru, the bugs are known as *chirimacha*; in Venezuela, as *chipo*; and in

---

* A similar invasion by *T. cruzi* can take place in the liver, lungs, spleen, brain, or bone marrow.

† *Rhodnius prolixus*, the second-most important vector for Chagas disease, is also found in northern South America.

Central America, as *chinche picuda*. In the Argentinian Andes that Charles Darwin visited in 1835, the bloodsucker was (and is) known as *vinchuca*, although Darwin mistakenly referred to it as "Benchuca" in his notes.

This brings us back to the 1959 study by Saul Adler. The eminent Israeli parasitologist suggested that Chagas disease was responsible for Darwin's chronic ill health and eventual death. "His symptoms can be fitted into the framework of Chagas disease at least as well as into any psychogenic theory for their origin," Adler wrote.*

The key to Adler's hypothesis was the fact that Darwin himself had described being set upon (and presumably pooped on) by Benchuca, a.k.a. *vinchuca*, a.k.a. *Triatoma infestans*, while visiting Argentina in 1835:

> At night I experienced an attack, & it deserves no less a name, of the Benchuca, the great black bug of the Pampas. It is most disgusting to feel soft wingless insects, about an inch long, cawling [sic] over ones body; before sucking they are quite thin, but afterwards round & bloated with blood, & in this state they are easily squashed. They are also found in the Northern part of Chili & in Peru: one which I caught at Iquiqui was very empty; being placed on the table & though surrounded by people, if a finger was presented, its sucker was withdrawn, & the bold insect began to draw blood. It was curious to watch the change in the size of the insects body in less than ten minutes. There was no pain felt. —This one meal kept the insect fat for four months; In a fortnight, however, it was ready, if allowed, to suck more blood.

The initial response to Adler's hypothesis was mixed. Two journal articles, one by Nobel Prize–winning biologist Peter Medawar in 1967, suggested that Darwin "had both Chagas infection and a neurosis."

---

* Psychogenic illnesses are those thought to arise from mental or emotional stressors.

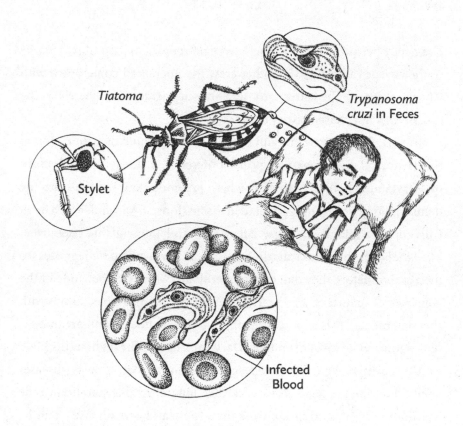

Others, however, weren't convinced. In 1977, Ralph Colp Jr. published a book on Darwin's illness in which he claimed that Darwin's health problems were purely stress-related. He followed up with a second volume in 2008 in which he refuted the Chagas disease hypothesis, writing, "Adler's theory of Chagas disease met with a series of receptions: acceptance, rejection, again acceptance, and then controversy."

Other scientists rejecting the Chagas hypothesis noted that Darwin had reported "palpitations and pain about the heart" even before the *Beagle* set sail, an indication of previously existing heart disease. Those poo-pooing (sorry) the possibility that Darwin had Chagas disease also mentioned the fact that there was no record of him having "the fever that characteristically accompanies the initial [Chagas] infection" (i.e., the acute phase of the disease that almost always presents before the dormancy period). Similarly, there were no reports in Admiralty records of the voyage indicating that other crew members had come down with Chagas disease. The latter isn't terribly surprising, though, since the disease was not characterized until 1909.

In 2011, Darwin's health and death became a topic of speculation at the Historical Clinicopathological Conference, held at the University of Maryland School of Medicine, where previous examinations into the demise of long dead patients included discussions of Alexander the Great, Christopher Columbus, Edgar Allan Poe, and Ludwig van Beethoven. The attendees had an extensive list of Darwin-related diagnoses to discuss, but before they did, gastroenterologist Sidney Cohen, one of the conference organizers, made sure to temper the media's expectations with this statement: "This is purely a symptom-based assessment, an analysis of this journey of invalidism that [Darwin] suffered throughout his life."

Ultimately, Cohen and his colleagues concluded that "Chagas [disease] would describe the heart disease, cardiac failure or "degeneration of the heart"—the term used in Darwin's time to mean heart disease—that he suffered from later in life and that eventually caused his death." They also observed that the onset of Darwin's chronic illness, which he described in his writings as beginning in 1840, occurred after a latent period of several years past the 1836 return of the *Beagle*. This latency would be expected after an initial exposure to *T. cruzi*.

Had that exposure occurred, the protozoan would have first invaded Darwin's circulatory system, then colonized his stomach, small intestine, and gallbladder, where damage to associated nerves would have resulted in gastrointestinal distress characterized by excessive vomiting, flatulence, and belching—which, indeed, Darwin had.* Finally, another complication from Chagas disease, chronic heart failure, would have occurred, killing the most famous naturalist of all time.

---

* Cohen also suggested that, absent Chagas disease, cyclic vomiting syndrome and gastric ulcers could explain Darwin's long-term gastrointestinal distress.

In an attempt to resolve once and for all the mystery of Darwin's demise, there have been requests to test his remains for the presence of *T. cruzi* DNA using modern polymerase chain reaction (PCR) techniques. This procedure has been used on samples from nine-thousand-year-old mummies from Chile and Peru, proving that Chagas disease was already well established in human populations in South America by then. One man's research project is, however, another man's desecration, and the request was turned down by the curator of Westminster Abbey, where Charles Darwin is buried.

As a result, the best that scientists can do is speculate. Though Chagas disease is consistent with Darwin's symptom patterns, we cannot be certain if his ailments and death were the result of the disease, the disease in combination with other ailments, or something else entirely. "Darwin's lifelong history does not fit neatly into a single disorder . . . I make the argument that Darwin had multiple illnesses in his lifetime," Cohen concluded.

Whether or not Chagas disease contributed to Darwin's ill health and eventual death, the naturalist would doubtless have been interested in the fact that the geographical territory of the bitey bug in question is expanding. As the climate warms, the range of *Triatoma infestans* is spreading north. Additionally, some species of reduviids not previously known to feed on humans are now making the switch—likely due to human encroachment into natural areas. Perhaps predictably, an explosion in the number of Chagas disease victims has followed. According to the World Health Organization (WHO), between six and seven million people are currently infected, mostly in Latin America, and the Centers for Disease Control and Prevention (CDC) estimates that more than three hundred thousand people in the United States have the disease.

According to Loyola University New Orleans professor Patricia Dorn, a Chagas disease expert, if there is a silver lining in all of this doom and gloom it's that the number of new infections in Latin America is actually dropping. This, she explained to me, is primarily due to the effectiveness of pesticide spraying programs. Additionally, Dorn said, although up to 50 percent of the North American *Triatoma* specimens tested were carrying the *T. cruzi* parasite, the bites of these species are relatively ineffective at transmitting Chagas disease. She explained that this was because, unlike their southern relatives, the North American insects do not defecate while feeding. Because of these fortunate table manners, Dorn estimates that only one in every two thousand bites transmits the disease to a victim. As a result, the vast majority of people carrying the disease in the United States were *not* infected in the States. Instead, they were bitten in Latin America and brought the infection back home with them.

It's not all good news, however. Even when they do not transmit *T. cruzi*, *Triatoma* bites are a primary cause of anaphylaxis, a serious and potentially life-threatening allergic reaction familiar to those with severe asthma or peanut allergies. Dorn also told me that a recent paradigm shift toward an antiparasitic treatment for chronic Chagas disease remains controversial, since it relies on the assumption that the *T. cruzi* parasite remains present in all chronic sufferers, rather than placing the blame (as previously thought) on the patient's overactive immune response.

Finally, although transmission of *T. cruzi* to humans is still rare in the United States, American dogs are showing postmortem signs of the disease, quite possibly because they eat the *T. cruzi*–infected insects and/or come into contact with their feces. Texas A&M's veterinary school associate professor Sarah Hamer led a study on government work dogs along the Texas-Mexico border, and found that 7.4 percent of the study animals tested positive for Chagas disease. In a similar study on shelter dogs across seven diverse ecoregions of Texas, Hamer and her colleagues found that "a conservative statewide average of 8.8%" of the dogs tested positive for the infection.

Arguably, the most infamous of blood-borne canine diseases is heartworm. Caused by the parasitic roundworm *Dirofilaria immitis*, it can also affect cats, coyotes, foxes, ferrets, bears, sea lions, and even

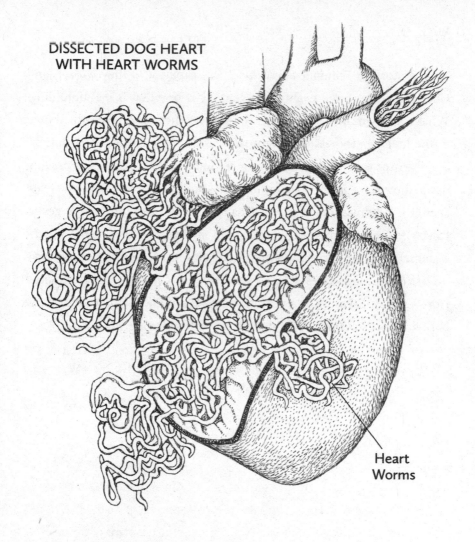

DISSECTED DOG HEART WITH HEART WORMS

Heart
Worms

humans. The only method of transmission is through the bite of an infected mosquito. If allowed to progress, masses of threadlike worms up to twelve inches in length will fill the right side of the heart and the great veins supplying it, giving them the appearance of being stuffed with angel-hair pasta. For dog owners, prevention (through monthly pills or biyearly shots) is far cheaper and easier than treatment, in which drugs are administered to kill the worms. After treatment, the dead parasites immediately begin to break down, necessitating that treated dogs avoid exercise for several months, lest pieces of the heartworms become lodged in the pulmonary vessels, causing death.

The story of reduviid bugs, the *T. cruzi* parasites they harbor, and Chagas disease in the twenty-first century is one that is still unfolding. What is known, however, is that the insects that may have contributed to the lifelong sickness and eventual death of "the Father of Evolution" are themselves evolving—adapting to habitat destruction by preying upon the very species that brought it about. Thus far, we have been spared widespread outbreaks of Chagas disease, but future changes in insect behavior, combined with climate change and the encroachment of humans (many of them living in conditions of poverty), could have disastrous consequences. And whether or not Charles Darwin suffered from Chagas disease, it is a near certainty that he would have found the barber bug's evolution as fascinating as it is frightening.

PART 3
〜〜〜〜〜〜

From Bad to Better

Then heard he through her frame the busy life-works ply,
But the sound was not of life; and he knew that she must die.

—Ebenezer Jones

[11]

Hear Here: From Stick to Stethoscope

IN 1816, ON a cool September morning in Paris, thirty-five-year-old physician René-Theophile-Hyacinthe Laënnec was walking past the Louvre palace when he saw two children playing with a long piece of wood. One child held an ear to the end of the stick while the other scratched the opposite end with a pin. Laënnec watched them for a few moments, their play providing him with a brief distraction from the harsh realities of his profession. For these were the days when consumption ran rampant in his beloved city, killing countless thousands of Parisians, including Laënnec's mother, brother, and two of his mentors.

Named for the observation that victims appeared to be slowly consumed by disease from within, the term "consumption" was originally used for several respiratory diseases, including lung cancer and bronchitis. The particular killer lurking in Paris, and the one that would become most associated with the term, had been marking the bones of its victims with characteristic knobby lesions* since the ancient Egyptians (where it showed up in the mummies), Greeks (who called it *phthisis*), and Romans (who referred to the wasting disease as *tabes*).

---

* "Lesions" are defined as tissue damage caused by a disease or trauma.

Like the ancient physicians, Laënnec, his colleagues, and their fellow Parisians, had no clue what they were dealing with. The cause of consumption was thought to be a combination of bad air ("miasmas") and heredity. What they did know was that consumption killed slowly, draining victims of their energy and their color, and producing significant weight loss.

In a phenomenon that would later be echoed by the mid-1990s style known as "heroin chic,"* the symptoms of consumption became highly romanticized in the days during which Laënnec practiced medicine. In Europe and its current and former colonies, pale skin and waifish waists (the latter enhanced by stiff corsets) became the beauty standard for early nineteenth-century women. Artists, writers, and poets almost glorified the deadly disease. American essayist Ralph Waldo Emerson wrote that his consumptive fiancée was "too lovely to live long," while English poet and landscape gardener William Shenstone said, "Poetry and consumption are the most flattering of diseases." But unlike the doomed hourglass-figured heroines of *La Bohème* and *La Traviata*, people suffering from consumption were likely feeling anything but romantic. The illness tore through and shredded the great cities of Europe. Thousands of victims were laid to waste after draining bouts of night sweats, chills, and violent, uncontrollable coughing.

One symptom of consumption was the presence of lesions called tubercles (from the Latin for "little hill") on the lungs and lymph nodes. In 1839, the German physician Johann Lukas Schönlein may have been the first to use the modern moniker for the disease, tuberculosis (TB), though it would be nearly a half century before consumption would become widely known by that far less romantic-sounding name. The rebrand followed Schönlein's fellow physician and countryman Robert Koch's 1882 discovery that the tubercles were caused by a bacterium, which he named *Mycobacterium tuberculosis*.

---

* Thin and pale-skinned, with dark under-eye circles, model Kate Moss became a sort of poster child for this, thankfully, brief pop-culture phenomenon.

Following this new knowledge came a dramatic revision in woman's fashion. Long, trailing skirts were abandoned, as they were thought to sweep bacteria into the home, and corset sales plummeted as well, since it was believed that in reducing the flow of blood, the undergarment would exacerbate the effects of TB. Men's styles were also affected, with beards and muttonchop sideburns falling out of favor, as they were thought to harbor hordes of microbes.

In the late 1800s, what had begun as instructions for sufferers of TB to find sunshine, fresh air, and altitude evolved into the massive sanatorium movement, with facilities opening across mountainous regions of Europe, like the Alps. In 1885, the first US sanatorium opened in upstate New York's Saranac Lake, followed by a facility in Denver.

It wasn't until 1943, though, that microbiologist Selman Waksman discovered a true cure for TB. He isolated the substance streptomycin from another bacterium and determined that it killed *M. tuberculosis*. The first human patient received the antibiotic in late 1949 and was cured of the disease. More new drugs followed, and by the early 1990s it seemed as if TB might be completely eradicated. Unfortunately, that was not to be. As for why, there was plenty of blame to go around: funding was pulled for worldwide TB drug distribution, many infected people did not stick with their treatment regimens, and cheaply produced antibiotics did not contain what they claimed to contain. As a result, mutated forms of TB began appearing on the scene, and these strains were resistant to antibiotics that had been effective until then. Multidrug-resistant tuberculosis (MDR-TB) has now clawed its way back to plague status across much of the world, with 1.4 million deaths from TB reported in 2019 by WHO.[*]

Though not spread by skirts or sideburns, as feared by people in the

---

[*] According to WHO, two-thirds of the approximately ten million cases of TB worldwide in 2019 originated in eight countries. In order of incidence they are: India, Indonesia, China, the Philippines, Pakistan, Nigeria, Bangladesh, and South Africa.

late 1800s, tuberculosis is in fact *highly* contagious. It travels through the air via coughing, sneezing, or spitting. Once internalized, the bacterium attacks the lungs primarily, but can also affect the kidneys, spine, brain, and even the heart, where it can cause inflammation, a thickening of the outer pericardium, and fluid buildup in the pericardial space. The heart infection known as tuberculous endocarditis (TBE) was first diagnosed in 1892. It can be especially deadly because even today it is typically diagnosed late, and often accidentally, during procedures like heart valve replacement or open-heart surgery. Sometimes it is discovered only during autopsy.

For physicians making determinations about TB in 1816, there were two principal modes of diagnosis. Both involved listening to the sounds made by the body, a practice known as auscultation (from the Latin verb *auscultare*, "to listen"). The first method was known as percussion, during which the physician would tap on the patient's chest (or abdomen) with a middle finger or a small hammer and listen as the sound resonated back to their presumably trained ear. The method was developed by physician Leopold Auenbrugger, the son of an Austrian innkeeper. He had seen his father tapping on wine barrels to determine the volume of wine they held, and he adapted the technique to determine whether a patient's chest was full of secretions. If so, like a full wine barrel, the tapping would produce a sound that Auenbrugger described as low-pitched and dull. For his patients, the sound almost invariably meant that they were suffering from consumption. Auenbrugger had honed his ability to differentiate between these tones while working at a Spanish military hospital in the 1750s. He checked his finding by performing autopsies to determine if there was TB-related fluid buildup in the chest cavity surrounding the heart and lungs.

The second sound-related technique available to early nineteenth-century doctors was called "immediate auscultation." In effect, this meant pressing an ear directly to the patient's chest in order to listen for lung and heart sounds. This came along with a wide-ranging set of issues. Many patients did not bathe, while others were infested with lice and other tiny vermin. Still others were too obese for chest sounds to be heard clearly, and the very idea of a male doctor pressing his head to a female's breast presented its own problems.

After encountering a particularly embarrassing situation with a plump female patient, Laënnec remembered his earlier encounter with the two children at play:

> I recalled a well known acoustic phenomenon: if you place your ear against one end of a wood beam the scratch of a pin at the other end is distinctly audible. It occurred to me that this physical property might serve a useful purpose in the case I was dealing with. I then tightly rolled a sheet of paper, one end of which I placed over the precordium (chest) and my ear to the other. I was surprised and elated to be able to hear the beating of her heart with far greater clearness than I ever had with direct application of my ear. I immediately saw that this might become an indispensable method for studying, not only the beating of the heart, but all movements able of producing sound in the chest cavity.

Laënnec had invented the stethoscope (from the Greek *stēthos*, for "chest," and *skopein*, meaning "to explore") and he spend the remainder of his life experimenting with a design. He settled on one that was nearly indistinguishable from the ear trumpets used by the hard of hearing. Laënnec also learned to differentiate the chest sounds heard through the stethoscope made by patients suffering from pleurisy, emphysema, pneumonia, and, of course, tuberculosis. The invention gave physicians another measurement, like pulse rate, that could be compared to a generalized "normal" reading, adding an important diagnostic tool to their black leather bags.

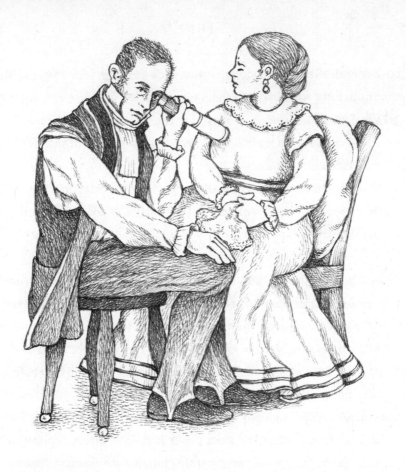

Stethoscopes also caused quite a stir among the Parisians who could afford them, or boast that their physicians used them. According to nineteenth-century specialist Kirstie Blair, along with this surge in popularity came "a growing fascination in medicine and in popular culture with the operation of the heart, pulses, and circulation."

Not long after getting married in 1824, Laënnec began to suffer a series of symptoms that included weakness, coughing, and shortness of breath. Leaving Paris for the improved climate of Brittany, he experienced some slight improvement, but soon his condition worsened. Perhaps unwilling to accept what he already knew, Laënnec gave his stethoscope to a nephew, asking him to auscultate his chest and to describe the sounds he heard. The resulting diagnosis was horrific. He was suffering from "consumption," the disease he himself had helped to elucidate with his invention. René Laënnec died of tuberculosis on August 13, 1826. He was forty-five years old.

Today, tuberculosis remains a serious health problem, especially in developing countries where socioeconomic conditions and poor healthcare infrastructure make it difficult to obtain the monthslong antibiotic treatment required to cure multidrug-resistant TB.

The stethoscope, meanwhile, has undergone tremendous innovation since its invention, but the basic concept remains the same. The two-eared (binaural) version was invented in 1851 by Irish physician Arthur Leared and went into production the following year. It remains a physician's go-to instrument for listening to a patient's heart and lungs. It is also used to check on the condition of blood vessels, like the carotid artery, where the sound of blood squeezing past a blockage has been well characterized.

And though the stethoscope no longer has quite the same high-tech appeal that it did in nineteenth-century Paris, in a 2012 study, it beat out surgical scrubs, the reflex hammer, the otoscope (the thing they stick into your ear), and the pen, as the medical equipment associated with the highest perception of trustworthiness of the doctor seen with it.

Dr. Laënnec would have been proud.

Surgery of the heart has probably reached the limits set by nature
to all surgery; no new method, and no new discovery, can overcome
the natural difficulties that attend a wound of the heart.
—STEPHEN PAGET (IN 1896)

[12]

Don't Try This at Home . . .
Unless Accompanied by a Very Special Nurse

A LITTLE MORE than a century after René Laënnec's discovery of the
stethoscope, another breakthrough would provide crucial inroads into
multiple aspects of modern cardiology. This new technique would
become an integral tool in the placement of cardiac pacemakers and the
replacement of heart valves, allowing physicians to reopen closed coronary arteries and deliver drugs directly into the heart—all without the
risk and trauma of cracking open a patient's rib cage, or blindly puncturing the heart wall with a syringe. What's more, the story behind it would
turn out to be as strange as anything a novelist could dream up.

Though Werner Forssmann was born into a family well entrenched
in Berlin's upper middle class, things became far more difficult for him
when his father was killed in action in 1916 during World War I. With
his mother forced to spend long hours at an office job, twelve-year-old
Werner was encouraged by his grandmother and his physician uncle
to continue his education. Bright and scientifically inquisitive, he took
their advice, graduating from one of Germany's best secondary schools

before entering medical school at the Friedrich-Wilhelm University in Berlin in 1922.

While studying to become a surgeon, Forssmann became intrigued with the concept of examining and treating the heart atraumatically—that is, without cutting into it. He realized that procedures like intracardiac injections, which were the only method to get medication directly into the heart at the time, were as dangerous as they were important. Blindly puncturing the walls of a beating heart could, for example, damage coronary vessels and lead to bleeding into the pericardial cavity. Forssmann reasoned that if a noninvasive technique yielding similar effects could be developed, it would become an important tool for cardiologists.

Forssmann graduated from medical school in 1928, and was a year into a surgical residency at a hospital near Berlin when he recalled seeing an old print of a researcher passing a tube down the jugular vein of a horse and into the right side of the horse's heart. This had been done in order to measure the pressure of the blood being pumped to the lungs. Forssmann tried in vain to convince his superiors that something similar could be done in humans, using the antecubital vein, a superficial vein of the upper arm, located near the crease of the elbow. Since the vessel also carries blood in a fairly direct route back to the heart, Forssmann reasoned that it could be used to gain access to the organ without surgery. Through the antecubital vein, doctors might be able to inject dyes that could be viewed with a fluoroscope, a form of x-ray machine that provided real-time images of the interior of an object.*

---

*   Although the danger of burns from fluoroscope-related exposure to ionizing radiation had been known since Wilhelm Röntgen's experiments in 1895, fluoroscopy was used for some seriously trivial reasons, none more so than the Foot-O-Scope. Installed in shoe stores beginning in the 1930s, customers would insert their feet into the boxy-looking contraption as a way to fit them with properly sized shoes. Amazingly, many of the estimated ten thousand units sold in the United States were still in use until the 1970s.

Forssmann's supervisors disagreed, and though they forbade him from attempting the procedure, the newly minted physician decided to carry on anyway. For a tube with a suitably small diameter, he settled on a urethral catheter, which also had the proper length. The problem Forssmann now faced would be getting his hands on the catheters and other surgical instruments he required, since they were kept in a locked closet to which he had no key. Undeterred, his solution was to chat up the operating room nurse, who *did* have a key. According to Forssmann, he "started to prowl around nurse Gerda Ditzen like a sweet-toothed cat around the cream jug." The young surgeon's pitch was apparently so effective that Ditzen not only gave up the keys to the closet, but she eventually volunteered to be his experimental subject.

On the appointed night, and well after the surgical theater had closed, the stealthy pair made their move, using Ditzen's keys to access a small operating room. The nurse expressed a desire to sit in a chair during the procedure, but Forssmann convinced her that it would be best if he strapped her down to a surgical table instead. She agreed, and Forssmann proceeded to prepare her left arm—or at least that is what she believed. Halfway through the preparations, Forssmann slipped away for several minutes, which must have greatly puzzled the strapped-down nurse. Unbeknownst to Ditzen, Forssmann had been busy locally anesthetizing his own arm, making an incision near the bend of his elbow, and slipping the well-oiled catheter into *his* antecubital vein. Only when he returned to her side did Nurse Ditzen realize that she had been duped.

After some well-earned grumbling, the nurse agreed to continue assisting Forssmann, which must have come as something of a relief to him since he had already fed the catheter in to a distance of twelve inches. He untied his partner in crime, and the pair set off for the x-ray room, where Ditzen was able to convince the x-ray nurse on duty to take a fluoroscopic image of Forssmann's shoulder and chest. Forssman then had to fend off a concerned fellow-physician friend, who had rushed into the x-ray room, threatening to yank out the catheter. Forssmann managed to do so, but to his disappointment, when he examined an initial radiograph he realized that the catheter tip had not yet reached his heart.

Undeterred, he advanced the tube to the twenty-four-inch mark, reportedly feeling no pain, only a sensation of warmth as he guided the catheter along its path. As the tip reached the base of his neck, Forssmann inadvertently stimulated the nearby vagus nerve and began coughing. After recovering, he stood behind the fluoroscope while Nurse Ditzen held up a mirror so that he could watch his own progress on the fluoroscopy machine. Then he pushed on, literally, until the end of the catheter finally entered the right auricle, an ear-shaped external extension of the right atrium. The radiography technician snapped several pictures, which provided Forssmann with the fluoroscopic evidence he needed, and which would subsequently be published in a scientific paper.

Although Forssmann took some serious flak from his supervisors, he was allowed to continue on in his position as a surgical resident, eventually transferring to the Berliner Charité Hospital, one of the largest university hospitals in Europe. But in November 1929, things fell apart quickly after the media descended on the esteemed facility and began writing about what Forssmann had done. Instead of congratulating the young physician, the medical community widely reacted with contempt. Bizarrely, the chair of surgery at another hospital charged Forssmann with plagiarism, claiming (with no substantiating evidence) that *he* had performed the first cardiac catheterization in 1912.

Meanwhile, derided by his colleagues for what they saw as a publicity stunt, Forssmann was dismissed on the grounds that he had not received permission to carry out the catheterization. Rehired in 1931 because of his surgical skills, he performed a total of nine catheterizations on himself over the course of a year before he was given the boot yet again. Taking a position at the municipal hospital in Mainz, Forssmann met and married Elsbet Engel, a resident working there in internal medicine. Soon, though, they were *both* let go, since married couples were forbidden to work together.

Perhaps taking the hint, Forssmann left the field of cardiology, becoming a urologist and opening up a practice with his wife (equipped, one must assume, with a full complement of catheters) near Dresden. He served as a medical officer for the German army during World War II but was captured in 1945 and served in an American POW camp for a short time until the end of the war. He returned home to find Dresden reduced to ashes but his family, miraculously, alive.

For the next three years, Forssmann was forbidden to practice medicine because of his affiliation with the Nazi Party, which he had joined in 1932, and so he worked as a lumberjack while his general practitioner wife became their growing family's principle means of support. In 1950, he was able to resume his urology practice, this time in the impressively named spa town of Bad Kreuznach.

As an outsider in the now fast-developing field of cardiology, Forssmann watched as cardiac catheterization labs opened in the United States and London, where his pioneering efforts were lauded. In Germany, however, he was turned down for a professorship at the University of Mainz because he had failed to complete his PhD dissertation.

"It was very painful," Forssmann said about his exile, many years later. "I felt that I had planted an apple orchard and other men who had gathered the harvest stood at the wall, laughing at me."

Despite his *serious* moral failings before and during the war, Forssmann was awarded a Nobel Prize in Physiology or Medicine in 1956 for his work on cardiac catheterization, which, after all, turned out to be a groundbreaking procedure. Upon learning that he'd won the award, he told a reporter, "I feel like a village parson, who has just learned that he has been made bishop."

Soon after, "Bishop" Forssmann was offered a position heading up a German cardiovascular institute. He turned the job down, explaining that it had been over twenty years since he had last experimented on himself, and that he lacked knowledge about recent advancements related to the cardiovascular system. Undeniably true, however, is the fact that many of these advancements had resulted from his own pioneering efforts in the field of cardiology.

Doctors now perform catheterizations for many reasons, and they do so by threading catheters through veins in the arm, groin, or neck to approach the heart or any of the four coronary arteries supplying it. One important use of catheters is balloon angioplasty. Picture inflating a balloon in a narrowed or blocked coronary blood vessel in order to widen it. After this, the catheter is used to place an arterial stent, a springlike device that holds back the walls of the newly widened vessels, preventing them from narrowing again. Cardiac catheters are also used to take pressure measurements in specific heart chambers, to snip off tiny bits of cardiac tissue for biopsies, to look for valve problems, and to repair or replace those valves if they're found to be defective.

After suffering from two myocardial infarctions, Werner Forssmann died in 1979, but not before he wrote an autobiography, the aptly titled *Experiments on Myself.* In it, he spent little time writing about the swastika-wearing elephant in the room—his Nazi Party affiliation. According to a journal article on the topic, he may have joined the party out of an early belief that National Socialism was better than the alternative, communism, but his attitude eventually changed to one that was more critical of Nazi ideology, a political trajectory typical of many German doctors during the era.

According to the support letters that Forssmann gathered in order to earn the "denazification certificates" that would allow him to work again, his mentors and associates described him as neither a militarist nor an activist, but as someone who detested the violence that was carried out

by the party he had joined. There is evidence to suggest that he refused
the opportunity to carry out unethical experiments and that he continued
to offer Jews medical care after it had been disallowed. Ultimately, he
was designated as a Category 4 Nazi (i.e., "a follower") by the French
Occupation Administration and fined 15 percent of his salary for three
years.

In the end, readers are left with the story of a groundbreaking
discovery—the unique aspects of which are tainted by the abominable
political affiliation of its discoverer.

Somewhere, far down, there was an itch in his heart,
but he made it a point not to scratch it.
He was afraid of what might come leaking out.
—MARKUS ZUSAK, THE BOOK THIEF

[13]

"Hearts and Minds" . . . Sort Of

THE IDEA OF a connection between the mind and the heart (as well as the blood running through it) remains firmly entrenched in our language, songs, and poetry. The words of William Shakespeare, John Lennon and Paul McCartney, Emily Dickinson, Tom Petty, and Stevie Nicks are bursting with cold hearts, broken hearts, and hearts given away in vain, while other tickers are lonely, get dragged around, or get chained up. Of course, there are also hearts that are bursting with joy or have messages coming straight from them. As for blood, take a minute or two and see how many blood-related idioms you can come up with for anger or lust. I'll wait.

Okay, that's enough (and sorry if that little exercise made your blood boil).

All of this is likely due, in large part, to roughly fifteen hundred years of adherence to the teachings and terminology of the Roman physician Galen and his influential followers, who believed that the heart was the seat of the soul as well as emotion. Additionally, some of these concepts, like "cold-blooded" and "hot-blooded," were themselves handed down from even more ancient philosophers like Hippocrates and Aristotle, who associated the heart with emotions, the soul, intelligence, and memory.

Traditional Chinese medicine, which has many present-day adherents, also has strong associations between the mind and heart. TCM has always held that the heart is primary among the organs.* In addition to its role as a pump, its practitioners believe it to be involved in emotional and mental processes, serving as the residing place of the mind and spirit, consciousness, and intelligence. TCM also holds that heart dysfunction leads to psychological and physiological problems, ranging from palpitations and restlessness to pallor, shortness of breath, and memory loss. It is noteworthy that all of these things are also acknowledged as symptoms of heart disease by Western medicine—albeit with alternative explanations for their causes.

Likewise, the holistic healing system of Ayurvedic medicine emphasizes the role of the heart as vital to the concept of mind/body/ spirit. It maintains that while Western medicine, which focuses on symptoms and diseases, is often a lifesaver, a healthy life also depends on keeping the bodily energies (*vata*, *pitta*, and *kapha*) in balance. Ayurvedic medicine proposes to maintain this balance through a combination of diet, herbs, meditation, and other relaxation techniques like yoga.

Although advances in the fields of medicine, psychology, and psychiatry, as well as modern research into behavioral physiology and neurophysiology, have conclusively proven that the heart is *not* the seat of the mind, it took centuries for that concept to firmly take hold in the West. The first hints of movement away from cardiocentrism came as early as the seventeenth century—but these claims were often based on poor or nonexistent science, and as a result, they were met with nothing that might be confused with acceptance. Perhaps the most influential of the early noncardiocentrists was the philosopher and mathematician René Descartes (1596–1650). Descartes was famous for his contributions

---

* A 2019 study published in the journal *Geriatrics* estimated that 14 percent of the population of mainland China over the age of fifty utilized a TCM practitioner.

to geometry and algebra, but he also had a keen interest in anatomy and physiology. In 1640, he claimed that the true "seat of the soul, and the place in which all our thoughts are formed" was . . . no, not the brain, but close—a tiny nubbin of endocrine system inside the brain known as the pineal gland.

Located between the two cerebral hemispheres, the pineal gland gets its name from its roughly pine-cone shape. It was the last of the endocrine (i.e., hormone-releasing) glands to be discovered, and it is now known to be involved in regulating circadian rhythms, like our internal twenty-four-hour clock, and some reproductive hormones. Unfortunately, Descartes's concept of the pineal gland, which he described as suspended within one of the brain's chamber-like ventricles and surrounded by "animal spirits," was nearly as faulty as his explanation of its function. Descartes reckoned that, "Since [the pineal gland] is the only solid part in the whole brain which is single, it must necessarily be the seat of common sense, i.e., of thought, and consequently of the soul; for one cannot be separated from the other." Descartes also reasoned that since the brain had both left and right sides, that fact somehow precluded it from being involved in mental activities.*

Things began to turn around for Team Craniocentrism with the work of English physician Thomas Willis (1621–1675), who became a pioneer in the modern understanding of the brain and neurophysiology. Willis took part in autopsies during which he studied the anatomy of the brain and its intricate supply of blood vessels, especially a circular convergence of arteries at the base of the brain which came to be known as the circle of Willis. A professor of natural philosophy at Oxford, he was expected to teach his students about the soul, but, informed by his research, he

---

* Even further anatomically from the facts was the Flemish physician Jan Baptista van Helmont (1580–1644), who also believed that soul did not reside in the heart. Instead, he claimed, it could be found within the folds (or rugae) of the stomach.

used the brain as a starting point rather than falling back on the standard cardiocentric explanations. In addition to his work on human bodies, Willis conducted experiments on animals, determining that different regions of the brain were specific to different functions. Willis also used his anatomical knowledge and medical observations to provide early insight into the craniocentric origin of intellectual disabilities and psychological disorders, including narcolepsy and myasthenia gravis (a neuromuscular disease that causes weakness in skeletal muscles). He also described some disabilities affecting the brain as being caused by so-called disorders of brain chemistry. He even coined the term "neurology."

While all of this is incredibly impressive, readers should remember that the mid-seventeenth century was a *very* different time for scientist types. Although Willis was certainly a game changer in the field of neurobiology, his writings are rife with rationales that were clearly meant to placate the Church of England. Additionally, his treatment for emotional disorders left some room for improvement, since it included beating patients with a stick.

But quibbling aside, the ball had been set rolling, and by the 1670s, the brain, which for millennia had been believed to be little more than a radiator to cool the heart, would begin to replace that organ as the seat of the mind, soul, intellect, consciousness, and emotion—at least in the West. Coinciding with this shift was a growing knowledge about nerves and the involuntary workings of the autonomic nervous system, which meant a new understanding of the connections between the heart, body, and mind. As we'll soon see, this transition also ushered an understanding of how emotional factors like stress, poverty, personal tragedy, and unhappiness can lead to heart disease. And eventually, the cessation of brain activity, not the stilling of the heart, became the determinant of life's end.

Through it all, poets, songwriters, and other storytellers ignored the adoption of craniocentrism by the scientific community to varying degrees—and given the alternatives a shift would've created (Janis Joplin's "Piece of My Brain" and Joseph Conrad's *Brain of Darkness* come to mind), this was quite likely a wise move. Actually, though, even as modern science has conclusively eliminated a link between the heart and emotion/cognition, many Westerners still turn to that imagery not only in their metaphors and music, but in their own quasireligious beliefs.

Some people believe that the heart contains the emotional characteristics of its owner, and that a transplanted heart can transfer these personality traits from donor to recipient. Most famous, perhaps, was the experience described by the late Claire Sylvia, who in 1988 became Massachusetts's first heart-lung transplant recipient. She later documented her experience in a best-selling memoir. In it, the previously health-conscious dancer describes her recovery from transplant surgery, during which she began experiencing serious changes in her habits, attitude, and tastes in both fashion and food. The latter was exemplified by a newly acquired taste for beer and a sudden craving for junk food, especially chicken nuggets of the KFC variety. As it turns out, she informed her readers, these were the very objects found in the jacket of her eighteen-year-old organ donor after he was killed in a motorcycle accident.

In her own book, cultural historian Fay Bound Alberti floats several possible explanations that could either dispute or support the phenomena experienced by Claire Sylvia and others. One is wishful thinking: the organ recipient may feel reassured to imagine that some part of the donor's personality lives on within them. It is also possible that the recipient may find the feeling of possessing someone else's heart traumatizing and may undergo genuine psychological alteration as a result. Alberti mentions the possibility that there is in fact some form of "systemic memory," in which the cells of the body are somehow imprinted with the experiences of the individual.

This last line of reasoning is unsupported by mainstream science, but it is often embraced by those who practice a previously mentioned type of alternative medicine called homeopathy. Homeopathic medicine holds that water retains the memory of the material dissolved in it, and that "like cures like"—in other words, that health benefits can be derived from ingesting pills, tinctures, or superdiluted solutions containing trace, or even undetectable, amounts of the substance that caused the malady in the first place.

As an example of the practice, one popular British homeopathy website recently suggested a new treatment for varicose veins, a circulatory problem in which weak or damaged venous valves allow blood to pool in the vessels, thereby twisting and distorting them. The condition is common in the feet and legs, where low-pressure blood must overcome gravity to return blood to the heart. Conventional treatments for varicose veins range from pressure socks (akin to the tight skin covering giraffe legs) to assorted therapies designed to close off affected veins and promote the growth of new vasculature in the area.

Meanwhile, the suggested homeopathic treatment is ingesting *Pulsatilla*, a genus of herbaceous perennials known as pasqueflower. Blooming in early spring, the plant's bell-shaped flowers are beautiful. Less beautiful is the fact that *Pulsatilla* is also *highly* toxic if consumed. In addition to inducing hypotension (blood pressure of less than 90/60 mm Hg) and reducing heart rate, *Pulsatilla* can cause diarrhea, vomiting, convulsions, and coma. Reportedly, the plant was used by indigenous Blackfoot people to induce abortions.

The British homeopathic website also notes that should *Pulsatilla* be unavailable, something called "Calc carb" is a good substitute, stating

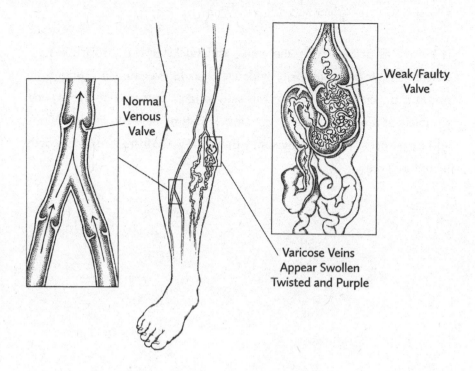

Normal Venous Valve

Weak/Faulty Valve

Varicose Veins Appear Swollen Twisted and Purple

that "People who do well with Pulsatilla often tend to be mild-mannered, avoiding arguments if they can . . . Whereas Pulsatilla tends to suit people who are generally warm-blooded and prefer to have fresh air in their homes, those requiring Calc carb are definitely chilly, with markedly sweaty feet. They hate damp conditions or damp weather but, like those needing Pulsatilla, tend to be mild in manner, perhaps verging more to the shy side or slightly nervous."

Personally, I would suggest that calc carb be referred to instead as "O ShAT" (oyster shells annihilated thoroughly) to reflect its true identity, as the main ingredient is bones, shells, and eggs. It is also used as a gastric antacid (e.g., Tums) and is found in household cleaners. If this is starting to sound familiar, that's because calc carb is often referred to as calcium carbonate ($CaCO_3$), or chalk.

Finally, although the phrase "There's a sucker born every minute" is closely associated with showman and huckster extraordinaire P. T. Barnum, there is actually no evidence that he ever said it. Though its origin is uncertain, the phrase was said to have become popular among gamblers and confidence men starting sometime between the late 1860s and early 1870s. For some reason, I thought this information was worth including here.

"I think you are wrong to want a heart.
It makes people unhappy."
—L. Frank Baum, *The Wonderful Wizard of Oz*

[14]

What Becomes of the Brokenhearted?

While most beliefs about the heart's emotional and spiritual importance exist outside the realm of modern scientific proof, recent research
into one particular form of coronary disease has found an indication that
hearts and minds are connected after all—if not in the way ancient or
alternative medicine suggest.

In 1990 cardiac researchers in Japan studied a group of thirty patients,
each of whom had entered the hospital complaining of chest pains and
shortness of breath. When tested initially, they all showed symptoms
resembling those of a heart attack: dysfunction of the left ventricle, as
well as abnormal electrocardiograms (ECGs), graphical representations
of the electrical activity of the heart. Upon examination, however, the
doctors found no signs of narrowing in the coronary arteries, a symptom
typically found in patients suffering from an infarct (i.e., tissue death due
to insufficient blood flow). In fact, the majority of the patients showed
no signs of heart disease at all. Even odder were the results of another
test performed to evaluate the condition of the left ventricle. After
inserting cardiac catheters to inject dye into the ventricle (thank you,
Werner Forssmann!), the physicians took x-rays as the patients' hearts

went through their cycles of filling and emptying. When examining the resulting ventriculograms, the researchers were struck by the fact that as the left ventricle finished contracting, it took on a weird shape: narrow at the top and ballooning out at the bottom. It reminded the Japanese researchers of the octopus traps, or *tako-tsubo* ("octopus" plus "pot"), that were used by local fishermen. Then, in yet another departure from the typical outcomes of a myocardial infarction, the majority of the patients in question saw their heart conditions resolve during the following three to six months. Apparently, whatever damage had occurred was completely reversible, which made this condition unique among diseases of the cardiac muscle, a.k.a. cardiomyopathies.

Since the initial studies of what has become known as Takotsubo syndrome, researchers have made considerable strides in their understanding of who experiences this strange malady and what triggers it. Interestingly, 90 percent of those who suffer from Takotsubo syndrome are postmenopausal women, and most of them have recently experienced acute physical or emotional stress, some of it as severe as surviving a recent suicide attempt. Many others had suffered grief over the death of a loved one. The relationship between bereavement and Takotsubo syndrome has led to an alternative name for the condition: broken heart syndrome.

The workings of Takotsubo syndrome actually make a great deal of sense. During highly emotional or stressful situations, the body's nervous system (specifically the sympathetic division of the autonomic nervous system, which regulates unconscious body systems) floods the circulatory system with stress hormones—the fight-or-flight response. These chemical messengers prepare the body to deal with real or perceived threats by managing physiological functions like heart rate, blood pressure, and breathing rate. In normal situations, this sympathetic response gets shut down when the threat passes or when emotions subside. But in patients with Takotsubo syndrome, researchers theorize that there is decreased

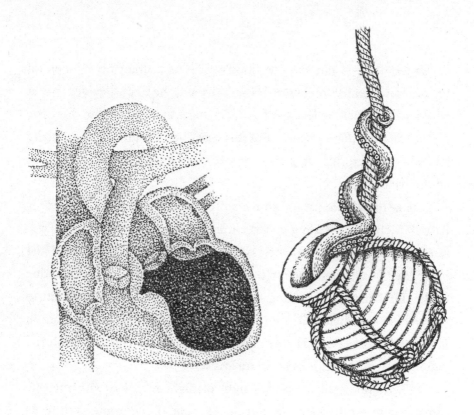

communication between the brain regions that process emotions and the autonomic nervous system. This causes the sympathetic nervous system to overrespond by continuing its outpouring of stress hormones—the overabundance of which leads to potentially serious cardiovascular problems. These can include spasming of the coronary arteries and their microscopic branches, a phenomenon that could explain the left ventricular dysfunction and chest pains observed in Takotsubo patients.

There are, however, unanswered questions regarding the condition. For example, it remains unclear why the left ventricle takes on its peculiar octopus-trap shape. It's also unknown whether the brain's overproduction of stress hormones is *caused* by the emotional trauma the patient suffered or whether the brain dysfunction responsible for the overstimulation of the sympathetic nervous system was already present, thus making that person more susceptible to Takotsubo syndrome.

Uncertainties aside, the condition serves as a dramatic example of the intimate connection between the heart and the brain—evidence that emotions like grief can lead to physical changes in the heart—in this case, changes of a temporary nature. But this heart/brain connection is actually a two-way street, since it is also clear that a damaged heart can lead to emotional dysfunction.

I spoke to cardiologist and University of Wisconsin professor emeritus Patrick McBride, a leading expert on cardiovascular risk factors. I was interested in learning why stress and depression negatively affect the heart and how the situation might be countered. He emphasized that due to the number of confounding factors, the link is extremely difficult to research. For example, when somebody's spouse dies, it is quite common for the survivor to end up in the hospital with a heart attack. But while the pattern is clear, the reasons for it are less so.

McBride reviewed the body's fight-or-flight response with me, the same system which kicks in during Takotsubo syndrome. While its cocktail of adrenaline, cortisol, and other stress-related chemicals is useful in managing physical threats, it can be counterproductive when it comes to emotions. When someone is under chronic stress, as they might be if a loved one had passed away after a long illness, those hormones may circulate so frequently that they can irritate the heart and blood vessels, damaging their inner lining, or endothelium. While this single layer of cells was until recently thought to be relatively inert, it actually has an endocrine function. Over the past two decades, researchers have shown that the endothelium releases its own set of hormones into the blood.

"Second to second and minute to minute, the endothelium is responding to our chemical environment," McBride told me. "If muscles need more oxygen, the chemicals released by the endothelium dilate the blood vessels that supply them, while contracting the blood vessels elsewhere."

When the endothelium becomes inflamed, the damaged cells also release chemicals like histamine, bradykinin, and cytokines (a broad category of small proteins also released by cells of the immune system). One result is that the blood vessels become more porous, and they leak plasma into the tissues surrounding them. This leads to the characteristic swelling, redness, and pain that we associate with inflammation. Meanwhile, the chemicals that have been released signal the body's repair team to show up and get to work.

This process is helpful when inflammation is acute, but not so much when the condition becomes chronic. McBride compared the constant presence of inflammatory chemicals to rubbing your skin until it becomes raw. What's more, as the lining of blood vessels becomes more porous during prolonged inflammation, chemicals in the blood can become modified, making them behave differently. One such change occurs when LDL cholesterol undergoes the process of oxidation, becoming Ox-LDL, a substance known to be involved in the formation of atherosclerotic plaques. MacBride likened Ox-LDL to bacon grease left in a pan.

Things can get even worse if the person in question already *has* atherosclerotic plaques, since chronic inflammation can cause the vessel's inner lining to crack open. As the body's repair crew rushes in to plug the damage, a blood clot forms. Usually clotting is a good thing—an intricate cascade of so-called hemostatic chemical reactions whose fibrous final product (the clot) can effectively halt blood loss from a ruptured vessel. Here, though, the feces hits the fan if a piece of the clot breaks off and flows downstream, where it can get stuck in an increasingly small vessel like a coronary artery or an artery supplying part of the brain. This can lead to a heart attack or a stroke, respectively.

Now, somewhat more well versed on the relationship between stress and the heart, I decided to change course with McBride, seeking to explore the methods currently being used to counteract the heart-unhealthy effects of stress.

Surprisingly, McBride led off with spirituality.

"It's very clear to me that people with spiritual lives do better—and that's research-driven." When people aren't afraid of their own mortality, he explained, they have better outcomes.

His claim, however, is controversial, since for every study pointing to the benefits of religious activity as it relates to health, there are critics who claim that even the best of these studies were faulty, because they lacked controls or did not consider covariables like age, sex, ethnicity, education, behavior (like smoking and alcohol consumption), and socioeconomic and health status, before reaching their conclusions.*

The fact remains, however, that patients who receive social support or are in strong relationships are more likely to have better outcomes. "People who are lonely or widowed have worse outcomes," McBride said.

Over the past four decades, McBride and his cardiac rehab colleagues have been working to address the high rate of depression that follows heart attacks. The reason for their efforts is that, like other types of acute stress, depression aggravates circulatory tissues. In combination with the heart disease that led to the event, this can have deadly consequences. McBride told me that currently, one in every two to three recovering patients is likely to suffer from the mood disorder. To address this problem, as part of their protocol McBride's team screens every patient that has had a heart-related event for depression, regardless of whether the event was the placing of a stent, bypass surgery, or a heart attack. As a result, the medical team at the University of Wisconsin's Preventive

---

* The shortcomings of these studies were reviewed in 1999 in Richard P. Sloan, Emilia Bagiella, and Tia Powell's article "Religion, Spirituality, and Medicine" in the *Lancet*.

Cardiology Clinic has had psychologists and therapists on staff since the 1980s and has been championing a mindfulness program since 1994.

Mindfulness is a therapeutic technique with its roots in Buddhist meditation. When practicing mindfulness, one attempts to focus one's awareness of thoughts, feelings, and bodily sensations on the present moment, rather than rehashing the past or worrying about the future. The technique also emphasizes acceptance of thoughts and feelings without judging them, helping practitioners understand that there is no "right way" or "wrong way" to feel at any given moment. Since the late 1970s, mindfulness has become a popular stress-management program, and it is commonly employed in prisons, hospitals, and recently, in schools, where anxiety among children has become a serious concern.

McBride told me that initially the professionals running his cardio rehab program referred to these mindfulness classes as "stress management" or "stress reduction."

"All the guys would show up," he said.

But then the staff started calling it "mindfulness meditation" and bringing in elements of yoga and tai chi. "And men would never show up. It was just too Eastern for these Western dudes."

I laughed. "How'd you fix that?"

"We went back to calling it 'stress management,' and the men showed up in droves."

McBride and his colleagues use mindfulness techniques to combat the "very real fear factor" that patients go through after surviving a cardiac event. That fear has always been present, but in recent years the internet has augmented it, allowing people to quickly access a vast amount of information. Some of this online info, regarding things like healthy diets and the need for exercise, is quite useful. But, as with all self-diagnosis, the danger exists that a patient might wander onto a website promoting untested dietary supplements or simplifying medical topics to the point of inaccuracy. The blanket claim that cholesterol harms your body serves as an example of the latter. Of course, these types of problems can be a counterproductive focus for patients recovering from cardiac-related issues like heart attacks or coronary bypass surgery. This emphasizes the need for rehab programs that provide information and instruction derived from peer-reviewed sources, like reputable medical journals. Note: Most hospitals currently have cardiac rehab programs, but these programs vary in scope—not an unimportant fact to consider when compiling information on the pros and cons of hospital options.

One constant found in many cardiac rehab programs (including the one run by McBride) is that that bringing partners, close relatives, or friends into the mix is an important part of the process. Among other things, cardiac rehab classes teach the partners and peers not to tiptoe around the person, wondering "When are they going to have the big one?"—a fear that the patient will invariably pick up on. The partner-focused classes also address related health problems, like erectile dysfunction, common in patients after a cardiac event, and provide resources about what to do should there be another heart attack, including teaching the partner CPR.

Ultimately, whether one believes in the power of meditation and yoga or just wants to make things easier for both partners during a difficult time, cardiac rehabilitation programs have been associated with a significant reduction in ten-year mortality after coronary bypass surgery, and a marked reduction of readmission and death after myocardial infarction.

The problem, though, according to McBride, is that while those who participate in cardiac rehab have better outcomes, only about one in four patients will enroll. A number of barriers to participation have been proposed. These include lack of health insurance, depression, a perception that the rehab program is inconvenient or unnecessary, and the travel time involved and transportation needed to get to and from the program.

After studying these barriers, researchers at the Mayo Clinic concluded that while factors like age (the older the patient, the less likely they'll participate) and sex (females are less likely to enroll) could not be modified, there were practices that *could* increase participation. These include in-hospital primary care by a cardiologist, referral to a cardiac rehab program while the patient is still in the hospital, in-hospital education about the importance of these programs, and a discussion with patients about transportation-related issues and how to overcome potential problems.

As off-putting as rehab programs might initially seem to the patient, they have undeniable benefits, especially, McBride emphasized, group programs. If a patient sees a peer running on a treadmill and knows that that person had their own bypass surgery only eight weeks ago, it might really hit home that now they're exercising and doing well. "They'll go, 'Hey, I just had bypass surgery. How did you get to the place where you are now?'"

"Social support of other patients becomes really important," McBride said. "They'll open up to other patients."

McBride also suggested that Chinese traditional medicine can be effective in preventing and treating cardiovascular disease, referring to it as part of a broader category of integrative medicine (IM). Roughly defined, IM seeks to understand the unique set of circumstances (physical, mental, spiritual, social, and environmental) affecting each individual's health. It then addresses them with a multidisciplinary approach tailored to the individual. McBride's group has been using IM as part of its cardiac rehabilitation for the past twenty-five or so years, ever since several physicians joined the group who were already combining Eastern and Western medical approaches.

In addition to ways to counteract stress, McBride's lab researches other methods of improving cardiac outcomes. The team has been testing the effects of an array of compounds on arterial function—specifically, whether or not unhealthy arteries would dilate when they were exposed to test chemicals. Among the compounds they have tested are vitamins A, C, D, and E, as well as substances like ginseng, resveratrol (a chemical produced by several plants in response to an attack by pathogens), grapes, red wine, and garlic.

"And what were the results?" I asked.

"Well, I can tell you that those vitamins didn't work at all."

"So, what *did* work?"

"Red wine did. Dark beer did. But the dietary supplements did not."

The most effective compounds that McBride's team have tested have been statins, which are a class of chemicals, which includes drugs like Lipitor, that lower blood cholesterol levels. There are two sources of blood cholesterol, diet and the liver, and statins work to block the enzyme responsible for the latter. "They remarkably reduce inflammation, improve endothelial function, and reduce atherosclerotic plaques," McBride told me.

Having been taking statins myself for the past fifteen years, I found it comforting to hear him mention them in nearly the same breath as dark beer and red wine—two medications I have always strongly supported.

McBride also mentioned flavonoids, referring to the antioxidant compounds found in foods like berries, apples, citrus fruit, legumes, and even tea. Antioxidants are compounds that scavenge or prevent the formation of unstable molecules called free radicals, which are involved in tissue damage. Other antioxidants include vitamins C and E and carotenoids—though McBride stressed that he did not believe that they work in dietary supplement form, where one is never quite sure what is in the pill, and thus cannot substitute it for a healthy diet containing these compounds.

He also stressed the anti-inflammatory effects of the Mediterranean diet, with its emphasis on lots of vegetables, olive oil, and garlic, along with reduced intake of saturated fats and increased monounsaturated fats.*

One more important takeaway from my conversation with McBride was an overall sense of the importance of moderation to a heart-healthy life.

"An Ironman [triathlon] is not the right amount of exercise, and doing nothing is not the right amount of exercise," he told me. "Taking a daily walk is the right amount of exercise. People say, 'Red wine is good for you, so I'll drink a whole bottle.' No, the right amount is three ounces."

Part of America's general cardiac unwellness is due to our dietary habits, which have been leaving moderation behind. In a trend that began in the late 1970s, portion sizes (especially in fast-food and chain restaurants) in the United States have increased at a rate mirrored by increases in obesity. According to Harvard Women's Health Watch, "a typical movie-theatre soda, once about 7 ounces, can now be 'supersized' to 32 or 42 ounces," while a 2- to 3-ounce bagel now weighs in at 4 to 7 ounces.

Our meat consumption has been on the rise as well, and global demand for meat has quadrupled over the past fifty years. One notable and related study examined death rates from "circulatory disease" during World War II, focusing particularly on Nazi-occupied Norway: It turns out that despite the increase in stress, between 1942 and 1945 around 20 percent fewer people died of cardiac events. The reason? With their livestock confiscated by the Germans and little or no access to meat, eggs, or dairy products, the native population was forced to survive on a

---

* McBride also hailed the benefits of the DASH eating plan, which stands for "dietary approaches to stop hypertension."

low-fat diet of vegetables, grains, and fruit. As a consequence, incidents of heart disease dropped.

I came away from my research into the topic with a list of recommendations for a heart-healthy lifestyle in the face of an often-stressful world. The list includes exercising, eating a diet with more fish and less fat, reaching or maintaining an appropriate body weight, getting enough sleep (seven hours per night seems to be the magic number), not smoking, drinking only a moderate amount of alcohol, using stress-reduction techniques, and getting regular medical checkups.

After running down the list I'd compiled as my interview drew to a close, I asked McBride if there was anything else he'd like to add.

"Moderation in all things," he replied. "I think that would be a beautiful message."

Hearts will never be practical until
they are made unbreakable.
—The Wizard of Oz, MGM, 1939

[15]

What's Snakes Got to Do,
Got to Do with It?

IN NATURE, HEARTS and circulatory systems have evolved into efficient forms of internal transportation whose primary functions allow organisms to exchange vital materials like nutrients and gases with the external environment. We humans, however, have outpaced the evolution of our cardiovascular system, testing the limits of its ability to adapt to junk food, toxins, pollutants, smoking, and stress.

Medical research has responded to the challenge. In recent decades, for example, we've seen the rise of low-fat diets and high-tech medical procedures like coronary bypass surgery, during which clogged coronary arteries are replaced with veins from a patient's arm or leg. Far more complex are mechanical hearts. In 1982, American cardiothoracic surgeon William DeVries successfully implanted the first total artificial heart (TAH), the Jarvik-7, into a sixty-one-year-old retired dentist named Barney Clark. Clark survived for 112 days, enduring a series of serious medical conditions, which included respiratory failure requiring a tracheotomy, as well as "fevers, stroke, seizures, delirium, renal failure, and bleeding related to anticoagulation [an inability to form blood clots]."

He eventually succumbed to colitis. The story's initially positive publicity was soon blunted by Clark's downward spiral, and the subsequent negative press did much to turn the goal of mechanical hearts from permanent destination to bridge therapy for patients awaiting transplants.

On the subject of heart transplants, Christiaan Barnard (1922–2001) performed the first successful one on December 3, 1967. During a five-hour operation, fifty-three-year-old Louis Washkansky received the heart of Denise Darvall, a twenty-five-year-old car accident victim. The organ functioned quite well, but, unfortunately, the immunosuppressant medications that prevented rejection of the heart also left Washkansky vulnerable to infection. He died of double pneumonia eighteen days later.

It is now estimated that something like five thousand heart transplants are performed worldwide each year, most in the United States. But even so, millions of lives are lost each year to heart disease, and thousands more die while sitting on long waiting lists for heart, liver, and kidney transplants. We've already looked at the past history of xenotransplants and current efforts to genetically design a strain of organ-donating pigs. But it also appears that nature can provide us new methods to treat faulty hearts that are animal-based but also animal-friendly. And a growing number of researchers are now turning back to nature and its amazing evolutionary modifications in their search for answers.

One particularly noteworthy adaptation found in the animal kingdom is the ability of certain hearts to repair themselves when they become damaged—a feature human hearts, tragically, lack. When someone suffers a heart attack, it's usually due at least in part to the obstruction of one or more of the coronary arteries, which cuts off blood both to the heart and to whatever region the vessel previously supplied. Deprived of oxygen, the cardiac muscle tissue located downstream of the blockage dies. If the patient survives, the dead muscle tissue is replaced by scar tissue, which isn't contractile and which prevents new cardiac muscle cells

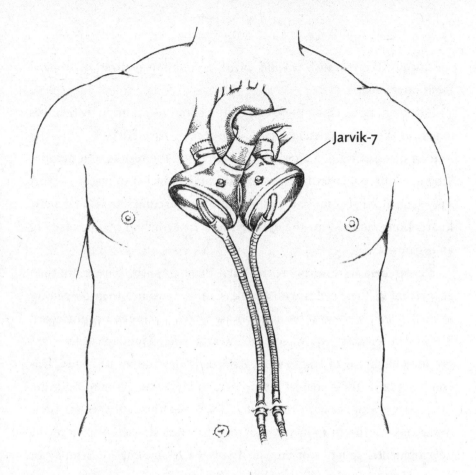

Jarvik-7

from forming. As a result, sections of this intricate pump are no longer functional, and the heart's beautifully coordinated workings are disrupted. Survivors are often left susceptible to future problems, including, but not limited to, additional heart attacks and, ultimately, heart failure.

But what if physicians could replace lost or dysfunctional cardiac tissue? Such a treatment would be transformative, especially given the fact that approximately half a million Americans are diagnosed with heart failure each year, and these patients have a one-year mortality rate of nearly 30 percent. Since the phenomenon of cardiac regeneration does not occur in humans (or any mammals, for that matter), researchers have turned to the most ancient of vertebrates for an answer, the fish:

specifically, the zebrafish (*Danio rerio*), a common denizen of tropical freshwater aquaria.

Although research on these egg-laying South Asian minnow relatives began in the 1960s, their popularity, especially as models for studying human diseases, surged after 2013. That's when the results of a decade-long quest to sequence their genome became available to researchers. A fully sequenced genome can be thought of as the complete set of genetic instructions required for the development, growth, and maintenance of an organism.

Researchers were surprised to learn that zebrafish share more than 70 percent of their genes with humans, and that more than *80 percent* of genes associated with human diseases have a zebrafish counterpart. They also possess equivalents to nearly every human organ and produce hundreds of transparent, externally developing embryos. This combination of traits enables researchers to model the human condition in a fast-growing, easy-to-maintain, easy-to-see-through species. Gene mutations can be introduced into zebrafish test strains to explore the developmental genetics of human diseases like muscular dystrophy or to model cardiac abnormalities, thus allowing drug researchers to test potential therapeutic compounds.

Most exciting, though, was the discovery that zebrafish hearts are able to fully regenerate after the amputation of up to 20 percent of their single ventricle. Admittedly, this type of wound has become somewhat less common in humans, with our recent pivot away from pillaging and gladiatorial combat, but the discovery has had huge reverberations for heart research. Scientists noted that in zebrafish experiencing such an amputation, a clot quickly forms to prevent catastrophic blood loss. The truly unique part, though, is that within thirty to sixty days of the injury, the clot is replaced by fully functional muscle cells.

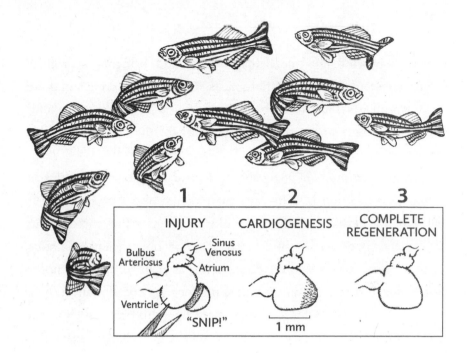

In a mature mammalian heart, cardiac muscle cells (a.k.a. cardiomyocytes) stop reproducing—and thus, stop producing new cells. Conversely, the zebrafish heart is not only able to produce new functional muscle cells, but it does so without the input of stem cells. There will be much more on stem cells, but for now we can think of them as a class of embryonic or adult cells with the capacity to develop into different types of cells, depending on how they are stimulated.

In adult zebrafish, new cardiac muscle cells originate from previously existing myocytes. When a specific region of the heart is damaged, undamaged myocytes in the area reenter the reproductive phase of their life cycle and begin to crank out new functional cardiomyocytes—muscle cells that are ready for action. These new myocytes migrate into the previously damaged area and replace the scar tissue that had initially formed in response to the wound. The zebrafish heart has, meanwhile, built a connective-tissue framework at warp speed, with blood vessels rapidly regrowing into the damaged area and bringing with them collagen-secreting cells called fibroblasts. The collagen frame laid down by the fibroblasts is referred to by researchers as a "regenerative scaffold," and it serves as a structural base of support for the new heart muscle being formed.

Given the positives of being able to regrow functional heart muscle, the obvious question is: Why can't mammalian hearts do it? From an evolutionary perspective, the most likely reason is that this inability is actually beneficial—or at least it was for our ancient ancestors. Since cardiomyocytes stop dividing shortly after birth, they are not as susceptible to cancer-causing genetic mutations. As a result, cancer of the heart is extremely rare.* Since all mammals share this trait, it is clearly an ancient mammalian adaptation—or more likely, an even older adaptation that developed in early vertebrates, since there's only a single species besides the zebrafish, a type of North American newt (*Notopthalamus viridescens*), known to have cardiac regenerative ability.

The inability of our cardiomyocytes to divide also makes perfect evolutionary sense if you consider that our distant ancestors were not burdened by crappy fast-food diets, obesity, smoking, and other recently developed heart-straining bad behavior. As such, this adaptation serves as a great example of how our organs evolved in a time that was very different from our own. As for why zebrafish buck this particular vertebrate rule— it is likely the result of a beneficial mutation, since it pays to be able to

---

* In brief, cells go through something akin to the life span of an organism. They're formed, then they grow and reproduce. As they move through maturity, they don't reproduce as much (if at all), and they often look very different than when they were younger. Then they work for a while, get worn out, and die. Think of cancer cells as cells that get locked in the reproductive stage. Because that's all cancer cells do: They reproduce—over and over and over again, never becoming functionally mature cells. Instead, they spread to other regions of the body (often via the circulatory or lymphatic systems). When they get there, they reproduce and reproduce and reproduce some more, ultimately screwing up the function of wherever it is they happened to have landed.

repair your own heart if you happen to be a tiny minnow-like fish (or a similarly pint-sized red-spotted newt) and thus a popular menu item for a host of bitey predator types.

But no matter how this trait evolved, the inability of the human heart to repair itself currently presents us with a serious problem, and science has presented us with the opportunity to confront it. Researchers are examining several approaches. These include identifying chemicals that would do one of the following: stimulate mature cardiomyocytes to divide; transform cells like fibroblasts into cardiomyocytes; or compel cardiac stem cells to differentiate into cardiomyocytes. Each of these is a seriously complex endeavor, especially considering that they each require modifying the behavior of cardiac blood vessels as well. After all, any potentially rejuvenated muscle tissue will need a fully functional blood supply to carry in the tissue repair team, nutrients, and oxygen.

Although removing large sections of the zebrafish heart elicits a significant regenerative response, researchers still need to develop zebrafish models demonstrating a similar response to more common human heart maladies. Because of this, scientists are now attempting to produce zebrafish models of human cardiac valve disorders, congenital heart defects, and lipid-related issues like high cholesterol.

The learning curve is steep, but researchers hope that what they have learned about the zebrafish hearts, as well as the hearts of other nonmammals, may one day usher in a new era in therapeutic heart regeneration.

GIVEN THAT HUMANS are even closer genetically to reptiles than they are to fish, it makes perfect sense that reptiles are also proving extremely valuable to the medical community. The Burmese python (*Python bivittatus*) is yet another example of how nonhuman hearts are helping scientists develop therapies for some very human diseases.

The snake in question, easily identified by the distinctive arrowhead markings atop its head, is native to the grassy marshes, forests, and caves of Southeast Asia. The Burmese python is usually ranked as the second- or third-largest species of snake in the world, and the female can attain lengths upwards of twenty feet, with a girth approximating that of a telephone pole. A python of that size would likely weigh in at around three hundred pounds.* Males are slightly smaller, maxing out at around fifteen feet.

I actually had one of these beautiful reptiles as a teen, although mine was only about four feet in length. Still, the presence of a snake in my home excited my friends and me to no end, especially at feeding time. But not everyone felt that way about *il serpente*, this including my mom and at least six of my beloved eight Aunt Roses.† I also vividly recall that, after being informed about Alice, the guys doing construction on our house on Long Island shunned my room like it was a plague ward. I, on the other hand, was fascinated by the constrictor's calm demeanor, how it shed periodically, and its ability to unhinge its jaws before swallowing a weekly parade of larger-than-python-head-sized mice. The medical community's interest in the snake, though, is related to something I never knew during my childhood. Nobody did. It was the observation, made in 2005 by researchers at the University of California, Irvine, that within

---

* The green anaconda (*Eunectes murinus*), native to South America, is generally believed to be the second *longest* snake in the world but the largest in terms of body mass. The largest specimen ever measured was twenty-eight feet in length and was estimated to weigh over five hundred pounds. For sheer length, the reticulated python (*Python reticulatus*) sits atop most lists, with one specimen measured at thirty-three feet.

† Readers of my previous books may recall that I had something like eight Aunt Roses while growing up, and that to differentiate them I set up a sort of field guide based on characteristics like height and facial mole placement.

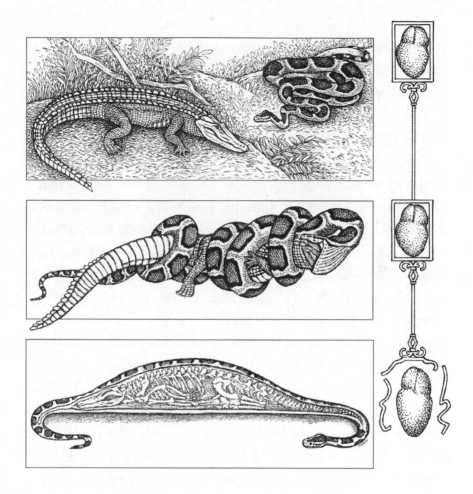

three days of consuming a meal, the heart of a Burmese python increases in size by 40 percent.

I spoke to University of Colorado Boulder researcher Leslie Leinwand, who has been studying this phenomenon in Burmese pythons for over a decade. She explained that the unusual adaptation is a byproduct of the python's odd meal schedule. In its native environment, a python can go for a year without eating any food, suffering very little in the way of negative effects—despite the fact that any mammal would perish should it try this feat. "And so these animals have adapted to do extreme things," Leinwand told me. "One of these is eating gigantic meals when they get the opportunity."

Like boa constrictors and anacondas, pythons are constrictors. They hunt by ambushing their prey, which can often outweigh them by up to 50 percent. The little guy I owned as a kid ate rodents, but in the wild, Burmese pythons prey on pigs, deer, and even small humans. The snakes subdue the victims with a nonlethal bite before quickly entrapping them within the thick coils of their muscular bodies. Contraction of these muscles causes the prey's chest cavity to compress, preventing it from expanding and thus filling the lungs during inspiration. Death by asphyxiation follows. Soon after, the snake uncoils, and unhinges its jaws (a truly interesting sight to behold). Then, beginning at the head, the constrictor "walks" its way down the victim's body, eventually swallowing its meal whole.

This type of feeding puts the snake in a fair amount of danger from predators. Imagine for a moment consuming something the size of a Great Dane in one bite and then waddling off to wait for digestion to run its course. On second thought, maybe you shouldn't imagine that. The bottom line is that it makes sense that constrictors evolved the ability to feed less frequently.

But pythons have also developed a phenomenal shortcut that enables them to get back on the move as fast as possible. Not only are they able to digest prey up to half of their own body weight in a period of only four to six days, but they are able to harness that digestion into tissue growth. With the exception of the Burmese python's brain, which is confined within the skull, "nearly every organ in the body undergoes an extremely rapid growth in size and mass," Leinwand told me.

This change is not merely due to an accumulation of fluid; it consists of actual tissue growth, usually taking place within twenty-four hours after consuming a meal. "Something that would never happen in a mammal," Leinwand added. I thought about mentioning one particular post-Thanksgiving weigh-in that I'd experienced, but thought better of it.

Leinwand's initial research interest had been in physiological human heart growth—namely, the growth that takes place in athletes' hearts. Most people think of heart enlargement in humans as purely a symptom of disease, and it's true that it can result from conditions like untreated high blood pressure or coronary artery disease. This type of heart growth is called pathological hypertrophy. "Hypertrophy" refers to growth in the *size* of specific cells—in this instance, the cardiac muscle cells. And "pathological" relates to an injury or disease state. To be clear, not all hypertrophy is bad, since, for example, it is a common result of weight training.*

"In some of these [disease-related] cases," Leinwand explained, "the muscle gets really, really big, but that happens at the expense of the heart chambers. So you end up with extremely thick heart walls but relatively small chambers. That's different than what happens in a highly conditioned athlete. They end up enlarging both the muscle *and* the chambers proportionately—larger amounts of muscle but also larger chambers for the blood to flow into and out of." This was the type of growth she was seeing in pythons.

Leinwand told me that the idea had come to her that if her team was able to understand how a python heart could grow so quickly, they might be able to reverse or prevent heart disease in humans, specifically by offering a lifesaving option for those whose hearts had become too unhealthy to benefit from exercise. (In hearts that *can* withstand exercise, the benefits include better circulation, increased oxygen supply to tissues, lower blood pressure, and a decrease in blood triglyceride levels.)

---

* Hypertrophy is different from hyperplasia, a form of growth in which cells remain the same size but increase in number. An example of hyperplastic growth would be the increase in body size that occurs throughout childhood.

Unfortunately, though, the logistics of the python-related experiments quickly turned into a nightmare. Sometime in the 1990s, people in southern Florida began releasing their unwanted pet pythons into the Everglades. One of the most species-diverse habitats in the United States, the Everglades are also a perfect environment for the tropical temperature–loving pythons, allowing them to survive the entire year with neither the threat of cold weather nor a seasonal lack of food. Worst of all, the menu available to the invasive serpents in the Everglades is an extensive one, which includes endangered species like bobcats, as well as other midsized mammals like raccoons, opossums, and foxes.* The scaly invaders also became the dominant predator of marsh rabbits.

Within three decades, the population of pythons in Florida had risen to somewhere between half a million and a million. It was an ecological nightmare that in 2012 led the US Department of the Interior to ban the sale of Burmese pythons, while another federal ban the same year prohibited the transportation of pythons across state lines. These bans have been a step in the right direction for the Everglades, although they have done little to curtail the rapidly growing snake population embedded there. But the crisis made it almost impossible for Leinwand and her colleagues to obtain specimens for their studies.

Still, Leinwand told me, after almost three years of struggling with the python-transportation roadblock, the scientists were able to maneuver around "a web of bureaucracy" to secure the specimens required.

They began their investigation of the python heart by focusing on those specimens that had recently eaten. Among their earliest discoveries was that when they drew blood from snakes that had recently consumed a large meal, it was white—"so filled with fat that it was basically opaque,"

---

* A 2012 paper in the *Proceedings of the National Academy of Sciences* reported that sightings of raccoons and opossums in the Everglades decreased by almost 99 percent between 2003 and 2011, while bobcat sightings dropped almost 88 percent. Pythons were identified as the culprits.

Leinwand told me. In humans, that would be bad news, since the fat would likely accumulate in their organs and lead to heart disease, in which fatty plaques form within the walls of the narrow heart-supplying coronary arteries.

"So, when we saw the snake blood looking like milk," she said, "I wondered why they didn't exhibit any symptoms of heart disease, since their hearts must be filled with fat."

But further examination revealed that the postmeal python hearts weren't filled with fat at all. In fact, there was actually *less* fat in their hearts than in snakes that had been fasted. Eventually, the research team figured out why.

"For you and me," Leinwand said, "or even a healthy rodent, fat is fuel, and it gets burned. When you begin to get heart disease, you stop burning fat and it accumulates in the heart."

But in pythons, something else occurs. A python's response to consuming a megameal is that its heart turns into a fat-burning machine. Simultaneously, the heart grows larger—but not in the pathological way, which given the animal's peculiar feeding habits would be maladaptive. Leinwand wondered what it was about the snake's physiology that enabled its heart to develop like that of an athlete rather than a couch potato.

The researchers determined that the trigger substance for the dramatic increase in heart size was the fat in the blood—more specifically, three fatty acids that occur naturally in food. These are myristic acid, palmitic acid, and palmitoleic acid. Humans take them as dietary supplements, like fish oil (though cardiologist Patrick McBride would likely make a face at this).

Leinwand and her team proved the role of the fatty acids by injecting the trio of substances into fasted snakes. In each case, their hearts increased in size, just as if they had recently eaten a meal. The same three substances also worked in mice, whose hearts responded by growing as large as the hearts of mice that had been exercising for weeks *without* the fatty-acid supplement. And, remarkably, in mice and fasted snakes alike, that growth maintained normal anatomical proportions, rather than mimicking the heart enlargement caused by disease. You'll recall that in pathological heart enlargement, myocardial growth is *not* reflected by an increase in the volumes of the atria and the ventricles. Finally, and hopefully, there has been no evidence that the fatty-acid cocktail employed by the researchers triggers any disease pathways.

Although the initial results were startling, it is clear that there remains a considerable amount of work to be done. The next phase of Leinwand's studies will test the fatty acid against larger animal models of heart disease. The hope is that each of these steps will bring them closer to their real goal.

"Ultimately," Leinwand told me, "my colleagues and I hope to use what we've learned not as a substitute for heart-healthy exercise but, for example, in cases where patients suffering from heart disease are unable to exercise, and where a therapeutic alternative could offer those patients a healthier heart and a longer life."

And having already established a biomedical company, Leslie Leinwand hopes that one day she'll turn snake oil sales into a positive career choice. It was a point made clear in 2017, when she won the American Heart Association's Distinguished Scientist Award for outstanding contributions to the field of heart health.

Someone has to stand up and say,
"The answer isn't another pill.
The answer is spinach."
—BILL MAHER

[16]

Grow Your Own

TO EXPLORE A very different approach to cardiac regeneration I visited Harald Ott, a researcher at the Harvard Stem Cell Institute. Ott and his colleagues are involved in an ambitious project: to grow human hearts, and potentially other organs, from stem cells.

When most of the roughly two hundred types of cells found in the human body reproduce, the outcome is an identical cell. Muscle cells produce more muscle cells, fat cells (adipocytes) produce more fat cells, and so on. Stem cells are different, though, because given the right conditions, they can be stimulated to produce different cell types. Most stem cells still have their limits. Stem cells in the blood, for example, can produce only other blood cells. But embryonic stem cells are special, because they can be stimulated to produce any type of cell (so they're described as "pluripotent"). They can be harvested from a few places—like the umbilical cord or from embryos, with the latter making their collection and use extremely controversial. Their pluripotency, however, has made them incredibly valuable for researchers involved in stem cell therapy, which seeks to address diseased or malfunctioning organs not by transplant but by growing those organs from stem cells.

"So why is it necessary to engineer human hearts?" I asked Ott, who is at the forefront of a very unique aspect of stem cell research.

He explained that the field of medicine had gotten really good at dealing with acute problems like traumatic injuries and diseases like pneumonia. As a result, more people are surviving those acute events, and many of them are living to a ripe old age—an age when their organs begin to break down.

"Some of the tissues like the liver or bones after fracture have built-in regeneration systems," Ott told me. "But many organs [like the heart] don't have the capability to regenerate themselves."

Initially, this might not be a huge issue, since some of these organs, like the lungs, have a reserve of extra cells. But that reserve can run out.

"End organ failure is a global epidemic affecting millions of people," he said. "So rather than dying from car accidents, pneumonia, and other issues, millions of people are getting older and older, and accumulating massive injuries that lead to deteriorating function."

As a result, there has recently been a serious shift in gears within the medical research community. Whereas for much of the twentieth century the goal was to repair damaged tissues and organs, considerable effort is now focused on engineering organs like the heart, kidney, and pancreas, in order to replace the patients' original but failing equipment.

Ott was first drawn to stem cell research because of the work of cardiologist Doris Taylor at the University of Minnesota in the mid-to-late 2000s. Her studies initially centered around rebuilding cardiac function by transplanting stem cells into the hearts of test rabbits that had suffered acute myocardial infarction. During Ott's time in Taylor's lab, they determined that simply injecting cells into a faulty heart wasn't effective enough, and that they would need to regenerate three-dimensional structures, not just repair them. Since then, Taylor has continued her work, eventually moving on to become the director of regenerative medicine at the Texas Heart Institute. Ott, meanwhile, took a fellowship in cardiothoracic surgery at Massachusetts General Hospital and a position as an instructor of surgery at Harvard Medical School.

Ott explained that his current experiments build off tissue-engineering research from the 1990s. In those studies, researchers showed that functional three-dimensional tissue could be generated by building cells onto a scaffolding of an extracellular matrix composed primarily of collagen.* The extracellular matrix of a tissue is secreted by its cells and gives tissues like bone and cartilage their form and their distinct physical characteristics. The characteristics of a matrix composed of collagen are that it can be stretched without breaking (i.e., it has tensile strength), that it does not elicit an immune response (i.e., it has low antigenicity), and that it readily allows other cells (like myocytes) to grow on it.

"I'm not an engineer by training," Ott told me. "So when I started working at this, instead of creating a scaffold from scratch I used organs from cadavers."

Ott and his colleagues put cadaver hearts through a process called decellularization, in which special detergents are used to dissolve away all of the cells. What they were left with was a flexible heart-shaped structure composed solely of a collagen-based extracellular matrix.

I examined one of his early decellularized heart specimens, which had

---

* If this sounds vaguely familiar, it's because the rapid regeneration of the zebrafish heart takes place upon a scaffold of collagenous connective tissue as well, in that case laid down by fibrocytes after a traumatic injury.

come from a pig. It was opaque and stark white in color, composed of solid components like collagen, elastin, and fibronectin—a cell adhesion molecule that binds cells to these substances (kind of like glue). Basically, though, it looked just like a pig heart. I found it fascinating that the complex structure before me had been produced by cells that no longer existed. What they had left behind was a heart whose architecture had been precisely preserved, thus making it perfect framework upon which Ott and his colleagues could build a new heart.

Because all of the cells had been removed, leaving behind only structural proteins, the scaffold would not be subject to the same immune response as a transplant heart. When the body recognizes cells as allogenic—that is, not of the self, and therefore immunologically incompatible—the immune system attacks those cells. This is the primary reason for the rejection of so-called allotransplants, those transplants that come from incompatible donors. Upon what was essentially a blank template, though, a research team could theoretically build a compatible organ with little fear of rejection.

A key question that remained, however, was: How would Ott and his team repopulate his heart-shaped framework with new cells that wouldn't be attacked? He explained that his research had gotten a huge boost from the 2012 Nobel Prize–winning discovery by John Gurdon and Shinya Yamanaka that mature cells could be genetically reprogrammed into stem cells. They did this by introducing four genes responsible for keeping stem cells immature into the mature cells. Even better news was that the resulting cells weren't just any stem cells—they were stem cells of the pluripotent variety. You'll recall that depending on how these stem cells are stimulated, they have the ability to differentiate into any of the approximately two hundred cell types that exist in the human body. As for where these mature pretreatment cells would come from, the easier the access, the better, and so it was exciting when researchers found that fibroblasts fit the bill.

Although they coinhabit the myocardium with cardiac muscle cells, fibroblasts are also the most common type of cell found in connective tissue, which includes the skin's dermal layer. Among other things, and as mentioned during the zebrafish discussion, they are responsible for producing structural proteins like collagen and elastin fibers and the extracellular matrix, the noncellular material that surrounds cells. Ott explained that the ease of accessibility to fibroblasts in the skin makes them far less trouble to obtain than cells from a cardiac biopsy.

Once the fibroblasts have successfully been turned into stem cells and then into cardiac muscle cells, they can then be seeded back onto the cell scaffolding. This, so far, has been Ott's sticking point. His team has been able to grow small patches of heart, and to get those cells to contract when stimulated. But they have not yet been able to create an entire pumping human heart.

Other labs working on the issue aren't attempting to build new hearts, but are exploring the use of similarly reprogrammed sections of contractile cells. Led by Sian Harding, a professor at Imperial College London, researchers in the United Kingdom and Germany were able to grow patches composed of human myocytes, which were then stitched on to the hearts of living rabbits, where they became fully functional cardiac muscle tissue. With human trials set to begin soon, it is hoped that this technique will enable cardiologists to replace the noncontractile scar tissue that results after a heart attack.

But patches of myocardial cells do not a heart make, and one major challenge that Ott and his colleagues face is getting his reprogrammed cells to form three-dimensional structures, like the coronary blood vessels that would be required to supply his newly built hearts. The cells themselves need to create these structures, serving not just as building blocks but as participants in the manufacturing process. Maddeningly, the blueprints for that behavior already exist within the cell, coded into the genetic portfolio—but so far, they are not accessible to scientists, who are still on the search for a way to switch the behavior on.

Until Ott and his colleagues can "flip this switch," they will improvise. Unable to create blood vessels purely from scratch, they decided to start at the same place they're starting with heart tissue: with scaffolding—in this case a section of decellularized blood vessel. Like the rest of the heart, the coronary blood vessels that supply it leave behind a framework of connective tissue once their cellular components have been dissolved away.

"We tell the cell, 'You're an immature blood vessel cell, and by the way here's a pipe. Can you just line that pipe for me?' And then the cells will do that," Ott said. "That's what's really unique about our scaffolding—in these decellularized organs, we actually have intact piping."

Creating three-dimensional structures to replace their faulty human counterparts remains a serious challenge. But using a previously existing scaffold, in this case the connective-tissue framework from a formerly functional blood vessel, isn't the only avenue of research being explored to address the problem.

Biomedical engineer Glenn Gaudette from Worcester Polytechnic Institute is also working on therapeutic heart regeneration, but he found himself using a rather different type of framework after one of his grad students returned from lunch with something amazing he had discovered in the cafeteria.

I met with Gaudette in his lab to discuss what happened next.

He began by explaining that anyone doing repair work on a damaged heart, or any damaged organ for that matter, knows the importance of blood vessels—many of them of the microdiameter variety.

"Heart muscle dies when it doesn't get enough blood flow," Gaudette told me.

This was, as Ott had previously pointed out to me, of particular concern in heart regeneration studies and was proving to be a sticking point in Gaudette's own research. While his team has been able to get heart cells to grow on the scaffolding of blood vessels around a decellularized heart, they have not been able to fully reproduce its structural and functional complexity.

"And that's why we came up with this," Gaudette said, offering me something small and green to peruse.

I held the object carefully, marveling at its veins and how it looked remarkably like a spinach leaf that he might have purchased at a grocery store. Gaudette assured me that it was, and that he had.

"Those veins transport water," he said. "Our veins transport blood. From an engineering standpoint, they both transport fluids. So my grad student at the time, Josh Gershlak, said, 'If we get rid of all the spinach, would we still have those vessels left behind?' And that's where the whole experiment started."

As Ott does with his donor hearts, Gaudette and Gershlak (now a postdoc) subject their spinach leaves to a chemical bath that strips them of their cells but keeps the extracellular framework. Similarly, this allows the vessels to retain their original structure and prevents that structure from being rejected by the eventual recipient's immune system.

Gaudette gave me a tour of his lab, during which I saw how the specimens are prepared. The spinach leaves he employs are hung individually in small bottles, each located about four feet below a gravity-fed supply of special detergent. As the detergent drips downward, it passes into a series of thin rubber tubes, each terminating in a large-gauge hypodermic needle, which is inserted into the tip of a leaf stem.

This gravity drip system provides a constant flow of detergent into the leaf. As the detergent encounters the plant's cells, it opens tiny holes in them and allows their contents to drain, so that when the detergent departs out the leaf tip, it carries with it the cellular contents. After a five-day perfusion period, what remains is a colorless, structurally perfect model of a leaf, albeit one with no plant cells remaining. The model is composed of the sturdy structural polysaccharide named cellulose.

If that substance rings a bell it's probably because plant cell walls are composed of cellulose, also known as the dietary fiber that passes undigested through our intestines, Roto-Rooter-style. In fact, no vertebrates can digest cellulose by themselves, although some do enlist the aid of endosymbiotic bacteria. Vast numbers of these microorganisms exist in digestive system organs, like the cecum of a horse or the rumen of a cow. The symbiosis aspect relates to the fact that the bacteria get a nice, warm place to live, while their four-footed cosymbionts get the benefit of the cellulase (the enzyme that breaks down cellulose).

Released into the digestive tract, the enzyme comes into contact with the herbivore's cellulose-rich diet, breaking the polysaccharide down into easily digestible compounds like simple sugars. This adaptation allows the herbivore's digestive system to extract nutrients and energy from previously undigestible stuff like grass. Among the invertebrates, even those infamous for their plant- and wood-munching prowess, many can't digest cellulose without help either. Some groups of termites, for example, require endosymbiotic bacteria in order to digest wood, and baby termites will starve to death if they don't obtain their own starter colony of flagella-waving gut microbes.

They do so by consuming a bit of feces from a parent or nestmate. Other termite species are flagellate-free, having evolved the ability to produce their own cellulase without the need to host fifty billion or so microbial gut-guests.

For Glenn Gaudette's purposes, though, the important thing is that cellulose is not only structurally sound, but it is close to being biologically inert, with the human body showing little or no immune response to the stuff. As such, it's a near-perfect biocompatible material, and has already been approved for use in certain medical devices. These include sheets composed of cellulose fibrils, spun by bacteria and applied to wounds, and implantable capsules for drug delivery.

Cellulose is a component of more than one attempt to engineer structures like the heart from scratch. Instead of using spinach leaves, researchers at Tel Aviv University are taking the 3D bioprinting approach. Their efforts, though early, seek to use a biopsy from a patient as the "ink" for their 3D printer. In April 2019, with much fanfare and media coverage, Tal Dvir and his team announced that they had indeed printed a small heart (about the size of a rabbit's heart). The obstacles confronting these scientists are many. Among them is the fact that while the cells of the printed structure can contract, the heart itself cannot yet pump. Additionally, Dvir's team will need to address the question of how to print the tiny blood vessels of the heart.

There is still much research to be done, and there are many obstacles to overcome. But the outlook for cellulose is exciting. Gaudette's lab has been able to get human heart cells to grow on the spinach scaffolding, and experiments are now underway to dissolve the cellulose after it has served its purpose. The hope is that one day, vessels formed on a structure of cellulose will be able to be stimulated to become surrogate blood vessels composed solely of human cells.

And while it's impossible to predict how much of this research will ever find practical applications, it is intriguing that scientists like Gaudette are looking toward the plant kingdom for a new and extremely novel way to benefit humans.

Given the complexity of regenerating a heart, or other organs like kidneys and lungs, I wondered why something so drastic was necessary. Why not search instead for better repair techniques or focus on disease prevention?

The answers relate to the fact that there are approximately 40,000 organ transplants performed in the United States each year (about 10 percent of them are heart transplants), and, as of September 2020, there were approximately 109,000 candidates on the US national waiting list. These patients are beyond disease prevention, and in many cases have organs so damaged that repair cannot offer a long-term solution. It's been estimated that every day about twenty people die while waiting.

Harald Ott explains it like this. "If the radiator in your car is broken, they don't *fix* your radiator anymore—they just swap it out for another one."

With this in mind, the ultimate goal of regenerative medicine is to come up with a replacement for those hearts (as well as kidneys, livers, lungs, and intestines) that isn't dependent on an often terminally long waiting list or the prospect of having transplant recipients spend the rest of their lives on immunosuppressant drugs. Other researchers continue to look for those replacements in the animal kingdom—for instance, genetically altering pigs to provide organs comparable to human organs, without the threat of tissue rejection.

I asked Ott to speculate on where he thinks organ regeneration therapies are heading: "Let's say it's twenty years from now and all of this research has worked out *really* well. Somebody has a damaged heart. What happens next?"

"They walk into a clinic, a skin biopsy is taken, and you grow them a heart," Ott said. "Once the patient reaches a point where their heart is just not doing well enough anymore, you just swap it out."

"And other organs as well?"

"And other organs as well," he repeated. "That's what I hope for."

Acknowledgments

I'D LIKE TO thank my agent, Gillian MacKenzie, for her hard work, great advice, perseverance, and patience. Thanks also to Kirsten Wolf and Renée Jarvis at the MacKenzie Wolf literary agency for their assistance, especially during the bumpy stretches.

I offer my sincere thanks to Amy Gash and Abby Muller, my incredibly talented editors at Algonquin Books, and to Elizabeth Johnson for her significant and outstanding editorial input. Thanks also go out to Amanda Dissinger, Brunson Hoole, and the entire production and marketing teams at Algonquin. It is an absolute pleasure working with you!

I was very lucky to have interviewed with or received assistance from a long list of experts who were very generous with their time. Extreme thanks and gratitude go out to Ken Angielczyk, Maria Brown, Mark Engstrom, Chris Chabot, Jon Costanzo, Patricia Dorn, Miranda Dunbar, Glenn Gaudette, Josh Gershlak, Dan Gibson, Hirofumi Hirakawa, Leslie Leinwand, Burton Lim, Patrick McBride, Jacqueline Miller, Kristin O'Brien, Harald Ott, DeeAnn Reeder, Mark Siddall, John Tanacredi, and Win Watson.

I owe a huge debt of gratitude to my friends and colleagues in the bat research community and at the American Museum of Natural History. They include Ricky Adams, Frank Bonaccorso, Betsy Dumont, Neil Duncan, Julie Faure-Lacroix, Mary Knight, Gary Kwiecinski, Ross MacPhee, Liam McGuire, Shahroukh Mistry, Mark Norell, Mike Novacek, Maria Sagot, Nancy Simmons (It's good to know the Queen), Ian Tattersall, Elizabeth Taylor, and Rob Voss.

I've been fortunate to have had several incredible mentors, none more important than John W. Hermanson (Cornell University's Program in

Zoology and Wildlife Conservation). Among many other things, John taught me to think like a scientist, as well as the value of figuring stuff out for myself.

A very special thanks to my great friend, collaborator, confidant, and coconspirator, Leslie Nesbitt Sittlow.

As usual, my dear friends Darrin Lunde and Patricia J. Wynne were instrumental in helping me develop this project from a vague idea into a finished book. A million thanks also to Patricia for all of the amazing figures she drew (not to mention the spot-on advice). As always, I can't wait for our next project together.

I was privileged to work with the talented team at TED-Ed on the *Pump*-related video "How Do Blood Transfusions Work?" (https://ed.ted .com/lessons/how-does-blood-transfusion-work-bill-schutt#watch). Shout outs and special thanks to Elizabeth Cox, Logan Smalley, Talia Soliman, and Gerta Xhelo.

A special thank-you goes out to my teachers, readers, and supporters at the Southampton College Writers Conference, especially Bob Reeves, Bharati Mukherjee (RIP), and Clark Blaise.

At Southampton College (RIP) and LIU Post, thanks and gratitude go out to Greg Arnold, Margaret Boorstein, Nate Bowditch, Ted Brummel, Kim Cline, Gina Famulare, Art Goldberg (RIP), Alan Hecht, Kent Hatch, Mary Lai (RIP), Karin Melkonian, Kathy Mendola, Glynis Pereyra, Howard Reisman, Beth Rondot, Jen Snekser, and Steve Tettelbach. Thanks also to my LIU Post teaching assistants, especially Bushra Azhar, Elsie Jasmin, Kelly Howlonia, Nelson LiCalsi, and Yurie Miranda.

Sincere thanks go out to my best friend Bob Adamo (RIP) and the Adamo family, Jeanne Bass, John Bodnar, Chris Chapin, Kitty Charde, Kristi Ashley Collom, Alice Cooper, Aza Derman, Suzanne Finnamore Luckenbach (who predicted it all), John Glusman, Tommie Keene (RIP), Kathy and Brian Kennedy, Christian Lennon and Erin Nicosia-Lennon,

Bob Lorzing, the legendary and wonderful literary agent Elaine Markson (RIP), Maceo Mitchell, Carrie McKenna, Val Montoya, the Pedersen family and various offshoots, Ashley, Kelly and Kyle Pellegrino, Don Peterson, Pirate Mike Whitney (Iggy's Keltic Lounge), Jerry Ruotolo (my great friend and favorite photographer), Laura Schlecker, Edwin J. Spicka (my mentor at the State University of New York at Geneseo), Carol Steinberg (who came through when the going got tough), Lynn Swisher, Frank Trezza, Katherine Turman (Nights with Alice Cooper), and Mindy Weisberger.

Finally, my eternal thanks and love go out to my family for their patience, love, encouragement, and unwavering support, especially to my wonderful wife, Janet Schutt, and my son, Billy Schutt, my cousins, nieces, and nephews, my grandparents (Angelo and Millie DiDonato), my aunts and uncles (including my Aunt Ei and all my Aunt Roses), and, of course, my parents, Bill and Marie G. Schutt.

Creatures like the icefish and the snow-burrowing tube-nosed bats would have been as fascinating to me as a child as they are today. Back in the 1960s, though, my exposure to such things came mostly through shows like The Undersea World of Jacques Cousteau and Mutual of Omaha's Wild Kingdom ("Jim, these icefishes have big hearts that help insure their survival in tough conditions. And you can insure your own family's survival with coverage from Mutual of Omaha").

But even if this type of information wasn't quite as exciting to others (my parents and assorted puzzled relatives, for example), I'm certain that having watched my childhood reaction to learning about the existence of giant calamari, they got a sense of the exhilaration that might drive an adult to climb into a rotting blue whale carcass, go diving in the Arctic to observe icefish, or, in my case, to spend twenty years chasing down and studying vampire bats.

My parents and that whole generation of funny, loving family members—many of them first-generation Italian Americans—are all gone now. Luckily, though, I'm left with the very real comfort that no matter how odd my behavior may have seemed to these people—always peering under stones and collecting creatures of every ilk—they knew.

They definitely knew.

Notes

Front Matter

vii   "Heart" definition 1   "Heart," Science Flashcards, Quizlet, https://quizlet.com
/213580838/science-flash-cards/.

vii   "Heart definitions" 2–7   "Heart," Cambridge Dictionary, https://dictionary
.cambridge.org/us/dictionary/english/heart.

Prologue: A Small Town with a Big Heart

3   more than 380,000 blue whales   T. A. Branch et al., Historical Catch Series for
Antarctic and Pygmy Blue Whales, Report (SC/60/SH9) to the International
Whaling Commission (2008).

Chapter 1: Size Matters I

17   a pair of carotid arteries   J. R. Miller et al., "The Challenges of Plastinating a Blue
Whale (Balaenoptera musculus) Heart," Journal of Plastination 29, no. 2 (2017):
22–29.

24   the only way to pump more blood   Knut Schmidt-Nielsen, Animal Physiology
(Cambridge: Cambridge University Press, 1983), 207.

24n   An average-sized man   Knut Schmidt-Nielsen, Scaling: Why Is Animal Size So
Important? (Cambridge: Cambridge University Press, 1984), 139.

25   Initial research indicates that this could be so   J. A. Goldbogen et al., "Extreme
Bradycardia and Tachycardia in the World's Largest Animal," Proceedings of the
National Academy of Sciences 116, no. 50 (December 2019): 25329-32.

Chapter 2: Size Matters II

29   the first multicellular life-forms   R. Monahan-Earley, A. M. Dvorak, and W. C. Aird.
"Evolutionary Origins of the Blood Vascular System and Endothelium," Journal of
Thrombosis and Haemostasis (June 2013): 46–66.

35   an arthropod called Fuxianhuia protensa   Xiaoya Ma et al., "An Exceptionally
Preserved Arthropod Cardiovascular System from the Early Cambrian," Nature
Communications 5, no. 3560 (2014).

Chapter 3: Blue Blood and Bad Sushi

45   When Europeans came to the New World   Gary Kreamer and Stewart Michels,
"History of Horseshoe Crab Harvest on Delaware Bay," in Biology and Conservation
of Horseshoe Crabs, eds. John T. Tanacredi, Mark L. Botton, and David Smith
(New York: Springer, 2009), 299–302.

45   whelks also enjoy   Kreamer and Michels, "Horseshoe Crab," 307–309.

46   a growing problem with poachers   Mark L. Botton et al., "Emerging Issues
         in Horseshoe Crab Conservation: A Perspective from the IUCN Species
         Specialist Group," in Changing Global Perspectives on Horseshoe Crab Biology,
         Conservation and Management, eds. Ruth Herrold Carmichael et al. (New York:
         Springer, 2015), 377–78.

46   its lethality is due   Thomas Zimmer, "Effects of Tetrodotoxin on the Mammalian
         Cardiovascular System," Marine Drugs 8, no. 3 (2010): 741–62.

48   "A significant hindrance to treating"   "Researchers Discover How Blood Vessels
         Protect the Brain during Inflammation," Medical Xpress. February 21, 2019,
         https://medicalxpress.com/news/2019-02-blood-vessels-brain-inflammation.html.

48   studies have shown a probable link   Stephen S. Dominy et al., "Porphyromonas
         gingivalis in Alzheimer's Disease Brains: Evidence for Disease Causation and
         Treatment with Small-Molecule Inhibitors," Science Advances 5, no. 1 (January 23,
         2019), https://advances.sciencemag.org/content/5/1/eaau3333.

48   a growing suspicion that these sticky masses   Dominy et al. "Porphyromonas
         gingivalis."

49   Because victims can remain conscious   Terence Hines, "Zombies and
         Tetrodotoxin," Skeptical Inquirer 32, no. 3 (May/June 2008).

54   functional morphologists   D. M. Bramble and D. R. Carrier, "Running and
         Breathing in Mammals," Science 219, no. 4582 (January 21, 1983): 251–56.

58   an ancient form of immune defense   F. B. Bang, "A Bacterial Disease of Limulus
         polyphemus," Bulletin of the Johns Hopkins Hospital 98, no. 5 (May 1956): 325–51.

58n   Amoebocytes occur in other invertebrates   S. P. Kapur and A. Sen Gupta,
         "The Role of Amoebocytes in the Regeneration of Shell in the Land Pulmonate,
         Euplecta indica (Pfieffer)," Biological Bulletin 139, no. 3 (1970): 502–09.

61   a colleague of Fred Bang's   Jack Levin, Peter A. Tomasulo, and Ronald . S. Oser,
         "Detection of Endotoxin in Human Blood and Demonstration of an Inhibitor,"
         Journal of Laboratory and Clinical Medicine 75, no. 6 (June 1, 1970): 903.

61   nearly half a million horseshoe crabs   "Horseshoe Crab," Atlantic States Marine
         Fisheries Commission, http://www.asmfc.org/species/horseshoe-crab.

62   various policies have hampered its ability   Michael J. Millard et al., "Assessment
         and Management of North American Horseshoe Crab Populations, with Emphasis
         on a Multispecies Framework for Delaware Bay, U.S.A. Populations," in Changing
         Global Perspectives on Horseshoe Crab Biology, Conservation and Management,
         eds. Ruth Herrold Carmichael et al. (New York: Springer, 2015), 416.

63   a decrease in the abundance of horseshoe crabs   "Horseshoe Crab," ASMFC.

67   around 104 beats per minute   A. D. Jose and D. Collison, "The Normal Range
         and Determinants of the Intrinsic Heart Rate in Man," Cardiovascular Research 4,
         no. 2 (April 1970): 160–67.

69   Singaporean biologist Jeak Ling Ding   Sarah Zhang, "The Last Days of the Blue-
         Blood Harvest," Atlantic, May 9, 2018, https://www.theatlantic.com/science
         /archive/2018/05/blood-in-the-water/559229/.

Chapter 4: Insects, Sump Pumps, Giraffes, and Mothra

71   This tracheal system explains why   Silke Hagner-Holler et al., "A Respiratory
      Hemocyanin from an Insect," Proceedings of the National Academy of
      Sciences 101, no. 3 (January 20, 2004): 871–74.

72   some ancient (a.k.a. basal) insects   Hagner-Holler et al., "Respiratory
      Hemocyanin."

75   In bristletails, though, bidirectional flow   Günther Pass et al., "Phylogenetic
      Relationships of the Orders of Hexapoda: Contributions from the Circulatory
      Organs for a Morphological Data Matrix," Arthropod Systematics and
      Phylogeny 64, no. 2 (2006): 165–203.

77   The deoxygenated blood eventually returns   Reinhold Hustert et al., "A New Kind
      of Auxiliary Heart in Insects: Functional Morphology and Neuronal Control of
      the Accessory Pulsatile Organs of the Cricket Ovipositor," Frontiers in Zoology 11,
      no. 43 (2014).

81n   A recent study also shows a clear link   SPRINT MIND Investigators for the
      SPRINT Research Group, "Effect of Intensive vs Standard Blood Pressure Control
      on Probable Dementia: A Randomized Clinical Trial," Journal of the American
      Medical Association 321, no. 6 (2019):553–61.

84   the arteries running through giraffe legs   Karin K. Petersen et al., "Protection
      against High Intravascular Pressure in Giraffe Legs," American Journal of
      Physiology: Regulatory, Integrative and Comparative Physiology 305, no. 9
      (November 1, 2013) R1021–30.

Chapter 5: On the Vertebrate Beat

90   "beats from one end to the other"   "Sea Squirt Pacemaker Gives New Insight into
      Evolution of the Human Heart," Healthcare-in-Europe.com, https://healthcare-in
      -europe.com/en/news/sea-squirt-pacemaker-gives-new-insight-into-evolution-of
      -the-human-heart.html.

Chapter 6: Out in the Cold

105   Cardiologist Parag Joshi   "Cholesterol Levels Vary by Season, Get Worse in
      Colder Months," American College of Cardiology, March 27, 2014, https://www.acc
      .org/about-acc/press-releases/2014/03/27/13/50/joshi-seasonal-cholesterol-pr.

105n more heart attacks occur in the morning   Salynn Boyles, "Heart Attacks in the
      Morning Are More Severe," WebMD, April 27, 1001, https://www.webmd.com
      /heart-disease/news/20110427/heart-attacks-in-the-morning-are-more-severe#1.

110   this characteristic led a European food company   Srinivasan Damodaran,
      "Inhibition of Ice Crystal Growth in Ice Cream Mix by Gelatin Hydrolysate,"
      Journal of Agricultural and Food Chemistry 55, no. 26 (November 29, 2007):
      10918–23.

111   The antifreeze proteins work by latching on   David Goodsell, "Molecule of the
      Month: Antifreeze Proteins," PBD-101, Protein Data Bank, December 2009,
      https://pdb101.rcsb.org/motm/120.

114   As with glucose, nitrogen is thought   James M. Wiebler et al., "Urea Hydrolysis by Gut Bacteria in a Hibernating Frog: Evidence for Urea-Nitrogen Recycling in Amphibia," Proceedings of the Royal Society B: Biological Sciences 285, no. 1878 (May 16, 2018).

117   they competed badly in a lab setting   Jon P. Costanzo, Jason T. Irwin, and Richard E. Lee Jr., "Freezing Impairment of Male Reproductive Behaviors of the Freeze-Tolerant Wood Frog, Rana sylvatica," Physiological Zoology 70, no. 2 (March–April 1997): 158–66.

122   Polar bears (Ursus maritimus) are the other   Hirofumi Hirakawa and Yu Nagasaka, "Evidence for Ussurian Tube-Nosed Bats (Murina ussuriensis) Hibernating in Snow," Scientific Reports 8, no. 12047 (2018).

125   sixty-one species of mammals with confirmed extinctions   Committee on Recently Extinct Organisms, American Museum for Natural History, http://creo.amnh.org.

Chapter 7: Ode to Baby Fae

129   the 2009 documentary   Stephanie's Heart: The Story of Baby Fae, LLUHealth, YouTube, 2009, https://www.youtube.com/watch?v=sQbJoWP-wn4.

130   "a tactical error with catastrophic consequences"   Sandra Blakeslee, "Baboon Heart Implant in Baby Fae in 1984 Assailed as 'Wishful Thinking,'" New York Times, December 20, 1985.

130   "If Baby Fae had the type AB blood group"   Robert Steinbrook, "Surgeon Tells of 'Catastrophic' Decision: Baby Fae's Death Traced to Blood Mismatch Error," Los Angeles Times, October 16, 1985.

130   a "revolution in transplantation"   Stephanie's Heart.

131   transplanting these genetically modified pig organs   Kelly Servick, "Eyeing Organs for Human Transplants, Companies Unveil the Most Extensively Gene-Edited Pigs Yet," Science, December 19, 2019, https://www.sciencemag.org/news/2019/12/eyeing-organs-human-transplants-companies-unveil-most-extensively-gene-edited-pigs-yet.

Chapter 8: Heart and Soul: The Ancient and Medieval Cardiovascular System

139   The heart, known as ab   John F. Nunn, Ancient Egyptian Medicine (London: British Museum Press, 1996), 54.

140   That movement was innate   R. K. French, "The Thorax in History 1: From Ancient Times to Aristotle," Thorax 33 (February 1978): 10–18.

140   "they carried the 'breath of life'"   French, "Thorax," 11.

140   Some modern translations   Bruno Halioua, Bernard Ziskind, and M. B. DeBevoise, Medicine in the Days of the Pharaohs (Cambridge, MA: Belknap Press, 2005), 100.

141   "silent killers"   "Aortic Aneurysms: The Silent Killer," UNC Health Talk, February 20, 2014, https://healthtalk.unchealthcare.org/aneurysms-the-silent-killer/.

141   "are difficult to interpret"   Nunn, Ancient Egyptian Medicine, 85.

141 "remarkably close to the truth" Nunn, 55.

142 He held, for instance, that the trachea French, 14.

143n One explanation for this may be French, 16.

144 It was then that two physicians H. von Staden, "The Discovery of the Body: Human Dissection and Its Cultural Contexts in Ancient Greece," Yale Journal of Biology and Medicine 65 (1992): 223–41.

145 The younger scientist also described von Staden, "Human Dissection," 224.

145 fumigation and confessions "Lustration," Encyclopaedia Britannica, https://www.britannica.com/topic/lustration.

145 Anyone, therefore, who performed von Staden, 225–26.

146 "a magical symbol of wholeness" von Staden, 227.

146 He did so by opening up a dog's artery Nunn, 11.

151 Ferdinand Peter Moog and Axel Karenberg F. P. Moog and A. Karenberg, "Between Horror and Hope: Gladiator's Blood as a Cure for Epileptics in Ancient Medicine," Journal of the History of the Neurosciences 12, no. 2 (2003), 137–43.

152 "blebs or blisters full of blood" Pierre de Brantôme, Lives of Fair and Gallant Ladies, trans. A. R. Allinson (Paris: Carrington, 1902).

153 In looking over the history and the credentials David M. Morens, "Death of a President," New England Journal of Medicine 341, no. 24 (December 9, 1999): 1845–49.

153 attached thirty leeches to every new patient Amelia Soth, "Why Did the Victorians Harbor Warm Feelings for Leeches?" JSTOR Daily, April 18, 2019, https://daily.jstor.org/why-did-the-victorians-harbor-warm-feelings-for-leeches/.

154 leech saliva Sarvesh Kumar Singh and Kshipra Rajoria, "Medical Leech Therapy in Ayurveda and Biomedicine—A Review," Journal of Ayurveda and Integrative Medicine (January 29, 2019), https://doi.org/10.1016/j.jaim.2018.09.003.

156 he had not known that large amounts of blood John B. West, "Ibn al-Nafis, the Pulmonary Circulation, and the Islamic Golden Age," Journal of Applied Physiology 105, no. 6 (2008): 1877–80.

156 "There is no passage as that part of the heart" S. I. Haddad and A. A. Khairallah, "A Forgotten Chapter in the History of the Circulation of Blood," Annals of Surgery 104, no. 1 (July 1936): 5.

156 These capillaries conclusively linked West, "Ibn al-Nafis."

156 largely forgotten until 1924 West.

158 "This communication is made not through the middle wall" West.

159 From a medical perspective "Michael Servetus," New World Encyclopedia, http://www.newworldencyclopedia.org/entry/Michael_Servetus.

160 blood "soaks plentifully through" M. Akmal, M. Zulkifle, and A. H. Ansari, "Ibn Nafis—a Forgotten Genius in the Discovery of Pulmonary Blood Circulation," Heart Views 11, no. 1 (March–May 2010): 26–30.

160 "I do not see how even the smallest" Arnold M. Katz, "Knowledge of Circulation Before William Harvey," Circulation XV (May 1957), https://www.ahajournals.org/doi/pdf/10.1161/01.CIR.15.5.726.

161 Citing a lack of evidence   C. D. O'Malley, Andreas Vesalius of Brussels, 1514–1564 (Berkeley: University of California Press, 1964).

161 Modern biographers list poor conditions   Michael J. North, "The Death of Andreas Vesalius," Circulating Now: From the Historical Collections of the National Library of Medicine, October 15, 2014, https://circulatingnow.nlm.nih .gov/2014/10/15/the-death-of-andreas-vesalius/.

161 "Between these ventricles there is a septum"   G. Eknoyan and N. G. DeSanto, "Realdo Colombo (1516–1559): A Reappraisal," American Journal of Nephrology 17, no. 3–4 (December 31, 1996): 265.

Chapter 9: What Goes In . . .

164 the German physician and chemist Andreas Libavius   M. T. Walton, "The First Blood Transfusion: French or English?" Medical History 18, no. 4 (October 1974): 360–64.

165 Several dubious accounts   S. C. Oré, "Études historiques et physiologiques sur la transfusion du sang," Paris, 1876; Villari, "La storia di Girolamo Savonarola, Firenze," 1859, 14; J. C. L. Simonde de Sismondi, "Histoire des républiques italiennes du moyen âge," Paris, 1840, vol. VII, 289.

165 "All the blood of the prostrate old man"   G. A. Lindeboom, "The Story of a Blood Transfusion to a Pope," Journal of the History of Medicine and Allied Sciences 9, no. 4 (October 1954): 456.

165 Dutch medical historian Gerrit Lindeboom   Lindeboom, "Blood Transfusion."

165 "It turns out his vivid imagination"   Lindeboom, 457.

165n Reviewed by A. Matthew Gottlieb   A. Matthew Gottlieb, "History of the First Blood Transfusion," Transfusion Medicine Reviews V, no. 3 (July 1991): 228–35.

166 "The most considerable experiment"   Frank B. Berry and H. Stoddert Parker, "Sir Christopher Wren: Compleat Philosopher," Journal of the American Medical Association 181, no. 9 (September 1, 1962).

166 "Filling their empty veins"   Christopher Marlowe, Tamburlaine the Great, part 2, scene 2, lines 107–108, ed. J. S. Cunningham (Manchester: Manchester University Press, 1981).

167 "a manic-depressive type"   Cyrus C. Sturgis, "The History of Blood Transfusion," Bulletin of the Medical Library Association 30, no. 2 (January 1942):107.

167n Denys's first patient   Kat Eschner, "350 Years Ago, a Doctor Performed the First Human Blood Transfusion. A Sheep Was Involved," Smithsonian, June 15, 2017, https://www.smithsonianmag.com/smart-news/350-years-ago-doctor-performed -first-human-blood-transfusion-sheep-was-involved-180963631/.

168 "a great glass full of urine"   Berry and Stoddert Parker, "Christopher Wren," 119.

168 "cracked a little in the head"   Samuel Pepys, The Diary of Samuel Pepys, November 30, 1667, https://www.pepysdiary.com/diary/1667/11/30/.

168 paid him twenty shillings   Samuel Pepys, Diary of Samuel Pepys, November 21, 1667, https://www.pepysdiary.com/diary/1667/11/21/.

169   "found himself very well"   Edmund King, "An Account of the Experiment of Transfusion, Practiced upon a Man in London," Proceedings of the Royal Society of London (December 9, 1667). https://publicdomainreview.org/collection/arthur-coga-s-blood-transfusion-1667.

170   "it is much more easy to bring"   Meldon, "Injection of Milk."

170   "white corpuscles" of milk   H. A. Oberman, "Early History of Blood Substitutes: Transfusion of Milk," Transfusion 9, no. 2 (March–April 1969): 74–77.

171   According to a short paper he published   Austin Meldon, "Intravenous Injection of Milk," British Medical Journal 1 (February 12, 1881): 228.

171   prevent "that depression"   Meldon.

171   "I look upon it as a much better"   Meldon.

171   when so-called normal saline was finally adopted   Rebecca Kreston, "The Origins of Intravenous Fluids," Discover, May 31, 2016, http://blogs.discovermagazine.com/bodyhorrors/2016/05/31/intravenous-fluids.

171   Latta's rehydration therapy   Kreston, "Intravenous Fluids."

Chapter 10: The Barber's Bite and the Strangled Heart

174   "survival of the fittest" was actually coined   Herbert Spencer, The Principles of Biology (London: Williams and Norgate, 1864), vol 1., 444.

174   "smoky dirty London"   Charles Darwin, "Second Note [July 1838]," "Darwin on Marriage," Darwin Correspondence Project, University of Cambridge (July 1838), https://www.darwinproject.ac.uk.

175   "We went a little into society"   Charles Darwin, The Autobiography of Charles Darwin, 1809–1882, ed. Nora Barlow (London: Collins, 1958), 115.

175n "a destructive malady"   James Clark, The Sanative Influence of Climate, 4th edition (London: John Murray, 1846), 2–4.

176   Gully contended that by applying cold water   Ralph Colp Jr., Darwin's Illness (Gainesville: University Press of Florida, 2008), 45.

176   "At no time must I take any sugar"   Darwin, "To Susan Darwin [19 March 1849]," Darwin Correspondence Project.

176   "oxyde of iron"   Darwin, "To Henry Bence Jones, 3 January [1866]," Darwin Correspondence Project.

177   "precarious" with "symptoms of myocardial degeneration"   A. S. MacNalty, "The Ill Health of Charles Darwin," Nursing Mirror, ii.

177   "Idleness is downright misery to me"   Charles Darwin, More Letters of Charles Darwin, vol. 2, eds. Francis Darwin and A. C. Seward, https://www.gutenberg.org/files/2740/2740-h/2740-h.htm.

177n After Origin's publication in 1859   Darwin, "From T. H. Huxley, 23 November 1859" and "To T. H. Huxley, 16 December [1859]," Darwin Correspondence Project.

180   increasing flow to the oxygen-starved heart   William Murrell, "Nitro-Glycerine as a Remedy for Angina Pectoris," Lancet 113, no. 2890 (January 18, 1879): 80-81.

180 an "irony of fate"   Nils Ringertz, "Alfred Nobel's Health and His Interest in Medicine," Nobel Media AB, December 6, 2020, https://www.nobelprize.org/alfred-nobel/alfred-nobels-health-and-his-interest-in-medicine/.

181 a drug usually given sublingually   Neha Narang and Jyoti Sharma, "Sublingual Mucosa as a Route for Systemic Drug Delivery," Supplement, International Journal of Pharmacy and Pharmaceutical Sciences 3, no. S2 (2011): 18–22.

181 "Angina Pectoris Syncope"   Janet Browne, Charles Darwin: The Power of Place (New York: Knopf, 2002), 495.

182n malaria killed 405,000 people   World Malaria Report 2019, World Health Organization, https://apps.who.int/iris/handle/10665/330011.

183 30 percent of them went on to develop   F. S. Machado et al., "Chagas Heart Disease: Report on Recent Developments," Cardiology in Review 20, no. 2 (March–April 2012): 53–65.

184 "Knowing the domiciliary habits of the insect"   "Triatominae," Le Parisien, http://dictionnaire.sensagent.leparisien.fr/Triatominae/en-en/.

185 Making matters even worse   Julie Clayton, "Chagas Disease 101," Nature 465, S4–5 (June 2010).

185 There, decades after the initial infection   "Chagas Disease 101"; E. M. Jones et al., "Amplification of a Trypanosoma cruzi DNA Sequence from Inflammatory Lesions in Human Chagasic Cardiomyopathy," American Journal of Tropical Medicine and Hygiene 48 (1993): 348–57.

186 "His symptoms can be fitted into the framework"   Saul Adler, "Darwin's Illness," Nature 184 (1959): 1103.

186 "At night I experienced an attack"   Charles Darwin, "Chili–Mendoza March 1835," Charles Darwin's Beagle Diary, ed. Richard Darwin Keynes (Cambridge: Cambridge University Press, 2001), 315, extracted from Darwin Online, http://darwin-online.org.uk/.

186 Darwin "had both Chagas infection"   Colp, Darwin's Illness, 143.

187 In 1977, Ralph Colp Jr. published   Ralph Colp Jr., To Be an Invalid: The Illness of Charles Darwin (Chicago: University of Chicago Press, 1977).

187 "Adler's theory of Chagas disease"   Colp.

187 "palpitations and pain about the heart"   Darwin, Autobiography, 79.

187 "the fever that characteristically accompanies"   Darwin, Beagle Diary, Darwin Online, 315.

188 "This is purely a symptom-based assessment"   "Historical Medical Conference Finds Darwin Suffered from Various Gastrointestinal Illnesses," University of Maryland School of Medicine, May 6, 2011, https://www.prnewswire.com/news-releases/historical-medical-conference-finds-darwin-suffered-from-various-gastrointestinal-illnesses-121366344.html.

189 Chagas disease was already well established   "A 9,000-Year Record of Chagas' Disease," Arthur C. Aufderheide et al., Proceedings of the National Academy of Sciences 101, no. 7 (February 17, 2004) 2034–39.

189   "Darwin's lifelong history does not fit"   "Historical Medical Conference."

189   between six and seven million people   Jasmine Garsd, "Kissing Bug Disease: Latin America's Silent Killer Makes U.S. Headlines," National Public Radio, December 8, 2015, https://www.npr.org/sections/goatsandsoda/2015/12/08/458781450/.

189   more than three hundred thousand people   Garsd, "Kissing Bug."

190   Dorn also told me that a recent paradigm shift   R. Viotti et al., "Towards a Paradigm Shift in the Treatment of Chronic Chagas Disease," Antimicrobial Agents and Chemotherapy 58, no. 2 (2014): 635–39.

190   7.4 percent of the study animals tested positive   Alyssa C. Meyers, Marvin Meinders, and Sarah A. Hamer, "Widespread Trypanosoma cruzi Infection in Government Working Dogs along the Texas-Mexico Border: Discordant Serology, Parasite Genotyping and Associated Vectors," PLOS Neglected Tropical Diseases 11, no. 8 (August 7, 2017).

Chapter 11: Hear Here: From Stick to Stethoscope

195   One child held an ear to the end of the stick   Ariel Roguin, "René Théophile Hyacinthe Laënnec (1781–1826): The Man behind the Stethoscope," Clinical Medicine & Research 4, no. 3 (September 2006): 230–35.

196   "too lovely to live long"   L. J. Moorman, "Tuberculosis and Genius: Ralph Waldo Emerson," Bulletin of the History of Medicine 18, no. 4 (1945): 361–70.

196   "Poetry and consumption"   William Shenstone, The Poetical Works of William Shenstone (New York: D. Appleton, 1854), xviii.

197   Men's styles were also affected   Emily Mullin, "How Tuberculosis Shaped Victorian Fashion," Smithsonian, May 10, 2016, https://www.smithsonianmag.com/science-nature/how-tuberculosis-shaped-victorian-fashion.

198   The heart infection known as tuberculous endocarditis   Alexander Liu et al., "Tuberculous Endocarditis," International Journal of Cardiology 167, no. 3 (August 10, 2013): 640–45.

198   like a full wine barrel   Roguin, "Laënnec."

199   "I recalled a well known acoustic phenomenon"   Roguin, trans. John Forbes.

200   "a growing fascination"   Kirstie Blair, Victorian Poetry and the Culture of the Heart (Oxford: Oxford University Press, 2006), 23–24.

201   And though the stethoscope no longer has   M. Jiwa et al., "Impact of the Presence of Medical Equipment in Images on Viewers' Perceptions of the Trustworthiness of an Individual On-Screen," Journal of Medical Internet Research 14, no. 4 (2012), e100.

Chapter 12: Don't Try This at Home . . . Unless Accompanied by a Very Special Nurse

202   "Surgery of the heart"   R. S. Litwak, "The Growth of Cardiac Surgery: Historical Notes," Cardiovascular Clinics 3 (1971): 5–50.

202　graduating from one of Germany's best　H. W. Heiss, "Werner Forssmann: A German Problem with the Nobel Prize," Clinical Cardiology 15 (1992): 547–49.

203　While studying to become a surgeon　Heiss, "Werner Forssmann."

203　a form of x-ray machine　"Shoe-Fitting Fluoroscope (ca. 1930–1940)," Oak Ridge Associated Universities, 1999, https://www.orau.org/ptp/collection/shoefittingfluor /shoe.htm.

204　he "started to prowl"　Werner Forssmann, Experiments on Myself: Memoirs of a Surgeon in Germany, trans. H. Davies (New York; St. Martin's Press, 1974): 84.

205　Bizarrely, the chair of surgery at another hospital　Ahmadreza Afshar, David P. Steensma, and Robert A. Kyle, "Werner Forssmann: A Pioneer of Interventional Cardiology and Auto-Experimentation," Mayo Clinic Proceedings 93, no. 9 (September 1, 2018): E97–98.

206　"It was very painful"　K. Agrawal, "The First Catheterization," Hospitalist 2006, no. 12 (December 2006).

206　"I feel like a village parson"　Forssmann, Experiments on Myself, xi.

207　According to a journal article　Lisa-Marie Packy, Matthis Krischel, and Dominik Gross, "Werner Forssmann—A Nobel Prize Winner and His Political Attitude Before and After 1945," Urologia Internationalis 96, no. 4 (2016): 379–85.

208　There is evidence to suggest　Afshar, Steensma, and Kyle, "Werner Forssmann."

208　designated as a Category 4 Nazi　Packy, Krischel, and Gross, "Werner Forssmann," 383.

Chapter 13: "Hearts and Minds" . . . Sort Of

210n　A 2019 study published in the journal Geriatrics　Jessica Yi Han Aw, Vasoontara Sbirakos Yiengprugsawan, and Cathy Honge Gong, "Utilization of Traditional Chinese Medicine Practitioners in Later Life in Mainland China," Geriatrics (Basel) 4, no. 3 (September 2019): 49.

211　In 1640, he claimed that the true "seat"　Gert-Jan Lokhorst, "Descartes and the Pineal Gland," Stanford Encyclopedia of Philosophy, 2013, https://plato.stanford. edu/entries/pineal-gland/.

211　"the only solid part in the whole brain"　Lokhorst, "Pineal Gland."

213　In her own book, cultural historian　Fay Bound Alberti, Matters of the Heart: History, Medicine and Emotion (Oxford: Oxford University Press, 2010), 2.

214　Reportedly, the plant was used　John C. Hellson, "Ethnobotany of the Blackfoot Indians, Ottawa," National Museums of Canada, Mercury Series, 60, Native American Ethnobotany DB, http://naeb.brit.org/uses/31593/.

215　"People who do well with"　Jennifer Worden, "Circulatory Problems," Homeopathy UK, https://www.britishhomeopathic.org/charity/how-we-can-help /articles/conditions/c/spotlight-on-circulation/.

Chapter 14: What Becomes of the Brokenhearted?

217   Upon examination, however, the doctors   Takeo Sato et al., "Takotsubo
      (Ampulla-Shaped) Cardiomyopathy Associated with Microscopic Polyangiitis,"
      Internal Medicine 44, no. 3 (2005): 251–55.

218   Apparently, whatever damage had occurred   Alexander R. Lyon et al., "Current
      State of Knowledge on Takotsubo Syndrome: A Position Statement from the
      Taskforce on Takotsubo Syndrome of the Heart Failure Association of the
      European Society of Cardiology," European Journal of Heart Failure 18, no. 1
      (January 2016): 8-27.

222   there are critics who claim that even the best   R. P. Sloan, E. Bagiella, and
      T. Powell, "Religion, Spirituality, and Medicine," Lancet 353, no. 9153
      (February 20, 1999).

223   Since the late 1970s, mindfulness   "What Is Mindfulness?" Greater Good
      Magazine, https://greatergood.berkeley.edu/topic/mindfulness/definition.

224   a significant reduction in ten-year mortality   Quinn R. Pack et al., "Participation
      in Cardiac Rehabilitation and Survival After Coronary Artery Bypass Graft
      Surgery: A Community-Based Study," Circulation 128, no. 6 (August 6, 2013):
      590–97.

224   a marked reduction of readmission and death   Shannon M. Dunlay et al.,
      "Participation in Cardiac Rehabilitation, Readmissions, and Death after Acute
      Myocardial Infarction," American Journal of Medicine 127, no 6 (June 2014):
      538–46.

225   After studying these barriers, researchers   Shannon M. Dunlay et al., "Barriers
      to Participation in Cardiac Rehabilitation," American Heart Journal 158, no. 5
      (November 2009): 852–59.

227   "a typical movie-theatre soda"   "Keeping Proportions in Proportion,"
      November 2007, Harvard Health Publishing, Harvard Medical School,
      https://www.health.harvard.edu/newsletter_article/Keeping_portions_in_proportion.

227   global demand for meat has quadrupled   Hannah Ritchie and Max Roser,
      "Meat and Dairy Production," November 2019, Our World in Data, https://
      ourworldindata.org/meat-production.

227   despite the increase in stress   A. Strom and R. A. Jensen, "Mortality from
      Circulatory Diseases in Norway 1940–1945," Lancet 1, no. 6647 (January 20, 1951):
      126–29.

Chapter 15: What's Snakes Got to Do, Got to Do with It?

229   "fevers, stroke, seizures, delirium, renal failure"   Jason A. Cook et al, "The Total
      Artificial Heart," Journal of Thoracic Disease 7, no. 12 (December 2015): 2172–80.

231   these patients have a one-year mortality rate   Ibadete Bytyçi and Gani Bajraktari,
      "Mortality in Heart Failure Patients," Anatolian Journal of Cardiology 15, no. 1
      (January 2015): 63–68.

232 Researchers were surprised to learn   "Why Use the Zebrafish in Research?" YourGenome, 2014, https://www.yourgenome.org/facts/why-use-the-zebrafish -in-research.

232 Gene mutations can be introduced into zebrafish   David I. Bassett and Peter D. Curry, "The Zebrafish as a Model for Muscular Dystrophy and Congenital Myopathy," Supplement, Human Molecular Genetics 12, no. S2 (October 15, 2003): R265–70.

232 thus allowing drug researchers to test   Federico Tessadori et al., "Effective CRISPR/Cas9-Based Nucleotide Editing in Zebrafish to Model Human Genetic Cardiovascular Disorders," Disease Models & Mechanisms 11 (2018), https://dmm .biologists.org/content/11/10/dmm035469#abstract-1.

232 The truly unique part   Kenneth D. Poss, Lindsay G. Wilson, and Mark T. Keating, "Heart Regeneration in Zebrafish," Science 298, no. 5601 (December 13, 2002): 2188–90.

232 the clot is replaced by fully functional   Angel Raya et al., "Activation of Notch Signaling Pathway Precedes Heart Regeneration in Zebrafish, Supplement," Proceedings of the National Academy of Sciences 100, no. S1 (2003): 11889–95.

233 with blood vessels rapidly regrowing   Fernandez, Bakovic, and Karra, "Zebrafish," 2018.

234 it serves as a structural base of support   Juan Manuel González-Rosa, Caroline E. Burns, and C. Geoffrey Burns, "Zebrafish Heart Regeneration: 15 Years of Discoveries," Regeneration (Oxford) 4, no. 3 (June 2017): 105–23.

235 scientists are now attempting to produce   Panagiota Giardoglou and Dimitris Beis, "On Zebrafish Disease Models and Matters of the Heart," Biomedicines 7, no. 1 (February 28, 2019): 15.

235 may one day usher in a new era   Tanner O. Monroe et al., "YAP Partially Reprograms Chromatin Accessibility to Directly Induce Adult Cardiogenesis In vivo," Developmental Cell 48, no. 6 (March 25, 2019): 765–79.

236 It was the observation, made in 2005   Johnnie B. Andersen et al., "Postprandial Cardiac Hypertrophy in Pythons," Nature 434 (March 3, 2005): 37.

240 A 2012 paper   Michael E. Dorcas et al., "Severe Mammal Declines Coincide with Proliferation of Invasive Burmese Pythons in Everglades National Park," Proceedings of the National Academy of Sciences 109, no. 7 (February 14, 2012): 2418–22.

Chapter 16: Grow Your Own

243 "'The answer is spinach.'"   Bill Maher, Real Time with Bill Maher, September 28, 2007, https://www.youtube.com/watch?v=rHXXTCc-IVg.

247 Led by Sian Harding   Leslie Mertz, "Heart to Heart," IEEE Engineering in Medicine & Biology Society (September/October 2019).

251   Other termite species are flagellate-free   Gaku Tokuda and Hirofumi Watanabe,
      "Hidden Cellulases in Termites: Revision of an Old Hypothesis," Biology Letters 3,
      no. 3 (March 20, 2007): 336-39.

251   Dvir's team will need to address   Nadav Noor et al., "Tissue Engineering:
      3D Printing of Personalized Thick and Perfusable Cardiac Patches and Hearts,"
      Advanced Science 6, no. 11 (June 2019).

252   there were approximately 109,000 candidates   Health Resources and Services
      Administration, "Organ Donation Statistics," https://www.organdonor.gov
      /.statistics-stories/statistics.html.

EPIC SURVIVAL

EPIC SURVIVAL

Extreme Adventure, Stone Age Wisdom, and Lessons in Living
from a Modern Hunter-Gatherer

MATT GRAHAM
and JOSH YOUNG

G

Gallery Books

New York   London   Toronto   Sydney   New Delhi

G

Gallery Books
An Imprint of Simon & Schuster, Inc.
1230 Avenue of the Americas
New York, NY 10020

First Gallery Books hardcover edition July 2015

GALLERY BOOKS and colophon are registered trademarks of Simon & Schuster, Inc.

For information about special discounts for bulk purchases, please contact Simon & Schuster Special Sales at 1-866-506-1949 or business@simonandschuster.com.

The Simon & Schuster Speakers Bureau can bring authors to your live event. For more information or to book an event contact the Simon & Schuster Speakers Bureau at 1-866-248-3049 or visit our website at www.simonspeakers.com.

Interior design by Robert E. Ettlin

Manufactured in the United States of America

1   3   5   7   9   10   8   6   4   2

Library of Congress Cataloging-in-Publication Data is available.

ISBN 978-1-4767-9465-5
ISBN 978-1-4767-9468-6 (ebook)

This book is dedicated to anyone who is willing to view the natural world as a gift and to appreciate, enjoy, preserve, and learn from it.

CONTENTS

CONTENTS

EPIC SURVIVAL

HEART OF THE WILD

I live a life that is far different from that of most people I know or have read or heard about. Yet it is a life upon which our society was built. While most people's lives are calibrated by ones and zeros, the daily beat of city streets, and the frenetic demands of modern-day living, mine is dictated by the wild. At first glance, mine is far more dangerous.

For the past twenty years, I have all but ignored the digital age and pursued a hunter-gatherer lifestyle, living as man did many thousands and even hundreds of thousands of years ago. This has taught me that the wilderness is a place of truth that will accept and enhance the life of any person who is ready to live on its terms.

Many people who try hard to enjoy life end up feeling empty and cannot figure out why. For me, the solution to happiness has lain on green meadows, rocky passes, and flowing streams. The wilderness needed to be explored and better understood. Once I made that decision, I felt released from the chains of society. My legs would take me to the places I needed to go. If they couldn't carry me there, I didn't need to be there.

I've walked into the wilderness wearing only a loincloth and a pair of handmade sandals and carrying a blanket, a stone knife, and a bag of chia seeds. Once I lived there alone for half a year and tested the limits of what a person can endure physically and mentally. My body went through amazing physical changes. The caked tartar flaked off my teeth and my breath became sweet, and my thinning hair filled back in. I also adapted physically. After my body virtually crashed while eliminating toxins, I bounced back and could run like a wild animal and spear a fish from twenty yards. All my senses were heightened, and I found that the more my body was tested, the more energy I was able to draw from the land to sustain me.

For most of my life, I have lived off the grid in traditional-style wickiups, pit houses, and primitive structures with no electricity or plumbing, remote places all, ranging in location from the mountains of Utah to the bottom of the snow-covered Grand Canyon to the jungle of Kauai in a hut built out of banana leaves. I have made fires out of bark, vines, sagebrush, tamarisk twigs, and a cow's anklebone to stay warm and cook my food. To survive, I became a proficient hunter using the atlatl, a device for throwing a spear that gives it greater velocity, and was accomplished with it to the point that I once beat the world atlatl champion in competition.

I have run thousands of miles exploring every piece of the wilderness in the western United States that called to me. On these treks, I wear handmade sandals constructed from yucca fibers to keep me closer to the earth. I've run through the Sierra Nevada mountains, the Mojave Desert, the Sonoran Desert, the Grand Canyon, Death Valley, and up the middle of California. In these explorations, I interacted with the plants and the animals. I was in no hurry. I didn't want to be just another alien visitor. If I sensed I had something to learn from a particular place, I would stay for days, or even weeks at a time, and listen to what the land had to offer.

But I don't mean to brag in the least, nor do I want to push my lifestyle on anyone. I am interested in sharing my explorations of the land.

I want to show what the human body and mind can do when pushed to its utmost limits, and how nature can help take care of it. I hope to add a dimension to people's lives through what I have experienced and what I have seen.

One of the most important things I've learned from living off the land is that it makes you more observant. Today in our lives—and I know because I have stepped out of the wilderness for periods of time—we can become desensitized to the point where we stop looking around and appreciating the simple things that are right in front of us. We drive home or walk through a city and can't even recall half the things we saw on the street. When you live off the land, you are forced to be hyper-aware of everything. I believe that achieving this level of hyperawareness makes people more complete because in addition to getting to know themselves better, they also develop a greater understanding of their relationships with other people.

While living on the land, everything becomes critical. If you don't pay attention, then you die. It sounds dramatic, but it's that simple. Being out in the wild strips away the artifice so that when you reenter society, you don't have the same distractions. I find that when I'm talking to a friend, I am immersed in their story and give them my full attention rather than worrying about who is texting me. (Yes, I do own a cell phone.)

Being in a survival situation teaches you what it takes to live. That connection is very powerful. All your senses become more developed. You experience heightened hearing and clearer eyesight. It can be highly addictive. When you live in such a way, you realize the potential of what you can be as a human being on a physical level, and it makes you want to return for more.

I see this pull in students I take on survival courses. We will go out for a month, and they will develop themselves in ways they have never experienced. Sometimes they push themselves so hard they feel enlightenment, but at the same time, they are craving a cheeseburger and an ice cream sandwich. They return to their city lives and get all those things they were dreaming about on the trail, but when they lose that wilderness boost, they return for more.

When you realize what the wild can do for you, there is a desire to mesh it with the modern world, even if you are not going to quit your job and become a hunter-gatherer. We know what nature can do for people who live in urban environments. We know it can build their awareness, their physicality, and their senses, and all of these can be adapted into their everyday lives. In our society today, the earth underneath us has become a distant thing. Most of us live in a concrete world removed from the true ground. I want to help people step off the concrete and onto the land.

So let's take a journey. Enjoy the land with me as I take you on a walkabout through the Southwest. See the celestial light bounce off the mountain. Relish the calm of nature's caves. Sweat it out when a mountain lion circles my tent in the night. Feel the pain of climbing out of the Grand Canyon in three feet of snow. Experience the rush of running the entire state of California. Stop to smell the roses and eat them, too—the rose hips, that is. Learn how nature mediates and sustains life.

Know all along that this journey is a true story. It may have happened yesterday, or it may have happened a thousand years ago. No matter. Being intimate with the land carries the same context. When we leave home and travel by foot, the soles of our feet find new treasures, as do the souls of our bodies.

EPIC SURVIVAL RULE #1:

SET YOUR OWN PATH

I am going to take the necessary steps to learn to fully live off the land. I want to live the way the American Indians did, and possibly even like our earliest ancestors in an even more primitive style. I am going to find a way to be closer to the earth and give the planet more than I take from it. On the surface, in a modern industrial society this doesn't seem possible, but I am going to try to show that it is.

The easiest starting point, it seems, is to find a way to slowly strip everything down to a bare minimum. This will allow me to reach the point where I can go out and see my direct impact on the land by foraging in an area without destroying the earth.

Most people are constantly beating up the earth and taking from the land. They are commuting to work in cars or on buses, producing reams of trash, and depleting natural resources just by living their lives. Almost no one is doing this maliciously. There is just simply no way around it.

To me, it appears that the earth is slowly dying and constantly, though fruitlessly, trying to replenish itself from man's impact on it. I am going to leave a gentler footprint, but I am going to do this for myself.

In this process, I am not going to judge other people who live differently than me. Not judging people is difficult, but it is a lesson that applies to anyone's life. You think you know something, you pass judgment on that, but then you later realize you didn't really know it.

I am attracted to the wilderness because it is a place where

I will be able to constantly learn how much I don't know. I am going to find a spiritual connection as well. I have no illusions that this will happen quickly. This is going to take years to understand. But I know I will succeed, because I am willing to risk my life in the process.

Chapter One

A GOLDEN SUNRISE

I was lying on a mountain ledge adjacent to Clouds Rest, a thin ridge of granite rising some four thousand feet above the Yosemite Valley and more than nine thousand feet above sea level. Darkness had set in. Snow was piling up. I hoped that the final fifty feet I had climbed to a small plateau the size of a park bench would be enough to keep the bears away from me during the night.

I was here because of a loosely related chain of events that made my path in life seem clear to me. At age seven, my parents separated. My mom and I moved around a lot, changing apartments once a year. When I was sixteen, we moved in with a man my mom was dating. His place was forty-five minutes away from my school. The distance, coupled with my lack of interest in classroom work, was the reason formal education began to fade from my life.

I never fit in with the traditional educational system. I was ridiculously motivated to learn, but I felt like school had me sitting in a chair eight hours a day, spending only one hour a day running around outside playing games. That was out of balance. I needed half my time to be outdoors, and though I didn't know it yet, I needed a large portion of the outdoor time to be nature based. Over time, it became not enough for me to be swinging a wooden club at a ball and running around a diamond and touching squares. I needed to experience the circles and ovals of nature.

Sitting at my desk in school, I would envision the natural world, just to keep my mind from feeling boxed in. I wasn't designed to sit and absorb information. In every school I attended, I was considered the best athlete. God gave me the gift to be fast and powerful, as well as light and swift. But when I sat in a classroom for eight hours a day, I was fighting against my natural gifts.

The turning point came early, in second grade. We opened up our history books to a chapter on Native Americans. To this day, I remember seeing a picture of a beautiful spear point. My heart immediately felt something. Though I couldn't articulate it then, I knew that the person who created that perfectly flinted spear point was somehow special. It was nothing like the tools I had seen in hardware stores; it had soul.

As I was marveling at this creation, my teacher told the class that we were also going to study their spiritual beliefs, but cautioned us that they "didn't know God like we do." Her telling the class that the Native Americans were lesser in the eyes of God made me instantly more attracted to them, because I knew there was more to that than the religious dogma they taught in parochial school.

The more the natural world came into view, the more I pulled away from school. Like most boys, I loved superheroes. Superman was cool, but it was Tarzan who captured my imagination, because he lived in the wild and protected both the people and the jungle itself. I was in awe of his physical attributes and how he lived as one with the world around him.

I wanted to be Tarzan, though I wasn't quite sure how to achieve that.

I felt a strong pull toward the lifestyle I was exposed to in my childhood at our cabin in Lake Arrowhead. My father was an avid amateur naturalist, and he taught me about plants and the smells of the trees and took me quail hunting. He took me on hikes. He would show me the land features and make me smell the plants. Most kids would probably think that was stupid and want to get on with the climb. I, however, immediately felt a connection to the natural world.

Between these experiences and my mother taking me to the beach on the weekends during the school year, I became captivated by how alive I felt in nature. Whether I was in the mountains or the city park, I felt different from when I was on pavement or in a building.

I often found people confusing, but nature always made sense to me.

When I turned seventeen, I took advantage of my parents' allowing me to make my own choices and moved to Yosemite. It was an ideal age to begin studying the wilderness. I had the strength of youth, no family relying on me, and years to explore on my terms. I arrived in the late fall. I took a job working at an ice cream store there. When it closed for the season, I worked in a cafeteria for another six weeks. I was a model employee, so I was chosen to work special functions and catered events. The best part about the job was that I only worked two days a week.

That freed me up to explore the land.

It also led to my being near the top of Clouds Rest during the biggest snowstorm of the winter, with only bears as my neighbors. But something unexpected happened on the mountain ledge: I didn't feel trapped. I felt free. As much as I was there because I wanted to be there, I was also there because I had to be there. I had climbed Clouds Rest because I knew there was something up there.

It was bigger than me, and I had to experience it.

In Yosemite at the time, backpacking was all the rage. On days off and weekends, young people who lived and worked in the area would go on overnight campouts. To me, those trips seemed like no more than tourist outings. I wanted my first backpacking trip to be off the charts and slightly insane, which is why I chose to climb Clouds Rest.

It was winter, and I was off work for a three-day weekend. There was snow on the ground and the temperature ranged from a high of forty in the day to the single digits at night. I put on shorts, socks and shoes, and a medium-weight fleece. To allow me to move at a rapid clip, I didn't bring a lot of supplies. I had a bivy sack—a lightweight, waterproof cover—to sleep inside of and help ward off the elements.

My goal was to climb Clouds Rest and then proceed across Half Dome. I wanted to spend the weekend up there with as few provisions as possible, and see what the land offered me.

The summit of Half Dome was once thought to be impossible to reach. In 1870, a report called it "probably the only one of all the prominent points about the Yosemite which has never been, and never will be, trodden by human foot." Five years later, climber George Anderson, who was laying the cable route that climbers now use, accessed it. More than a hundred years after that, any climber in good physical condition is able to reach the summit.

People I worked with had cautioned me that Clouds Rest was too treacherous a climb and the wintry conditions were too brutal for my first long backpacking trip, particularly since I was going it alone—and in shorts! But I had to see it.

I pretty much ran up the mountain. As I was going up, a snowstorm was starting to blanket the land. About seven miles up, I reached a halfway point where people climbing Half Dome set up their base camps. The area was emptying out. Hikers were coming off the mountain and heading back to town. Every person I passed warned me that the storm was already outrageous and getting worse, which only made me want to go even higher.

Something was driving me to get up there and see what it looked like in the thick of the storm.

I continued past Little Yosemite Valley toward Clouds Rest. The trees started to thin out, and soon I was above the tree line. I reached a point about two miles from the peak. The land was pure white. The snow was getting thicker and thicker, and the visibility was about five feet. Because I was moving fast, I wasn't cold, despite the fact that I was wearing shorts.

Still, I had to continue climbing. I was feeling drawn. Though most people would feel that I was entering a dangerous situation, I wasn't scared or concerned for my safety. This was not a foolish ego trip. In fact, there was no ego involved. I was doing this for myself. I knew that as long as I made the right decisions along the way, I would not get hurt. I would not cross the line to where things became too dangerous. Fear would not become a factor.

Everybody has a certain way they approach and handle fear. I believe that people who are on a genuine path and are being true to themselves as well as everything around them can ward off fear. The situation is never just about us. There is a greater force that is pulling us to what we should be doing.

Knowing you are on the right life path can mitigate fear.

Some degree of fear can be appropriate and healthy. It can compel you to make sane decisions rather than rash ones. But fear creeps in and becomes dangerous when you are doing something you should not be doing. Uncontrollable fear occurs when you are in the wrong place at the wrong time and you are caught off guard. You have either arrived in the wrong place to begin with, or you have undertaken the wrong course of action. In that situation, we lose the ability to control our fear.

So there I was: The storm was packed with energy. With every step, the snow became thicker. It soon reached the point of a whiteout. Even though the peak of Half Dome was directly ahead of me, I couldn't see it. I was moving by simply feeling the land. If I detected a cliff, I turned and went in a different direction.

Eventually I hit a rock wall that stopped me. I stood still and tried to let the land tell me what to do. Every once in a while, a beam of light would break through the thick snow. From what I could see, it looked like there was a small peak directly in front of me, about fifty feet up a rock wall.

The sheer newness of the area was captivating. What was up there? For the excitement of what I might find, I decided to climb the snow-covered face of the wall. The climb turned out to be easier than it appeared to be because the wall was smaller than I had imagined.

The plateau at the top of the wall was roughly three by six feet. I tried to look down the other side, which would normally provide a sweeping view of Yosemite Valley, but the visibility was zero. I was walled in by blowing snow. I assumed the cliff in front of me was probably steeper than the cliff I had climbed, but there was no way to know for sure until the snow stopped.

I decided to spend the night there. Though the area was known for bears, I wasn't worried about bears coming for my food. I knew—or rather I had been told—that a bear couldn't climb a near-ninety-degree rock wall.

The snow was falling in thick sheets and the wind was picking up. I pulled out my body bag and my bivy sack and crawled into them. I lay there, not afraid but intensely curious about what the storm would do to me.

It was my first test of patience in the wild. I would have to wait out a night—or maybe longer—on this peak. I would have to appreciate and respect the energy of the massive storm. I didn't sleep a wink. The wind was way too intense. Throughout the night, as huge gusts blasted me, I curled up tighter and tighter into a ball. I was completely drenched. The body bag seemed to do nothing to protect me.

Because I had a Christian upbringing, I found myself praying throughout the night, not for my safety but for a gift.

During the night, during the storm, I received it.

Though I was lying near the top of Clouds Rest, wet and cold, covered by what felt like an iced towel, the excitement kept me warm. I had a feeling something was going to happen. I didn't know what. I wasn't the least bit afraid, because I was waiting for something extraordinary. I thought maybe the peak would break loose and float away.

At some point just before dawn, after being awake all night, I managed to fall asleep.

When I opened my eyes, I learned how Clouds Rest had gotten its name. The sunlight was shining on my face. I lifted my head and looked out over a crisp blue sky. At the exact same level of the peak where I was lying was a layer of clouds that went on forever. I was just above the cloud line, so I couldn't see anything below. It looked like I could step out and walk across the layer of clouds.

Actually, I thought about trying.

Still to this day, I have never seen such a sight in the mountains, or in an airplane descending through clouds, for that matter. Those clouds provided a perfect floor for the sky, much the way the land does for mountains.

I sat up and watched the sun rise. It was a magical sight. As the sun crested the horizon, the entire cloud layer turned orange. It took about an hour for the sun to fully rise and the white color to return to the clouds. The sky was now clear. There was no sign of snow.

I imagined that this must be God's view of us and the sunrise.

I packed my wet belongings and descended from the clouds. At the base of the rock, where I had climbed up to sleep, there were fresh bear tracks. Clearly, a bear had been trying to get up onto the peak to eat my food. I never saw the bear, and I never feared that I might have been in danger. The bear couldn't have climbed the rock, because it was too steep for an animal with claws.

I came down the mountain without incident and without seeing any hikers, all of whom had been discouraged by the storm.

For my first climbing experience, I took what I had seen as a sign. I had the feeling of being given a gift. I had gone to a place where everybody told me not to go; yet despite their warnings, everything inside me told me that I needed to go there to see something. And I had.

I had been blessed with the sight of a magic carpet of clouds and an immediate sunrise that few humans ever see. It was a transformative experience that validated the unusual, nonacademic path I had chosen, and showed me that the natural world held unlimited possibilities.

Coming off that experience, I was elated. I felt so full of light. When I returned to the valley floor and tried to talk to people about what I had seen, I felt as though I was talking to a wall. In comparison, everything else seemed dull.

The contrast was so stark that I began to separate life in society, the artificial world, from life in the wilderness, the real world. I began to understand that if I made nature my sole study, I would always see the world through a different lens than people in society.

But I was ready and willing to take that journey.

Chapter Two

STRADDLING TWO WORLDS

There was a lateral drift to my life from my late teens leading into the precipice of adulthood at age twenty-one. I felt that the flexibility of moving to different areas and exploring them would lead to an education of the wild, and ultimately to a grounding. It was my version of backpacking through Europe to discover myself and decide what kind of life I would lead.

I lived one winter in Mammoth Lakes. I found a job at a ski shop and spent my spare time hiking in the backcountry. I entertained the thought of touring all the national parks and working for several months in each one. I even called around to the parks to see which jobs were available. I was offered a job as a tour boat guide in the Everglades National Park in South Florida. I almost accepted the job, but I ended up turning it down because in my mind I hadn't seen enough of the Sierras yet. The Sierras felt like my home, and I wanted to explore them.

I had arrived in Yosemite with preconceived judgments on people's relationship to the environment. I thought that people who didn't use paper bags, and instead brought reusable ones, when they went to the market were pretentious. In Yosemite, there was a lot of acting "green." It was the first time I was exposed to that mind-set.

But the land kicked me in the butt. It seemed to be saying that whether or not this behavior was fake, to that person they were taking a step to be a better steward of the land. It also taught me that I needed to find a way to practice the ideals of a more natural lifestyle in my own way.

At the moment I made that commitment, I felt a change inside me taking hold. As soon as I started studying primitive skills, I looked for everyday ways to stop destroying the earth. At the time, I was living in the Sequoia National Park. My diet was conducive to healthy living. I ate mostly rice, lentils, and wild greens. I never ate processed food of any kind, or anything with sugar in it.

I took a pledge not to set foot in a motor vehicle for eight months. I had always been rebellious toward automobiles and supported the view that many of our environmental problems would not exist without cars. Without motor vehicles, people would be forced to have a greater respect for the place they live, as there would be no quick escapes. Most people love their cars and all the voice-commanded gadgets in them, but how many people are interested in an interaction with their car's exhaust?

I still strongly believe that cars and motorized transportation, in addition to causing environmental harm, are responsible for a universal loss of health and fitness, as well as our society's overall sanity. Anytime we fight against the grain of nature, there is a price to be paid. As we balance the needs of our lives, we need to be wary of how much compromise we are willing to accept. Going back to nature is an attempt to reestablish ourselves, and people who don't do that end up unhappy with their lives. When people search and find that balance, it provides a certain amount of peace and sanity.

I actually tried to do something about the overuse of motor vehicles. When Bill Clinton was president, I wrote him a letter to advocate building more bike and running trails that actually go places. I proposed that these trails parallel roads and interstates and run through cities. To me, it seemed that we built these tight highways with no shoulder and no place for a person to ride a bike or want to walk, let alone even safely walk. If we could just lay woodchips along the roads, we would start creating more happiness in our lives. To add an element, we could make those trails cut off the highway so that bikers and pedestrians could pick up a bit of scenery.

Part of the problem is that in order to make everything practical and fast, society has sacrificed our artistry and our connection with nature. To me, those words go hand in hand. Nature, without even trying, is an artistic masterpiece, but when we interrupt that flow with skyscrapers and freeways, it disconnects us. If we can figure out how to interject things in our life that make us feel happy and connected, like walking through a beautiful spot, there will be a way to maintain that artistry in our life that breeds feelings of passion. Working within the confines of nature and not mowing down every tree acknowledges that we want to be part of that artistry.

As soon as I made my pledge to stay out of cars, I went on a three-day run. But instead of covering my typical twenty-five miles a day, I ran sixty-five miles a day. It was intense. I had never run that far before, or even close to that far. When I finally stopped, I actually sat down and prayed. I thanked God for giving me the strength and the body to cover such a distance. I thought that maybe he had given me such power because of the commitment I had made to improve the earth.

That thought resonated. Right after I had committed to not riding in a vehicle, I was given the stamina and strength to do something that most people could not, even if they were extremely fit. There had to be a connection, though I couldn't quantify it at the time. The one thing I did begin to feel was that running heightened my instincts for the terrain.

Socially, I must admit, living on foot was a challenge. When friends invited me to a party that was eight miles from my house, I jogged there. It also presented problems when I asked a girl out on a date. Even the most adventuresome women weren't willing to walk five miles to dinner. Though my friends admired me, they didn't fully understand why anyone would purposefully make their lives harder. I felt I was making it better.

I gained a reputation among the year-rounders in Sequoia as a distance runner. The Park Service became aware of my ability to cover long distances of terrain very quickly. After I worked a short stint in the reservations office, the Park Service asked me to join the search-and-rescue team. They used me for urgent distress calls that required an immediate reconnaissance on foot.

One afternoon, there was an extensive search-and-rescue mission for a lost teenager. He and a group of friends had been out for a day hike. He had left the group early, but he never made it back to town. The search was concentrated in the area where he had last been seen. Day one ended with no clues.

After the second day ended with no sign of his whereabouts, I studied maps of the area. I asked the ranger in charge if they had covered a certain area that was several miles from their primary search radius. He told me that the young man couldn't have made it that far in two days. I told him that I believed he could have made it much farther.

Generally, people who are lost first walk in circles trying to find their way out. But it was clear to me after the search team had spent two days covering the circle route that he had chosen a different path.

The route I had projected was downhill. Given a choice, most people will walk downhill rather than uphill when they are lost. As he traveled, the terrain would have become lusher and moister, meaning he would have thought water was nearby. Though this was logical, part of my instinct about his location came from my time running in the area, spending time covering that land, and understanding where he could naturally wander.

Four and a half days later, the teenager popped out on a road and was picked up by a passerby in the area where I had suggested they search.

The head ranger was impressed that I had pinpointed the missing teenager's location. My knowledge of the terrain, combined with my ability to run and my interest in tracking, resulted in him putting me on a "hasty crew." They are the first responders in a search, and they generally hike or run to the site. I was teamed with another physically fit ranger.

Weeks later, a report of another missing teenager came in. The teenager had been backpacking with a friend who had left the hike to return to work. The teenager was supposed to stay another two nights in the backcountry. Three days had passed with no word from him.

My colleague and I took off running up the trail. We zigzagged, covering the valley floor on our way to his last known location. En route, we came upon a waterfall that was about six miles away. I felt something as we passed through the area. I felt that the teenager was nearby. Part of my sense of his location came from reading the land, but just as much of it was a strong gut feeling.

The ranger was understandably skeptical. The location I pinpointed was far from the kid's last known location. He agreed to stop and let me have a few minutes to explore my hunch.

I didn't find any foot tracks, so I stopped reading the dirt and started reading the water. I stared at the nearby waterfall. In my head I began to form a picture of what felt like actual events.

Okay . . . if somebody came down that waterfall a quarter mile upstream and they were unconscious, where would they end up? Downstream somewhere. Perhaps caught under that large rock.

I walked downstream and stood on the rock. Instantly, I had a sharp pain in my gut that felt like my appendix had burst. I made my way across the boulders to the other side to try to look under the rock. The other ranger was watching me, somewhat bewildered.

I was trying to establish a position to look under the rock. There was a strong current flowing downstream that obstructed my ability to see below the surface of the water. I looked from several different angles, but I couldn't make out anything that looked like a body. Then I started to take off my shirt.

"Can I go in?" I yelled up to my colleague.

"That's not your job," he said. "If you see something, we need to call in the dive crew."

"I don't see anything," I said. "I just *feel* like something is under there."

We exchanged a few more words, and he insisted that we call in the divers.

I again cautioned him that it was just a hunch. I was twenty-one. I wasn't an expert tracker by any means. I had been on the job for only a few weeks, and I didn't want to be responsible for wasting resources on a hunch.

Even if he doubted me, the ranger had watched me closely enough to know there was a possibility, remote as it might have been, that I might be right. We also had nothing else to go on. He called for the dive crew.

Hours later, the divers found the young man's body under the rock I had pinpointed.

The debrief was a solemn affair, but the head ranger was very firm about his belief in my abilities. "I've been on a lot of searches," he told the group. "I'm a trained tracker. I have no idea what led Matt to think there was a body under there. The only possible way is that he has a psychic ability. Without Matt, we never would have found this body. For that we are grateful. We would have spent two weeks with no results and probably a million dollars on the search."

The fact is, they never would have found the body because they were searching six miles away. The land was untrackable. Where we started out, it was pine-needled forest with a little dirt here and there. But above the waterfall, it was 100 percent granite. It takes a highly skilled tracker to track on rock, and the ranger service didn't have such a tracker.

I offered to speak to the teenager's parents, but the head ranger told me it was best to handle things through formal channels. "You've helped bring closure for the family," he said. "Remember that."

The season ended in October without any further serious incidents. We had two more fairly basic rescues in the ensuing seven months, but nothing near a casualty. I finished out the season on the search-and-rescue team—on foot.

I felt like it was time to move on. I wanted to explore the backcountry. I owned only about twenty pounds of gear. I had a sleeping bag, a bivy sack, a couple changes of clothes, a few pairs of handmade sandals, and some maps. I modified a pair of running shoes into sandals to alternate wearing with my sandals, because they would provide more cover for my feet in the snow. I loaded everything into my backpack and took off running to the east toward the Sierra Nevada mountain range.

I didn't have a destination in my mind.

The day I set out there was an early-season, blinding snowstorm. The visibility was less than ten feet. But when I reached the mountains, a path of blue sky emerged and seemed to guide me as I made my way over the pass.

I first stopped in Kern River Canyon, a wonderful valley that the naturalist John Muir loved. I ran into the north fork. This particular valley remains one of my favorite spots on earth. From there, I could see Mount Whitney, the largest peak in the lower States, which towers some 14,505 feet above the valley. The entire bowl surrounds the upper part of the valley and drains into it, so there are magical waterfalls bordering all sides.

One of Muir's favorite spots was the hot springs. I had previously been there a couple times. For a hiker, it is reachable with a decent amount of endurance and some know-how. But because it's a forty-mile walk from the closest drop-off point, you almost always have the place to yourself.

I stayed there for four days, recovering and regrouping. I relaxed in the hot springs, made teas, collected plants, and fished. I loved making tea out of the needles from the pine trees. I am convinced that it's one of the healthiest teas there is. You take the pine needles and smash them up into a pulp and drop that into boiling water. If you drink it often, the antioxidants in this tea will add years to your life.

One day while walking through the area, I met a random guy. He said he was camping in the lower valley and had hiked to the hot springs. I asked him how long he had been in the area. When he told me he had been out for four weeks, I only half believed him. He didn't look like the type who could stay out there long. He looked like an average city businessman.

It turned out he was a producer of Budweiser commercials—or at least that's what he told me. He said the reason he was able to stay out so long was that he was having a helicopter bring him supplies. Where I was camped that wouldn't have been possible, but based on where he was camped, the story seemed legitimate. He was on the border of the park and the forest, so a helicopter would be able to land there.

I spent a night there. We shared a campfire and talked a little bit. It was mostly small talk. He kept rambling on about the Bud commercials he had done and the different location shoots. He saw the land from a completely different point of view than me. We were both looking at the same water falling down the granite face of the mountain, but he was interpreting its value differently. I saw the small things the waterfall produced and fed, like the streams and the wildlife. To him, it was a backdrop for a beer commercial.

He told me I should be in a commercial because I had a "rough and worn look." I gave him my parents' phone number. Of course, after we parted company the following morning, I never heard from him or saw him again.

I stayed in the hot springs area for about a week. I then started hiking south, following the length of the range of the Sierras.

The Sierras are a monocline that runs north–south. The east side is steep, while the west side has a more gradual descent. On the east side, if you were to take a hiking trail up, it would be about fifteen miles to the ridge. If you were to take a hiking trail from the west side, you would be looking at more like fifty miles to that same ridge. I followed the north–south ridge for eighty miles, all the way until the Sierras tapered out at the foot of the Mojave Desert, which almost curls around them.

When I came down out of the mountains, I stopped in a small town called Tehachapi. In total, my journey had been about 130 miles. The walk over the Sierras was a nice way to cap off the months I had spent living on foot and not riding in a motorized vehicle.

There, I ended my goal of not setting foot in a car the old-fashioned way: I put out my thumb and hitchhiked back to my mom's house in Huntington Beach.

Chapter Three

ON MY TERMS

I found myself torn. I had set myself on an unusual path in life. But I was still young, and I felt as if I needed to reground myself in the real world. It wasn't about making money. In my life, money was for wants; it wasn't for needs. It was more a sense of trying to feel like I belonged to something concrete, so to speak.

During my walk through the Sierras, I admit that I felt lonely at times. I loved what I was learning in the wilderness, and how the wilds responded to me and I to them. But I also missed the ease of the city and being with my friends. I didn't want to be forced to choose one life or the other. I wanted to take the best of both worlds, but I wasn't really sure how to, or even if it was possible.

I ended up staying in the Huntington Beach area that winter and working for my dad. He owned a construction company that had a contract with the aerospace manufacturing company McDonnell Douglas. Life there wasn't all steel and concrete. I was near the ocean, where I could surf again. I was also back where I had grown up.

The working life wasn't completely foreign to me. I never had any trouble finding a job. People liked me, and I had a determination and dedication that showed. Employers were attracted to that, but I wanted to push away from normal jobs.

As I tried to relate to the people I had grown up with who had gotten married and had full-time jobs, I recognized that everybody's life is different. The phases are what allow someone to be a loner and not be alone. At that age, I was still developing and maturing as a person. I wanted to interact in a community and have friends, which made it harder to be alone.

Being alone is rewarding if you connect to the land, because what you get out of it is amazing. The plants and animals don't literally talk to you, but they almost do. There is nothing you can do to replace those experiences. When you are alone in the wilderness, you are struggling hard to understand something deeper, but nobody is there to help you. Ultimately, even the strongest people are going to feel a sense of loneliness.

The mental fortitude and patience it takes to be alone are way beyond what most people are capable of. Though I knew I was able to live alone, I didn't have the patience or mental strength at that age to be alone for significant periods of time.

I found that when I talked about being alone, most people could only conceptualize that as sitting on their couch watching TV and eating food. That has nothing to do with being alone. Even reading a book is not being alone because you inhabit someone else's illusionary world.

But being alone in the wilderness is something else entirely. There is no escape. You must be immersed and present with everything that is going on around you. Being back in society, I saw that people had created the ability to escape from the moment in different ways. While that makes life easier for that moment because it allows people to check out of reality, it also destroys our ability to be patient and present.

It helped that I didn't own anything, so I had nothing to come back to. All my belongings could fit in a backpack or tote bag. If it did not fit into my backpack, I gave it away. My only real connections were with friends or directly with the land, so for me to wander off on my own was easy. Yet I was still conflicted.

My experiences in the wilderness were giving me some deeper perspectives that others could not understand, nor could I fully. I appreciated the newfound and profound wisdom I was gaining, but I also began to feel a distance from people. To those around me, it seemed that I wanted to be a "hermit rather than a cool guy," as one guy put it. I also noticed that I had fewer close friendships and a harder time relating to people on everyday issues.

I found that for some, the phrase *living off the land* meant living "simple" and "lacking intelligence." I had discovered that this could not be farther from the truth. There is nothing simple about living off the land. It requires innate and learned intelligence to adjust to the natural changes around you—otherwise you will not survive. There is also evidence that the hunter-gatherer lifestyle stimulates psychological and physical well-being by returning us to the land, making it one of the most complex and invigorating ways to live.

It has been suggested that the cures to many modern ills can be found in natural living. The renowned naturalist and ecologist Paul Shepard conducted some of the definitive studies on early nature-based cultures. He concluded that humans, who over the millennia have spent nearly all of their social history in hunting-and-gathering environments, must be close to nature to properly grow emotionally and psychologically.

Shepard determined that the shift over our hundreds of thousands of years from hunter-gatherers to settled agriculturalists destroyed not only our surroundings but also us. Bucking our innate desire for the wilderness has taken us out of our natural element and as a result created many of the psychological and physical problems that exist today.

The one thing I knew was that I did not want to give up my enriching experiences in the wild just to be more relatable at a party. It was at that point that I decided I should not turn my back on the wisdom that nature was offering that could help others; rather I should further explore this call at whatever personal cost. I reached the conclusion that I would have to become a better communicator with the forest in order to explain and teach others the language of the earth. I made it a goal to figure out a way to bring friends and others to terms with the wilderness.

Before returning to Huntington Beach, I had walked virtually everywhere I went. I covered not just the length of California but also many of its nooks and crannies. I would meet people everywhere I went. But when I looked at them, I felt that they were boxed in and had lost their freedom. I didn't want to be like that. I didn't want to lose my freedom. I regarded these people as unhealthy, because they were trapped in a system that degraded their minds and their health. It scared me. I promised myself I would never fall into that trap.

For me, a large part of living off the land alone in the wilderness was not destroying or taking anything from the earth. Growing up, Christianity had taught me about being kind to people. But I had concluded that it wasn't enough to be kind to people while beating up the earth. To avoid falling into that trap, I kept my possessions to a minimum. I had seen friends in my field with ties to material things, and I could see it was pulling them away from the wilderness. That aspect scared me as well.

The other struggle I had was trying to share what I had learned. I felt that I was getting to a point where I could live off the land and not destroy the earth. But being back in society, it looked like everybody else was destroying the earth. I began to question what good I, as one person, could do if I slipped away and back into that lifestyle. Asking these questions resulted in me acknowledging that there was more to my life than just my relationship with the wilderness. I had to include other people's relationships with the wilderness.

One of the most important things I had learned from living off the land is to be more observant. I saw that my friends would drive home or walk around Huntington Beach and not even recall what they saw. Living off the land made me aware of everything. In addition to helping me know myself better, it also led me to pay closer attention to people.

I was careful not to be preachy. I tried to live by example. If asked, I would gladly share my love for the wilderness and my stories with people who wanted to hear them. But I focused on living by example. I knew that I could never convince somebody in a dialogue. I never dictated to others. Even though I was searching for answers, I never judged anyone else's life. I was arriving at my own conclusion on what was a kinder way to live. It was working for me. I was hopeful it would work for others if they chose to look for it. But at the same time, I felt a certain amount of confusion that so few people chose to live that way. Because more people weren't pursuing my chosen lifestyle, it made me constantly question what was wrong with it.

At some point, as I was rounding out my skills and filling out the full picture, I started leaning harder on the life of a survivalist. I was constantly pushing myself and riding a fine line. I would run sixty-five miles in a day from Death Valley to the Sierras. Every few months, I would sit on a mountaintop and fast for one full cycle of a day and two nights. I would run barefoot in the snow. Instead of sleeping on a bed when I was in civilization, I chose to sleep on a wooden board to harden my bones and my body. I would purposefully bathe under icy water in the wintertime.

Ultimately, that is what created the deeper connection for me and made me come to the realization that the wilderness would be my life's path. My intentions were pure. I believed in what I was doing so strongly that I was willing to die for it. But at the same time, I believed that the wilderness saw my intentions and therefore wasn't going to kill me.

EPIC SURVIVAL RULE #2:

BE PHYSICALLY IN CONTROL

Because I don't own a car, I am going to focus on three physical elements to cover the landscape: running, climbing, and swimming. I believe that if I can master those three natural modes of transportation, I can travel anywhere I want go.

Swimming is important to me because I grew up near the ocean, and I love to go on long-distance ocean swims. I use swimming as a relief from the heat. I also know that there will be times when crossing to safety requires me to swim across a body of water.

Climbing is critical because the earth is not flat. If I do not learn to scale a steep rock properly, I am putting myself in danger every time I move across the land. I will need to climb to cross mountain ranges. I will also need to climb to find a perch where I can avoid animals coming after my food at night.

Running is the most important means of transporting myself from one place to the next. It is my primary means of exploring new areas. It also allows me to speed up the basic tasks of living in the wild, like hunting for food or seeking shelter in threatening weather.

At the same time that I need to be strong in these areas, I am also starting to realize that no matter how physically developed I become, I cannot compete against nature. The wild will destroy me if I try to compete against it. I can't throw myself off a cliff and expect to pop up like a rubber ball, nor can I jump into a flash flood and survive.

I also see that it is futile to compete against the wild. Why would I want to go out and do battle with something that is so real, so beautiful, and so much a part of me? That defeats the point of being out in the wild.

However, the dilemma remains that at the same time nature is beautiful, it is fierce. That scares people. That ferocity keeps our awareness intact. I know that if I use my fitness to exist in that fierceness rather than fighting against it, I will be protected.

Chapter Four

"MUSCLE-HEAD MATT"

I was terrified of heights. That fear had been instilled in me by vague
memories of working in a circus and falling to my death. I don't know
what those memories mean, or where they came from. Some people at-
tribute such things to reincarnation. I'm not sure that is the case, but in
any event they were recurring thoughts. I knew that to be able to fully
connect with the land and be able to live as a hunter-gatherer I would
need to overcome my fear of heights and become a proficient climber.

The physical aspects of climbing appealed to me. I hoped to use
those to placate my fears. I grew up as an athlete and led a very physical
life. When I saw climbers, I noticed they always had good physiques. I
thought, *I want to look that fit.*

While those were the initial reasons I invested myself in climbing,
they would soon fade into the background. Climbing became less about
me and more about the art of climbing itself. It became about a dance
or a meditation with the rock, about how to listen to and move with the
land in the way it dictates, and about how not to force my physicality
on the rock.

Learning about climbing was more of a mental process than a physical one. The only formal instruction I had was a beginner-level climbing
class at a mountaineer school. I was taught basic rope work and a few

safety precautions. Following that one class, I would just show up in the valley, where there were enough climbers to help show me the ropes, so to speak.

When I first started climbing, I was such an accomplished athlete that I could just muscle my way up the rock. I attacked the rock like it was a physical challenge, such as bench pressing my body weight or sprinting a lap around a track. I was a *yarder*, the term used for people who pull overly hard with their arms. My ferocity and overall lack of grace on the rock caused the climbers to give me the embarrassing nickname "Muscle-Head Matt."

As challenging as making it up the rock was, I now had another obstacle to overcome.

In my early climbing days, I had several white-knuckle incidents. Most of them came when I wasn't climbing with a rope. They were soloing or scrambling experiences.

My first close brush with death was when my friend Jake and I hiked up to the top of El Capitan, the largest monolith of granite rock in the world. Experienced climbers from all over the world travel to Yosemite to tackle the three-thousand-plus-foot climb. After we made it up the trail, we decided not to hike back down the same way. Instead, we opted for a steep gully on the right side the trail.

The gully descended the full three thousand feet. From a technical standpoint, it was extremely dangerous. The rock was also very chalky and loose, which made it over-the-top dangerous. We also both had backpacks, giving us extra weight that increased the difficulty.

At one point, I was climbing down a big flake—a rock sheet that juts out on its own. I was holding on tightly. Jake was below me. His climbing technique was far more accomplished than mine. I was still muscling everything. Jake was using his feet to work his way down the rock, while I was using my arms, yarding on the flake.

Without a sound or any other warning, the flake snapped. A thin chunk of stone broke off in my hands, leaving me in a free fall. I went flying through the air, still grasping the stone. Instinctively, I let go of the stone to free my hands.

I continued to fall straight down the vertical rock about twenty-five feet until my backpack slammed down on a rock. I hit so hard it felt like I scrambled my internal organs. The impact flipped me. I hit the ground in a slightly sloped area and began sliding down the hill.

Frantically, I grabbed at rocks and trees to stop my slide. I managed to latch on to the base of a tree seconds before I fell off a two-hundred-foot drop. I hung there with my arms wrapped desperately around the tree, half my body on the rock and the other half dangling over the edge. I was a hundred feet below Jake.

"Are you okay?" Jake yelled. "Are you stable?"

I told him I was.

"Man, I saw you flip off that rock and I thought you were done," he said, summing up the fall.

So did I. I pulled myself back up onto the rock. It was my first real climbing lesson. The rock was warning me that I needed to become a more graceful climber. Jake was below me, and I could have very easily hit him with part of the broken flake.

And if I don't grab that tree, I die.

The near-death experience rattled me, but it also motivated me. My first thought was that I wouldn't climb chalky rock. Unfortunately, where

I lived, there was always going to be chalky rock. I needed to learn to climb it properly.

I fully realized that I would not be a true climber until I improved my technique and established some finesse. I also knew that I needed to find them on my own, not in climbing school.

I began studying the rocks. I would go out to the base and stare up at a rock for hours and hours. I wanted to establish an appreciation for the rock. These cliffs were an integral part of the earth, not climbing walls with plastic notches dragged behind pickup trucks to carnivals. The rocks had been carved and honed by Mother Nature over millions of years. I had to learn to respect them.

I also decided to climb easier rocks without a rope to learn how to move on the stone. This allowed me to stop being afraid that I wouldn't make it up the rock. During this time, I started to enjoy how the rock carried me over it. Once I felt that connection, I began to see climbing as a graceful dance that nature was teaching me, rather than as a man battling the elements to survive.

My technique needed improvement. I had to learn to distribute my weight more evenly and move the way the rock wanted me to move on it. Once I altered my technique, I started to feel like I was connecting with the rock. I started moving with the rock in a way that the land wanted me to move. It felt as if I were dancing on the rock. I could almost envision the rock gods and the tree gods looking up and saying, "I really appreciate how that guy is climbing."

Oddly, when you reach that point, in some ways it feels like you are giving something back to nature. Every aspect of the land is giving and receiving energy—all the rocks, all the animals, everything.

Let's say a climber makes his way up a rock crack. There are chipmunks frolicking and plants flourishing. The chipmunks communicate with their chirping and movements as they jockey for position for food. Plants have different ways of communicating that we don't fully understand.

Now let's say a climber brings a boom box and plays some heavy-metal music. Then he starts yarding on the rock. His rough climbing pulls a bunch of loose rock down, but he doesn't care. He keeps going. Maybe he steps on a loose boulder that rolls down the crack. During this process, he might have disrupted a squirrel house or even an unseen endangered falcon's nest. Even if there is no visible damage, he is certainly freaking out all of the animals and disrupting that whole area. Of course, he didn't realize that because he was not paying attention.

But if a climber goes up there and climbs with focus, attention, and purpose, even a hawk or a golden eagle that has lived on the land its entire life will fly by and appreciate the way he is moving on that piece of rock. Seeing that climber might just make the hawk's day. It lives there. It is part of the land, so in essence, the climber is giving something back simply by climbing well, as strange as that may sound.

I found this to be true. When I started climbing with finesse, hawks always came to watch me. When I sucked, I never saw a bird anywhere near me.

Of course, there's no way to know when you reach that level. There's not a light that goes off, or a special call you hear from the hawks. The only tangible way would be if a spectator were to look up and say, "That guy is moving gracefully up the rock, rather than fighting to get to the top without falling." What it comes down to is the difference between a person who has a strong intention and one who does not.

Intent is a powerful driver. Whenever I travel across the land, I will always connect with the native people. They will look at me and say, "Strong heart." What they are seeing is not a guy who can know everything and live off their land. The native people can feel when somebody has an awareness of and an intention to learn their land. That intent can be extended to anything in the wild.

About a year after I began climbing, one of the world's greatest climbers Peter Croft, showed up to watch me on several occasions. From high up on the rock, I would look down and see Peter staring at me. I wondered if he thought I was going to die because I was a hundred feet up without a rope. He would spend hours watching me, but by the time I descended, he was always gone.

I knew from other climbers that Peter was shy. One day, I bumped into his wife in town and asked her why he was watching me. "He loves to watch you climb because you climb like you appreciate and love the rock," she said.

That was huge for me. I had gone from "Muscle-Head Matt" to being appreciated by a world-renowned free climber. It validated that I had a connection to the rock, and that I could make the connection stronger if I looked and listened, instead of trying to battle the land.

I had come to realize that when you put up your defenses and try to battle something, you don't get very far, but if you work with something, be it the land or people, you get a better experience—even if that inter-action is with an immovable force. When you watch the best climbers on the hardest climbs, you won't notice how difficult it is unless they fall. They maintain their grace and cool until the moment they pop off the rock. It's the inexperienced climber that is grunting, grasping, battling, not moving well, and losing all technique and focus—and probably not getting up the rock. It looks like a schizophrenic spider trying to figure out where to go, darting in one direction, then another.

The process of climbing is like a jigsaw puzzle. You look up the rock ten to thirteen feet and mentally measure your moves. You think, what's the sequence? Where do I put my hand? I see a geological pocket in the stone caused by some type of organic vegetation that was trapped and has decayed. Does it look too soft? If so, then you shorten your moves until you can test the spots.

But the danger never disappears. Even after I became a confident climber, I had close calls. The first was when I was free soloing—climbing without a rope—Half Dome in Yosemite. Having been there on my first backpacking expedition, I knew there was an easy route on the south side that wasn't directly on the face. It was a big crystal dyke that went up for a couple thousand feet. I climbed the first four hundred feet or so and hit the critical juncture.

I knew from reading the guidebook that this move was the crux of the whole climb. You had to move along a polish (a rock ledge) for about eight feet before you reconnected with the dyke. When you use a rope, you clip in at that point so if you fall, you can swing over to the dyke. Without a rope, that move was dicier than most five- to six-rated moves, which was the rating of the climb.

I stopped and worked out the move in my head. Mentally, I saw it. But then I checked the footing. It was very unstable. I concluded the risk was not worth it. I climbed back down the dyke, descending the four hundred feet I'd climbed up, to the bottom.

Unsatisfied with my climb, I decided to take a more challenging way to the base. Instead of hiking back to the trail, I went down the face on an edge where there was a ravine. I had to climb down the ravine backward because it was so steep. The ground was loose. In some regard, it was reminiscent of the El Cap adventure, but technically much harder.

When I got about halfway down, I felt that I was in the clear. I then spotted a rope coming off the face. The long, static line was about a thousand feet. It came down off the face and disappeared into the ravine. The terrain was becoming increasingly steeper so I decided to use the rope as backup, as a sort of guide line. I put my hand on the rope and continued the downward moves.

After a few steps, the granite under my feet turned to loose ball bearings. The footing was so poor I started sliding. I grabbed the rope firmly with my right hand. I was now using it for support. Trouble was, a thousand feet of rope has a lot of stretch in it, so I was holding on but continuing to slide down the ravine.

After about twenty-five feet I could feel the rope getting tighter and tighter. Then it became so taut that I gripped it with my other hand. When it tightened fully, I had all my weight on it. But the rope was so long and I put so much force on it that it turned into a rubber band and popped me up out of the ravine.

I was literally flying across the face of Half Dome, Tarzan-style, some five hundred feet off the floor. To avoid crashing into the rock, I pushed off with my feet. Now I was free swinging out over the rock, doing a hundred-foot pendulum. I was desperately looking around, trying to determine where I could land. There wasn't a person in sight.

As I was swinging back and forth, I realized that the rope wasn't long enough for me to reach the bottom. I was stuck on a sheer face of Half Dome.

I spotted a dihedral, which is an inside corner of rock. I could see that it had a plateau on top and thought that maybe I could make it up to the plateau to get my bearings. But that escape plan was fleeting. I quickly realized the rope wasn't swinging me to the plateau; it was swinging me into a rock crack about thirty feet below the plateau.

I hit the crack hard, did a quick hand jam into it, and wedged myself in. I lifted my feet and stuck them in the crack so I could regroup and assess the easiest way down. I checked the rope's length again, but determined it wouldn't reach the bottom.

I looked up and saw that the plateau was only about thirty feet above me, in what appeared to be an easy climb. As a backup plan, I wedged the rope into the crack in case I needed it later. I then started free soloing. I was four hundred feet off the deck, doing this very difficult crack (rated 5.8, a fairly challenging climb for somebody without experience using a rope and a very difficult climb with no rope).

When I reached the plateau, I pulled myself up and walked over to the other side to find an easier route down. But there wasn't one. The other routes off the plateau were sheer rock, straight down. The crack where I had left the rope was the easiest way down. The rope, as I had already determined, was useless because it didn't reach the floor.

I climbed off the plateau and back down into the crack to where the rope was wedged. I left it there, and proceeded down. The climb was over. I just needed to make it down. I took it slowly, felt my way, and reached the bottom safely.

Once I became an experienced climber and shed "Muscle-Head Matt" for good, I established a cardinal rule: never free solo anything that I had already done with a rope. I broke that rule only once.

It happened on a route I had done a bunch of times with a rope. Each time I was up there, I kept thinking it would be a perfect solo route. It was 800 feet, 5 pitches. (Pitch is climber-speak for length; a pitch is 165 feet of rope.) It started out as a 5.8- or 5.9-rated climb. But in order to not repeat myself, I decided to do a 5.10 variation.

The variation was cool. I climbed up the first 150 feet of a thin finger crack and felt confident that I would be fine free soloing the route. I passed two climbers who were on ropes. It's kind of freaky for a climber who is roped up to watch somebody without a rope climb by them.

After I carefully passed the climbers, I reached the tricky part. There was a bulge right at a crack that required two hand jams followed by an unorthodox friction move. I climbed the bulge and executed the move. I stayed in control, but as soon as I stood up, my foot was just inches from popping out.

My heart started racing. I held fast and thought, *Slow down your heart.* My next thought was that it would've been awful if those climbers I passed had to watch me plummet past them.

There was still one bigger bulge above me that I had to climb. At that point, I was questioning my commitment. I had broken my rule not to free solo something I had roped. I made that rule because it kept me focused and in control. But I went ahead and did it.

I reached the next bulge. My heart was beating in triple time, and I kept trying to slow it down. There was too much dangerous exposure on the route, because at that point I was seven hundred feet off the deck of the steep, sloping face where the ledges were.

Though I made it without incident, that was the last time I broke my cardinal rule.

Any responsible climber on a new route should always be thinking, *Can I downclimb this route?* As you go up, you must focus on every move. However, you also have to work in both directions because there may come a point in a route, even as far as a thousand feet up, where you determine that you can't continue and need to downclimb the route. When that happens, you set aside ego for safety. No matter how well you know the rock, you have to know yourself better.

Climbing was an important phase of my education about nature. It was something I immensely enjoyed. Being on the rock was blissful. It was one of the few things I found that completely emptied my head of all thoughts and distractions. All of my attention had to be focused on one thing.

At that point in my life, I didn't know how to meditate. Climbing took the place of that. It channeled my head to a point that forced me to relax and surrender to nature.

If a climber doesn't surrender to the rock, he is not going to climb well. When someone watches an accomplished climber, they marvel at the way their back muscles ripple and their legs and arms attach to the rock. Most people think, *That person must be in amazing physical shape and have strong hands.* That's all true, but that is not even the half of it.

For a climber, it is about letting go of those thoughts. The climber cannot think about the physicality, because there is so much technique involved in the way you must balance your feet and your hands to take pressure off them. The climber must take a soft approach, rather than an aggressive one, or he will not be a climber for long.

When I first started climbing, fear was pervasive. I was scared of heights and had thoughts of falling. But once I learned to move over the rock the way nature wanted me to move over it, I felt like I was doing the right thing. Although I was free soloing routes two thousand feet up, those climbs weren't a death wish. I wasn't doing anything that was out of place to the degree that it sparked an uncontrollable fear. I was doing what I was supposed to be doing.

Chapter Five

THE ANCIENT FOOT MESSENGERS

From the age of six, running was my pathway to feeling free and the vehicle for my exploration of the land. My father would take me out jogging with him. I loved when he would vary our route so I could see different things. Even at that age, I realized that running was a way to put a smile on my face and to explore my surroundings.

Throughout high school, I was an avid runner. I could run a fifty-two-second quarter mile at a time when adults were winning gold in the Olympics with a forty-nine. But I wanted to run across the plains and up mountains to see life, not around a track. I didn't want to run for fitness; I wanted to run for awareness, as my forebearers had.

When I moved to the Sierras at age seventeen, I ran every day. I started out going on short runs of six miles or so. I then worked my way to running up Half Dome, which was twenty miles round-trip. But during that time, running remained a social pursuit. It was somehow ingrained in me that twenty-six miles—the distance of a marathon—was the farthest you were supposed to run. But as I began to study running and what it meant to the lifestyle I was choosing, that would all change.

At the time I became interested in running, America was in the middle of a jogging craze. People were running in greater numbers than at any time in the past. This led to the creation of specialty shoes for running. Recreational runners bought the shoes in droves, and eventually created a multibillion-dollar market.

Nike, then a nascent company, led the charge in creating different styles of running shoes. In my teens, I ran in a Nike shoe called the Air Tierra. The shoe was like a rubber moccasin, with no cushioning in the forefront and just a half inch of padding in the heel. I loved it. You could corner the edges of the rocks and maintain full control.

After I ran the heck out of my Air Tierras, I bought a second pair at a thrift shop and beat those up as well. But when I went to buy another pair, I couldn't find the shoe anywhere. I called Nike and asked why they stopped producing it. The customer service representative told me that too many people were getting stone bruises, so the company discontinued the shoe. The shoe that replaced it was beefed up in the heel and heavily padded in the front.

Soon the market was flooded with similar shoes. I didn't like training in the newer shoes that had come on the market. I remember looking at a pair of gel-filled shoes and thinking, *If I put these big cushions on my feet, I'm just going to destroy my body because my feet will not be able to hit the ground naturally.* That was when I started to experiment with different footwear.

I began running in my kung fu shoes. The thin sole allowed me to feel the impact of the ground. I actually came up with the idea from my martial arts training, which was based on the concept of impact training.

As a martial artist, I did impact training with my hands. I would punch wood constantly to build up cartilage and tissues. I knew from that type of training that the impact actually built strong, healthy joints. I decided to translate that to my feet.

Studies of impact training in martial arts have shown that martial artists who train on boards build cartilage, whereas martial artists who train on cushy, foam bags break down their bodies. The way to keep the body the healthiest is to connect it to the impact. The same principle applies to the feet and legs. You can't have a lot of cushioning and not expect that you are going to deteriorate something—which is the running-shoe myth.

The running-shoe myth, sold to consumers by manufacturers, is that cushioned shoes give a runner bounce and prevent injuries. But we now know from numerous long-term studies that running shoes have likely created more injuries than they have prevented.

Dr. Irene Davis, a sports scientist at the University of Delaware, tested runners in shoes and in bare feet. Her studies showed that when you are wearing a running shoe, your heel hits first, which is called a heel strike. This creates a hard ground-reaction force, or upward vibration, that has an impact on the legs. However, in bare feet, there is less ground-reaction force. The runner lands more on the midfoot and less on the heel, causing a reduced impact on the legs.

Dr. Davis and others have shown that rather than protecting our feet, running shoes actually harm them by causing the foot to overpronate. In normal pronation, the outside part of the heel hits the ground first. The foot then rolls inward at fifteen degrees, causing the weight of the body to be distributed evenly over the foot. When a foot overpronates, the foot rolls farther inward, which throws off the weight distribution. In this situation, the ankle does not properly stabilize the body. Normal pronation occurs best in bare feet.

When running barefoot, the feet receive continuous information from the ground. This allows the runner to respond to changes in what's underfoot much faster. You can use every part to your foot, which prevents overusing any one part and creating adverse shock in the ankle and leg. You can use your toes to cradle a rock for better balance and traction. When you hit the corner of a rock, your foot takes the impact and molds around the rock, whereas in a running shoe, the stiff edge of the shoe can catch the rock and roll your ankle.

Barefoot running was a concept I began to grasp in my teens. After I moved to Yosemite, I often ran barefoot, even in the winter. I found that running barefoot in the snow was good for conditioning. I also found that running barefoot in the snow increased the circulation in my capillaries. My feet never got cold, because I was constantly stimulating my capillaries and increasing the circulation in my feet.

Barefoot running led to my exploration of sandals as a far superior alternative to funkified Nikes. I began experimenting with sandals made of many different materials, including yucca, rawhide, leather, worn-out flip-flops, and old tires. My goal was to find footwear as close to a moccasin as possible so I could maintain a natural running position and be in touch with the ground.

Rawhide sandals are ideal for running. They are made from the hides of large animals such as elk. They work well in colder weather, as I can leave the fur on for a tiny bit of added comfort. I can also use the coarse hair grain to provide added traction in certain types of soil, such as damp clay. Rawhide sandals are stiff at first. It feels like you are running on plastic, but as you break them in, they conform to your feet.

How long a pair of rawhide sandals last depends on the quality of the hide. Provided that the hide is cut from the upper part of the neck, elk can last a maximum of three hundred miles. Buffalo hide, which is thicker, can take you six hundred to seven hundred miles. Tanning the leather slightly can increase the durability. This makes the hide last longer because it forms around the ground as you move and therefore wears more evenly.

Two-Ply Rawhide Sandals
WITH FUR LEFT ON

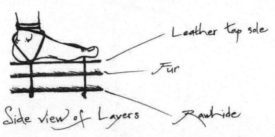

Leather top sole

Fur

Side view of Layers

Rawhide

Leaving the fur on adds more contour to the foot, provides better insulation, takes less time to produce, and the fur does not wear out before the rawhide does.

Yucca Fiber Sandals

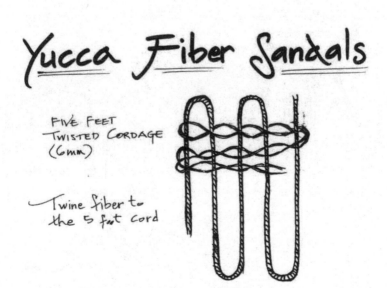

FIVE FEET
TWISTED CORDAGE
(6mm)

Twine fiber to
the 5 foot cord

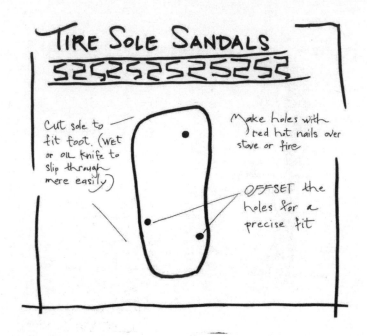

TIRE SOLE SANDALS

Cut sole to fit foot. (wet or oil knife to slip through mere easily)

Make holes with red hot nails over stove or fire

OFFSET the holes for a precise fit

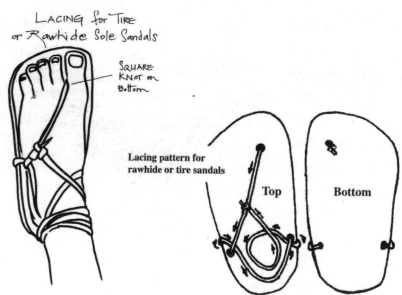

LACING for TIRE or Rawhide Sole Sandals

SQUARE KNOT on Bottom

Lacing pattern for rawhide or tire sandals

Top

Bottom

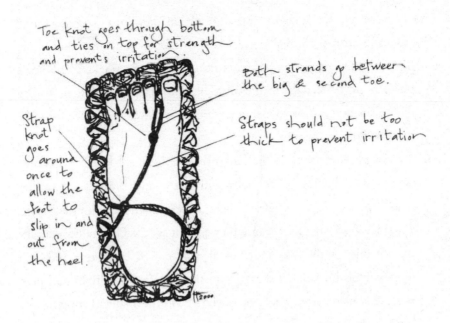

Toe knot goes through bottom and ties on top for strength and prevents irritation.

Both strands go between the big & second toe.

Strap knot goes around once to allow the foot to slip in and out from the heel.

Straps should not be too thick to prevent irritation

Yucca sandals have several advantages. They can be woven anywhere yucca grows, eliminating the need for killing an animal to make the sandals. I found that the Yucca sandals allow my feet to grip the ground closely. If they are woven tightly, the Yucca grabs the soil and rock well, even when the rock is wet. This helps when I'm hunting. With my feet holding firm, I'm able to throw more accurately. They also breathe nicely in both cold and heat. The downside is that they take longer to make and they wear out fast. I can get only about forty-five miles per pair.

Another advantage to sandals is that they are easy to make and that other fibers work just as well as yucca. No matter where I am, if I lose a sandal in a flash flood, I can make another pair in an hour. On some islands, there are different types of bark or cottonwood that will work. Agave plants can also be used—so, as one of my friends joked, if you really know what you are doing, you can have a tequila while making your sandals.

Fiber sandals also work well for tracking. They don't make any sound from step to step, so they are very effective for stalking animals. On the flip side, they are also good if you don't want to be tracked. The fibers break up the pattern left by the sole. These work even better than placing felt on the bottoms of your shoes, which is a standard practice for military avoiding the enemy.

For extreme survival situations, tire sandals are the best because they are the most durable. Three-quarter-inch tire sandals can last up to ten years. They can be made with old belted tires, which are readily available. The sidewalls work best because they are not as stiff. Generally, sidewalls are not lined with steel, and they also have a natural curve and arch support built in. Though the grip on the foot is not as tight and pure as sandals made from yucca or rawhide, the tire material imitates the cartilage in the body. This will make a first-time wearer sore until their cartilage toughens up.

Sandals provide a more natural way of running than $250 midstrike shoes with custom orthotics. After I switched to sandals and established a more natural stride, I didn't suffer from any hip, ankle, or knee injuries.

I also found that running in sandals puts my feet in better touch with the ground. I love the feeling of being closer to the earth rather than plodding along in a pair of funkified Nikes. Scott Jurek, a professional distance runner, once looked at my sandals and commented, "You must feel really connected to the earth running in sandals like that."

Indeed I do.

Even as I explored barefoot running and running in sandals, I still did not have a profound connection to running. My entire relationship with running changed in my late teens. I was in a public library rummaging through the free box of magazines when I stumbled upon an article about the Tarahumara Indians in the May 1976 issue of *National Geographic*. The story focused on their ability to effortlessly run long distances. In a two-day period, the best runners could cover 435 miles, which is equal to sixteen marathons. Even an average Tarahumara could run more than 100 miles without stopping or suffering sore joints.

To this day, the Tarahumara are considered the greatest runners in the world. Running has been part of the Tarahumara culture since the 1500s. It was a means of both athletic expression and survival. They delivered mail and intervillage communications and hunted for food on foot. While the article was inspirational for someone interested in long-distance running, what was most interesting to me was that the Tarahumara ran in handmade sandals rather than shoes.

At the time, I was spending half the money I earned on running shoes. I was also suffering from a variety of running-related injuries such as shin splints, persistent pain in my right knee, and hamstring injuries. I had already begun searching for an alternative.

To become the most efficient runner possible, I needed to see the Tarahumara Indians and study them in action. I was eager to examine the Tarahumara's sandals, as well as to study their running technique and to gain knowledge, inspiration, and wisdom.

When I set out on the trip, I didn't have much information to go on. I knew the Tarahumara lived in Copper Canyon, located in the southwestern part of the Mexican state of Chihuahua, about two hours south of the Arizona border. Copper Canyon is a vast system of six canyons, so I needed to find a starting point. The consensus from what I read was that I should go to a small town called Creel, which I did. When I reached Creel, I hopped off the bus and starting running.

I searched for the Tarahumara on their own terms—on foot. The land was very dry, and the canyon walls were green, like oxidized copper (hence the name). I found small village settlements and met several Indians, but I didn't feel comfortable invading their sacred space. Instead, I spent most of my time watching the men run and observing how they interacted. One thing was true: they were always running. The trails were constantly in use, and I seldom saw the men walking.

I spent ten days running, seeing the land, getting in shape, and orienting myself mentally to the running techniques of the Tarahumara. Watching them was a beautiful sight. It looked like the way a human should run. Their backs were ramrod straight, and they pointed their toes down. This caused them to land midfoot rather than striking with their heel first, which is a result of having the toes up. What was most noticeable was that the looks on their faces showed that they loved to run. There was no grimacing or grunting. They were actually smiling as they traversed the land.

The main thing I learned from them was how to strike the foot on the ground. They recognize that the nerves in the feet are as sensitive as the nerves in the genitals. Treating those nerves kindly proves to be the key to running distances. They also understand that running involves a "mass spring" that transfers potential energy into kinetic energy through the tendons and ligaments.

The tribe showed me how running could be a spiritual part of life. Running to them is like water flowing downhill, feeding the crops as it goes. The images of them running are beautiful. They sew their own clothes out of bright fabrics, and when you see them in the distance, they look like moving flowers dotting the trail.

I also observed their sandals closely. I was somewhat surprised when I realized the sandals were simply hard, heavy chunks of rubber. It was astounding to see how well they could run on something so hard. People think of rubber as soft and bouncy. But these are not wealthy people with resources, so they use very heavy semitruck tires for the soles. Used semi tires are inexpensive, and when made into sandal soles, they last fifteen years, even with the amount of miles the Tarahumara run.

When I finally held one of their sandals, I was even more shocked. The sandals weighed *three pounds* each. I wondered how they ran in them at all. Not only did they feel like leg weights but they were very stiff. Healthwise, the thing the sandals had going for them was that they were hard. They weren't cushions breaking the runners' bodies down like superlight running shoes stuffed with foam and gel.

I had brought my leather huaraches on the trip, which were the type of sandals the Tarahumara made before they used tires. My footwear intrigued the Indians I met. Many of them made comments about how fast I could run in light shoes made of leather. On a future visit when I showed up in thin tire sandals, I was amused to find the Indians were slightly jealous. They requested that I bring them some thin tires when I next visited.

It wasn't surprising to hear the positive health effects that so much running had on the Tarahumara. The Tarahumara were practically immortal compared to Americans. In addition to the miles covered, the Tarahumara's diet also contributed to their overall health. Diabetes and heart disease were virtually nonexistent.

Many articles have been written about tesguino, the corn beer that the Tarahumara consume at festivals. It was also detailed in the book *Born to Run*, by Christopher McDougall, which talked about the bacchanalian-style feasts they had the night before long runs. The truth is, they don't have those festivals very often. Americans like to think about drunken Indians, but they don't drink beer and rarely drink fermented corn.

I found that they drink more tea than anything. Whenever I stopped in a village, they always had a pot of pinole, which is hot tea made with pine needles. They keep it over a fire in big clay pots. They are such a corn-based culture that sometimes on top of the pinole tea they serve corn tortillas. They live and breathe corn, especially the poorer people, because they can't afford to grow or buy most other foods.

The traditional, Western view is that the corn eaten by the Tarahumara is stored as glycogen and then converted to energy. This serves as the explanation why the Tarahumara have such endurance on long runs.

Nutritionists teach us to do calculations where you burn x amount of energy and then replace that with y amount of calories. Under this scenario, in a 26.2-mile marathon, an average runner will burn about 2,600 calories. Following this logic, if a Tarahumara were to run 435 miles in two days, he would burn 43,500 calories.

But that's an American concept that really doesn't hold validity in ancient and Native American cultures. It also doesn't make sense that someone can store up or consume 43,500 calories in a few days. In fact, what happens in traditional cultures is that the runners burn so many calories and replace them with so few that over time their metabolisms become increasingly more efficient.

Take a Native American tribe such as the Navajo, whom I have seen up close. If they ate the same amount of calories as we do in our normal lives, they would gain fifteen pounds in a month. It wouldn't be entirely their fault they got fat. Genetically, they would still have the same superefficient metabolism. In our Western and European culture, we have entered a phase where our metabolisms are much more rapid because we haven't been in starvation mode the way traditional people have been.

The same is true in Kenya. There is an abundance of great distance runners from Kenya, all of whom are rail thin and do not consume an abundance of calories. This is because they have lowered their metabolism to the point where they don't require the same number of calories as an average American for energy.

Seeing how effective this was in Native American cultures, I went on a lifelong pursuit to make my body more efficient as a runner and use fewer calories. After a few weeks of going into starvation mode, I found that I needed half the calories to function at the same level. Of course, as soon as you return to three hearty meals a day, you've blown all that. You have to maintain yourself at close to the same level.

Obviously, we don't have the genetic predisposition to go as far as a Native American, but we do have the ability to readjust our metabolisms for short periods of time in training. I sought to find a balance that would allow me to live for long periods of time in the wild.

I knew that running had been central to the lifestyle of early hunter-gatherers. They ran to hunt food. They ran to avoid being attacked by animals. They ran to move from place to place quickly. And likely, the men ran to impress the ladies. I knew that to live in the wilderness and be able to hunt effectively I would need to understand running and be able to explain it to others.

The transition to natural running requires special training that should be undertaken in stages. You can't just take off your running shoes, hit the trail in bare feet or sandals, and expect to feel anything but pain. You don't want to race—pun intended—into anything full-on, or you will get injured.

To run in bare feet or sandals requires an understanding of the larger picture of running and a concept of what keeps the body strong. If a runner has worn cushioned shoes for most his life, he will have a weak body. Cushioned shoes compromise the body. His cartilage and his joints through his knees and hips will be weak. For somebody who has used that kind of compromising support, if they put on hard rubber sandals and think their body will feel better, they are in for a world of hurt.

Any transition to sandals must be undertaken gradually. You must first strengthen the feet before moving to sandals. The first step in the transition is to go barefoot as much as you can without hurting yourself. The best way is to start out walking in bare feet and then slowly build up to running. After walking around barefoot as much as possible for a week, try running a half mile.

Depending on where you live, you can try running barefoot on different surfaces. The best surfaces are grass, dirt, or sand. However, someone who lives in a big city should run on pavement once in a while. The idea is to have that feeling where your toes are grabbing the ground. Reaching that feeling will strengthen the underside of your feet and cause you to land more softly.

Over time, barefoot running will build a quarter inch of cartilage on the bottom of the feet. At that point, when you run on pavement, you will not feel that pounding slap of the feet. Once you have that feeling, you can start experimenting with traditional, thinner-soled sandals designed for rougher terrains.

I can run barefoot on any landscape with the exception of extremely hot sand. But it is a matter of the pace and how much I want to be able to look around and observe without worrying about what's on the ground. Sometimes going barefoot in rough terrain causes me to miss everything around me, because I'm looking down at my feet the entire time. When I run on new land, I wear sandals so I can observe.

In addition to strengthening the underside feet muscles, barefoot running adds more spring to the step and ultimately provides for a softer landing. Running on harder soles, such as rawhide or tire sandals, also builds up cartilage. A runner who wants to maintain a healthy body needs a balance of all those elements. The Tarahumara, for instance, know how important barefoot running is, so they occasionally take off their sandals and run barefoot just to keep their bodies healthy.

I do half my runs barefoot and half in thin-soled sandals. The exception is wintertime. When it is really cold, I will now wear a low-cut, barefoot-style running shoe with a tiny bit of padding. If I did that all the time, it would start to weaken my body. But because I do it sparingly, it doesn't.

Ultimately, a transition to natural running will prevent joint pain. Consider the Tarahumara and natural distance runners from African countries. When they are old men, they are still running constantly. They don't experience joint problems. This is not because they have über-genetics. It is simply because from the time they were young, they put their bodies on a different track from Western runners and did not attempt to overprotect their limbs with cushions on their feet.

Instead of running in natural shoes and surroundings, can you use a StairMaster, treadmill, or elliptical machine? Sure. It's not the best solution, but sometimes it may be the only one if you want to exercise. Inherently, anytime we go against the grain of nature and use technology, it will compromise something in our health and well-being. However, sometimes you have to use technology because of the circumstances.

Of course, it would be far better for the body to hike up a trail. You are outside, breathing fresh air. You also receive the benefits of walking down the trail, which creates a different neuromuscular sensation.

Running teaches you to have spring. The body is taking in the force, rather than pushing out of it. Bikers who never run are terrible runners. Despite the fact that they have muscular legs, they don't know how to take in impact; they only know how to push out. When they run it is awkward, because they have the strength to push, but they don't have the strength to land.

The ability to be able to run fluidly is a constant thought process that eventually becomes a natural one. Running starts with a conscious awareness of how you walk and move throughout the day, not just when you put on your running shoes. Most people don't even think about the way they walk. They just put one foot in front of another, and that is the same feeling they should try and achieve when they run.

I love to run for two reasons. The first is that it allows me to see new terrain. The other is that it feels good to be moving.

When I looked at the Tarahumara men running down a cliff side, they always had big smiles on their faces. Kids are the same way. When they have that spontaneous thought to run down a sand dune, it puts a smile on their face. You can take huge strides, and because it's soft, you won't hurt yourself if you face-plant. Happiness is a large part of running.

With the knowledge I had gathered from observing the Tarahumara and my experimentation with barefoot running and with sandals, I began to focus on running lightly, easily, and smoothly, all while searching for a harmony with the land. One Native American I met summed up the perfect feeling of running. He said: "When you run, you never run over the earth. What happens is that the earth just starts moving under you, and you move your legs with it."

That is the ultimate feeling of running, as you no longer feel like you are running. It is the point you reach when you feel like running is something you are supposed to be doing, not any sort of physical struggle. If you want to run all day, that is a great place to set your mind. It prevents you from using excess energy and allows the earth to dictate your movements. You feel like the earth is pushing you along as it rotates under you, and rather than exerting energy, you must figure out how to dance as varying terrain and obstacles come at you.

Confirmation that I had achieved that feeling came later, on my second visit to the Tarahumara. In addition to what I had learned, one of the most satisfying aspects of my time with the tribe was that I made a lasting connection. They branded me "Rara Miri El Blonco," which translates to "the White Tarahumara" or "white foot runner." I wear that as a badge of honor.

Chapter Six

PREPARING FOR AN EXTREME RUN

E veryone talks about finding his or her identity. For most people, there are experiences that become the building blocks of this process. These experiences begin to add up, and the sum total of them becomes the person's identity. Whether consciously or unconsciously, everyone goes through the process in some form or another.

When I turned nineteen, I was just like any other human trying to separate myself from other humans. I was working in a mountaineering shop in Mammoth Lakes, California. Though I performed the tasks to the best of my abilities, it was still just a job. My real focus was working to perfect my climbing and becoming a better distance runner. I felt that by somehow combining climbing and running and using those to explore the land, I could end my lateral drift and establish my own unique identity.

One day, I happened upon a book titled *The Pacific Crest Trail, Volume 1: California*. As the book detailed, the Pacific Crest Trail stretches 2,638 miles from the Mexico/California border to the Washington/Canada border. The California section, by far the longest section at some 1,700 miles, covers the length of the state. Therein lies its majesty.

The trail passes through every climate and nature zone the United States has to offer. As its name states, it runs along the crest of some of the most stunning mountains in the world. It climbs at points to fourteen thousand feet, dips down into the hottest, driest desert, and crosses rivers, towns, and even our busiest highways.

The book's focus was on hiking the Pacific Crest Trail. In the beginning, it talked about allowing five to six months to complete the trail. It mentioned that a couple named Ray and Jenny Jardine completed the journey in three months and three weeks. Another man, Bob Holtel, ran the trail in 110 days at the pace of a marathon a day, though he rested 46 additional days, meaning that he averaged about 16.5 miles a day. With my training, I was pretty sure I could blow those times away.

I had first heard about the Pacific Crest Trail when I was eighteen. Several of my climbing friends talked about hiking the trail, but running the trail seemed like an elusive dream. Even though the guidebooks said it took several months to complete, we were all certain we could do it faster. But no one actually tried.

I had friends who were in great shape and were accomplished runners. One of my friends, Bruce Davis, used to tell a story about running out into the wild for a day, spending the night, and running back the next day. He had worn a fanny pack and carried a bivy sack and a change of clothes. He was elated by the experience, and I could see by the light in his eyes that something had hit him really hard. It was a simple way to connect with the backcountry carrying minimal gear.

Ideas for long runs circulated in my head for years. I considered entering the Western States 100-Mile race. I contemplated running the John Muir Trail, which stretches 211 miles through the Sierra Nevada mountains. The dilemma was whether I wanted to start entering long races or to keep to the purity of running by going out and finding my own backcountry trails.

The more I read *The Pacific Crest Trail* and the more I thought about running the PCT, the stronger the pull became. It was a pursuit that would help define me, as well as teach me things about the land that I couldn't learn on long day runs. Rather than entering competitive races, I decided that running for the sake of exploration was the route I would take.

At first glance, running 1,700 miles might seem impossible. But from my research I was learning that running great distances was actually more natural for humans than most people realize. Several scientists who undertook lengthy studies concluded that humans are actually built to run extreme distances.

In one study, the noted anthropologists Daniel Lieberman of Harvard University and Campbell Rolian of the University of Calgary determined that short toes make humans better suited to running than to walking. They tested fifteen people both running and walking on a pressure-sensitive treadmill. They discovered that an increased toe length of just 20 percent doubled the amount of energy required to run and produced additional shock on the foot.

Their conclusion was human bodies are ideally equipped to run long distances.

In doing so, we are using our bodies the way our hominid forebearers did millions of years ago. In what is known as the "endurance running hypothesis," scientists and anthropologists believe that running for extended lengths of time is an adapted trait. The evidence suggests that it was the catalyst that forced *Homo erectus* to evolve from its apelike ancestors because they needed to obtain food.

For me, such information only fed my desire to take a run few had taken before, and it suggested that not only would my body hold up but it was actually built for the journey.

When I broke the news to friends that I was going to run the Pacific Crest Trail, nearly everyone I told was supportive and offered encouragement. However, several ultra-distance runners discouraged me, which came as a surprise. I expected more support from people who ran more than I did. I wasn't bragging that I was setting out to break the record. I explained that I was just planning to run the trail and see what happened. Despite knowing how physically fit I was, they were skeptical that my body would hold up.

At that time, the prevailing wisdom among ultra-distance runners was that they didn't peak until age thirty. Runners who regularly won races were closer to forty. The reason for this was mostly pacing and tactical smarts. It wasn't that younger runners weren't capable of winning those races; rather it was just that they would start too hard and not know how to pace themselves over fifty or a hundred miles.

I was fairly certain that my body would hold up. I was running like a madman. It was wintertime. I would wake up in the morning and run twenty miles before 10 a.m.—my morning jog in the snow. My weeks were filled with one hundred and fifty total miles over varying terrain. There were times I pushed myself extra-hard for consecutive days. I would run sixty-five miles one day, fifty the next, and maybe forty the third day. But I had never run those distances consecutively for weeks. I had little concept of how many miles I could do for days on end.

I did have other advantages. At that point in my life, I had never owned a car. I was always on my feet. I didn't even have a bicycle. I made a concerted effort to stay off wheels of any kind. My primary mode of transportation was on foot, which allowed me to stay connected with the earth every step of the day.

The thought of running the PCT brought on a rush of tremendous joy. Not only was it a historic trail, it was a journey that represented freedom in all regards. As much as I loved running across the land, those runs required brainpower. The beauty about getting on a trail is that you don't have to think. You hop on the trail and put one foot in front of the other. The next day, you do the same thing. There is a meditative feeling to being on a trail versus running cross-country.

It was a perfect time to run the PCT. The trail had just been officially completed. Previously, there were sections where people had to improvise the route because the trail was not continuous. Depending on the route taken, there were shortcuts in the uncompleted areas, particularly in one twelve-mile section that had not been finished because it was overgrown with brush and trees.

In May 1996, I felt I was ready. I headed to Seal Beach in Southern California to stay with family. The city was close to the Mexican border where I would start. I had done a lot of mountain training, so I reasoned that a couple weeks of flat training would be beneficial. I also began putting together the supplies I would need. Some supplies—very few, actually—I would carry rolled up in a cloth and tied around my waist. Others I would ship to post office drops along the trail.

I was trying to capture the feeling of what it means to take off running with only the essential gear I needed tied around my waist. To me that offered the ultimate freedom. It showed that I had developed my skills to the point where I was confident I didn't need a backpack and could travel a great distance with only the absolute essentials.

I had only about four hundred dollars to my name, so I decided to ask energy bar companies for products in exchange for publicity. I called PowerBar and Clif Bar. PowerBar sent me a hundred bars. Clif Bar was a little less generous. They first offered me a wholesale price. Then they ended up sending me sixty bars complimentary. I understood their reluctance because nobody knew who I was. I was just some kid saying he was going to run the state of California.

At that point, I thought that once I finished the trail and broke the record, they would all be calling me with endorsement offers. I wasn't convinced I would accept them, but the prospect of the offers was somewhat motivating.

I also stocked up on sunflower seeds and chia seeds. I had learned about the power of chia seeds from my research on Native Americans. Legend had it that when Geronimo was on the warpath and nearly starving, chia seeds sustained him. Many native tribes throughout the Southwest claimed that chia was the superfood of the land and that you could go hundreds of miles on just a few teaspoons of the seed, often referred to as the "food of the gods." At the time, chia cost only about ninety-nine cents a pound—though once it became a fad, the price spiked to thirty-two dollars a pound.

I boxed up supplies to be shipped to the mail drops I chose from the PCT book. For the California section of the PCT, the book listed eighteen drops that were on the trail itself, rather than in nearby towns. I could have done the trail without any drops by purchasing food in the towns the trail runs through, but using drops would allow me to have the foods I wanted rather than relying on what was available. I also didn't want to be forced to detour for supplies. I ended up choosing five drops based on the terrain in the area and what I thought my needs would be.

I planned to travel light. The fact that I was running and not hiking forced the issue even further. I took a five-by-five-foot piece of cloth that would serve as a blanket and laid it out on the ground. I placed my bivy sack, my water filter, my maps, extra shorts, and sandals, as well as my food, on the cloth and then rolled it up. I tied it around my waist. I slung a water bladder over my shoulder. Everything I carried had to serve a purpose.

I had very little margin for error.

Chapter Seven

RUNNING CALIFORNIA

June 1, 1996, was a hot, dry day on the California-Mexico border. The ground was singed from lack of rain and looked like a rippled potato chip. I was wearing shorts and regular running shoes to protect my feet. My cloth lined with supplies was tied around my waist. When I arrived in the small town of Campo, I walked south about a mile into the middle of the desert until I reached a barbed wire fence.

Staked in the ground next to the fence was a sign that read "Mexico" with an arrow pointing south. Another sign pointing north to a trailhead read "Pacific Crest Trail." I signed my name on the PCT register that was inside a small metal box. I then turned north toward the trail that was faintly carved out of the brush, and I started running.

There was no adrenaline burst when I started. I had a sense of bliss that I was about to see new grounds and that I was making the journey unencumbered. I felt like I had been to one of those spiritual retreats where everyone gives up their possessions.

I began thinking ahead. I knew there would be no water for at least twenty miles. Any groundwater in the area that came down the mountains from the winter snow would have long since dried up. I immediately focused on conserving water.

The first day was harder than I expected. It was so hot and humid that my feet began swelling almost immediately. After just one day, it was clear that traditional shoes weren't going to work. That night, I turned my running shoes into sandals. I cut the toe box out of the shoes and removed the tongue, giving my feet space to move around and spread out properly when they hit ground.

On the second day, I came upon two men who were hiking down the trail I was running up. Judging by all the gear they were carrying, they were day-trippers. They were moving at a good clip. As I approached, I greeted them with a wave.

"Wetbacks!" one of them said, pointing up the trail. "There's some wetbacks up there!"

I stopped. "Are they dangerous?" I asked in a concerned tone.

"I mean, I don't think so," the other said. "But they're Mexicans."

I cracked a smile.

"Be careful," the first guy said.

"Will do," I said, giving them a salute.

I continued up the hill. Sure enough, when I came around over the top of the hill, I spotted five guys hiking through the brush. They were wearing beaten-up blue jeans, dirty white T-shirts, and monstrous smiles. Clearly, they were happy to be in California starting their new lives. They all waved at me, and I waved back.

I was surprised by how much I was struggling in the first few days. The heat was affecting me, and my body hadn't adjusted to the task at hand. On the first day, despite being fresh, I covered only twenty miles. I was disappointed that my body wasn't responding better.

To improve blood flow, in the first few nights, I found a spot on a slight slope and lay with my head downhill and my feet uphill. I crawled into my bivy sack and put the piece of cloth I had brought over my body. This covered my body while exposing my legs to air. Keeping my legs uphill kept the blood out of them. I felt that this recovered them faster and left me feeling fresher the next morning. I ended up repeating this sleeping process every night of the trip.

Over the next few days, I covered roughly twenty-seven miles a day. For me, that was not much. I kept thinking, *This isn't right. I've run sixty-five miles in a day. Why can't I produce more? Is my body failing me?*

Water was at a premium early on, as I knew it would be. In addition to the year being unusually dry, it was summertime, with highs reaching one hundred degrees. Most hikers attempting the entire PCT start out in the spring because the land is still moist from the winter snow. Many of the springs and water sources mentioned in guidebooks were dried up. There were several days when I wouldn't find any water the entire day.

But even in a dehydrated state, with each passing day I seemed to become more a part of the land. I stopped feeling like a runner or a hiker, and I started to feel like I belonged. I started to feel like the land was moving under me, rather than me pushing over it. Achieving that feeling allowed me to increase my daily distance without any pushback from my body, and I hoped to establish a rhythm.

I zipped through the Laguna Mountains. Once the elevation increased, the shady oaks kept the sun off me. As I came down into the Anza-Borrego Desert, the climate shifted dramatically. The contrast was jarring. I was fully exposed to the sun, and it was torturously hot and dry.

One evening at sunset, I stopped to sleep for the night. I was out of water and feeling dry-mouthed. I reasoned that if I slept, then I would be starting up again in the hot sun, not knowing where I would be able to fill my water bladder. Despite the fact that I was tired, I decided to continue running through the night until I found water.

This was the type of decision I often had to make in survival situations. I had to ignore my immediate needs and look ahead. Sleep was important but water was more critical.

It turned out to be the right choice. The terrain was very open and there were no snakes, which would have made night running dangerous. The night sky was so clear that the trail was illuminated by a three-quarter moon that lit my way over the open terrain.

Early the next morning I found water. I pressed on across the desert. Soon, I rose into the San Jacinto Mountains. For most of the way, chaparral, a wiry shrubbery, brushed at my legs and arms. The climb was nothing compared to the descent. Leaving San Jacinto, the plunge was some eight thousand feet, passing through nearly every life zone in the state.

The down run was intense. The trail was steep with switchbacks. The drop was greater than hiking from the top to the bottom of the Grand Canyon—twice. The previous year I had trained in the Grand Canyon, so I was used to hitting the four-thousand-feet elevation mark and being done. But this time when I hit it, there was yet another Grand Canyon to go.

I made it down in about three hours, but I was completely out of food. I hadn't started out with a lot of food, to keep the weight of my pack down. I had finished all my food except for a tiny bit of chia seeds. Those would have to sustain me for several days until I picked up my first shipment or reached a town.

The San Jacinto Mountains were twenty miles from the top to the bottom. As depleted as I was, I still had to cross the desert flats and then run back up the other side to reach the town of Big Bear, which is basically at the same elevation. In Big Bear, I could find food.

I gutted out the ten miles across the desert, passing underneath the

I-10 freeway, and then began the climb to Big Bear. I could feel my blood sugar crashing. I was having a hard time even lifting my arms to keep them in sync with my legs. However, my legs were in such good shape that I was able to run without pumping my arms. I let my arms dangle at my sides like an ape and continued running.

A couple miles before I reached Big Bear, about six thousand feet up, I spotted a car. I was so depleted that at first I thought it was a mirage. But as I drew closer, I could see the outlines of a person hunching over the trunk. This was the first person I had seen since the Mexican immigrants several days earlier.

It was a lady organizing her belongings. I stopped and told her what I was doing. She didn't seem impressed or unimpressed. Actually, I wasn't sure she believed me. I asked if she had any chips or other food that could help me balance my blood sugar. She dug into her pocket and produced some hard candy, for which I was grateful.

When I reached Big Bear, I was happy to see civilization and food. I decided to treat myself—and also to put some much-needed calories in my body. I had twenty dollars on me. My first drop was in Agua Dulce, another 110 miles away. I had shipped both money and supplies, so I wasn't worried about spending money. I went into a diner and downed a burger and a milk shake.

After my all-American dinner, I camped outside of Big Bear. Because of the elevation, the night was bitter cold. When I lay down to go to sleep, my legs were cold. As uncomfortable as I was, my thoughts turned positive. The cold would heal my aching legs. It was the first time during my run that I was fully and willingly surrendering to the elements.

The next morning, I got up at 6 a.m. to get moving. I dropped into the Deep Creek Hot Springs canyon and headed toward the Mojave Desert. I was running down the trail when I came upon a guy sitting in his camp. He looked like he had just woken up. The guy was very fit, unlike some of the hikers I had seen in town.

He had reams of expensive gear strewn out all over the place. There must have been forty pounds of everything from tents and sleeping bags to cooking equipment and hiking boots. It looked like he had bought out the REI store.

Because he was almost directly in my path, I stopped. The guy gestured at the gear. "I bought all this fuckin' gear to hike the trail," he said. "I have a zero-degree sleeping bag, and I was cold last night."

"Really," I said, trying to sound sympathetic.

He shook his head in disgust. "I was going to do the whole trail, but this gear isn't working out for me," he said. "I'm done. If your car is around here, I'd like to get a ride out of here if it's not a problem."

"I'm heading north up the PCT to the border," I said. "I'm on foot."

A perplexed look came across his face. He pointed to my wrap. "With just that?" he said.

I nodded. He was speechless. He didn't know if I was messing with him or if I was some kind of badass. He went back to shuffling his gear, and I went on my way.

As I ran, I thought about people who try to re-create creature comforts in the wild with expensive camping gear. I didn't want to be judgmental. It is better than sitting at home and not trying to enjoy the wilderness. But the true experience of hiking the PCT or camping out in the mountains requires a modicum of separation from your living room.

I couldn't help but see the irony. Ultra-light sleeping bags weighed less than a pound. I had wanted to purchase one for my trip, but they were $250, well above what I could spend. Here was a guy who had an even more expensive zero-degree sleeping bag, and he was cold. In contrast, my system was to use the cold to regenerate my legs.

I smiled. While this guy was anchored to his gear, which would take him an hour to pack up, I had slept better and packed in thirty seconds. I had lived with the land, and it had taken care of me.

Emotionally, it was very Zen-like to wake up every day and know that I had one task: to run north. Most of the trail was two feet wide, with sand, pebbles, mud, or pine needles underfoot, and an endless sky in front of me. Every single step I took, every breath I drew, there was something captivating to behold—from plants and trees to creatures of all shapes, sizes, and sounds. I felt a duality with my body and spirit, as if they were working in unison while my feet danced over the trail.

Not knowing what was over the next hill or around the next curve was exciting. With each amazing vista I took in, I was compelled to push forward to see if there was an even more amazing one over the next hill. Being constantly surprised by the beauty of the terrain was the greatest joy of the journey.

There were obstacles in different parts. As I was coming down out of the San Bernardino Mountains into the Mojave Desert, ticks blanketed me. I had to stop every hundred yards and brush swarms of ticks off my legs. Then I'd run another hundred yards and repeat the process.

I did this for several miles through the area near Silverwood Lake until I reached the desert. Luckily, only a few bit me—on my testicles. That's where they end up when you can't find them. The problem is that you don't notice when they bite you, but when you find them and pull them off, the spot is tender for a couple days.

The Mojave Desert had one bizarre spot. I was running along and hadn't seen anything for eighty miles. Then all of a sudden, the trail crossed a highway right next to a convenience store. It was literally in the middle of nowhere.

I went inside. The store was empty, except for the cashier. I picked out an orange juice and a large bean burrito. I walked up to the counter to pay. The cashier quipped, "Don't eat much, do you?"

I thought he was making fun of me for buying the overstuffed burrito. Preparing to explain to him that I was running the entire PCT, I asked him what he meant. He told me that two hikers who had come in the previous day bought ten Big Ed's ice cream sandwiches (which are as big as they sound), sat down in front of the store, and polished them all off. I thanked him and headed out.

I began to think about the odd human interactions I was having. The people I encountered seemed to be using the trail for their own ends. The Mexican immigrants were using it to escape and rebirth themselves. The day hikers were trying to escape their daily lives. The guy with all the gear didn't seem to be trying to escape anything. But I seemed to be using the trail to find something.

After covering roughly 450 miles using only the supplies I had brought—plus the burger and shake, and burrito and orange juice—I arrived in the town of Agua Dulce to pick up my first drop. Unfortunately, the box was not there. I had mailed the first two boxes the day before I left. I then planned to call my mom at certain intervals so she could mail the other two boxes.

I calculated that I was several days ahead of schedule. I was totally famished, and it was a hundred miles until my next drop in Tehachapi. I had just $1.10 on me. I needed to figure out how to get the most nutrition out of my very limited budget. I ended up buying a bag of corn tortillas, three rotten bananas for half price, and a few limes. That was all I ate for the next hundred miles.

The naturalist John Muir called the High Sierras the "Range of Light." The range has some of the most majestic scenery in the world. There are piney forests interspersed with haphazard grassy meadows. Lakes connected by small fords dot the landscape. Immense conical peaks of granite that reach thirteen thousand feet in parts wall in the entire two-hundred-mile stretch. Even in the summer, the higher elevations are coated with snow and ice. Other sections have glaciers that never fully melt.

Kennedy Meadows is regarded as the gateway to the High Sierras. The area, at an elevation of 6,100 feet, has a small resort, several camp-grounds, and a resupply station for hikers heading up into the High Sierras.

I had shipped my third drop to Kennedy Meadows. Inside was my sleeping bag, a necessity in the High Sierras, as the temperature would drop to the freezing mark. Fortunately, the box was there. I unpacked my coat, the food, and some money. I rerolled my wrap and set back out as fast as I could.

I didn't like stopping for too long in the towns and talking to people. I wanted my interactions to be with the land and its natural elements. Almost on cue, I spotted my first mountain lion.

I was near Walker Pass heading up into the Sierras. The trail I was running on was an open ridge. Immediately adjacent was a ridge where the mountain lion was walking parallel to me.

I slowed down. I wasn't afraid, but I also wasn't sure what the lion would do. The lion kept moving and periodically looked over its shoulder at me. We were paralleling for about a mile. I hoped the mountain lion didn't regard me as an invader of his land, but rather as someone who wanted to learn about it. After a mile, the lion rolled his head at me and split off in a different direction.

In the nearly hundred miles from Kennedy Meadows to Trail Pass, the elevation rises from 6,100 to 11,000 feet. With every thousand feet, the temperature dips a few degrees. I soon realized that the north sides of the peaks had snow and ice in the mornings, while the south sides were slushy because the sun was baking them. I was hitting the north peaks in the morning and the south peaks late in the day. Ideally, I wanted this to be the other way around so that the north peaks would be slushy, allowing me to make my way down. Early in the morning, the north peaks were still frozen solid, leaving me to figure out a way to descend safely.

Depending on the ice and the steepness, I used a host of techniques. I sharpened sticks or used sharp rocks to chisel my way down. Occasionally, I would take sticks and wedge them in the ice below me and slide down. Other times, I would use longer sticks as brakes. I knew that if I started to slide without a braking mechanism I would end up pinballing off the trees.

For a two-hundred-mile stretch, from Mount Whitney to Tuolumne Meadows, the terrain was composed of undulating passes of thirteen thousand feet that quickly dipped to eight thousand feet in a valley and then rose back up to thirteen thousand feet.

The highest point on the PCT is Forester Pass. Officially, the elevation is 13,160, though the sign marks it at 13,200. The most surprising thing about being that high on almost all rocky terrain was the amount and variety of flowers.

Like a driver on a road trip marking towns as destinations, I set out to sleep on the top of every mountain. I had slept on mountain peaks many times before the PCT run. To avoid any nighttime drama, I always found an isolated peak so the bears didn't pay me an unexpected visit in their search for a midnight snack.

Most days, I went to sleep when darkness took hold and woke up just after first light. Some nights provided a welcome deep sleep. Others were too magical to miss.

Before reaching the High Sierras, I had camped on the very top of Mount Baldy at ten thousand feet. The city lights of the Los Angeles Basin glowed in the distance below as far as the eye could see. I knew the hustle and bustle of life was occurring. I could see it, but I was at such a distance that I didn't feel it.

In contrast, the volcanic peaks of the Sierras were lit only by the moonlight. There were no signs of any people. Within a matter of a hundred miles, ten million people had vanished.

In *Zen and the Art of Motorcycle Maintenance*, Robert Pirsig talks about how it is better to travel than to arrive. That is exactly how I felt when I hit the final stretch of the PCT in the Cascade Range. Unlike the Sierras, the Cascades were mostly flat with gently undulating terrain that was a pleasure to run. My legs felt like wheels gently rolling across the land.

The most beautiful area was the Lassen Volcanic National Park. The trail takes you to the east side of Lassen Peak, away from the tourist area located on the west side. Lassen Peak is still an active volcano. The heat from the ground warms the azure geothermal lakes that dominate the area. I swam in one lake that must have been eighty degrees.

The day I arrived in Seiad Valley, the final stopping point before the Oregon border, I ran fifty-five miles through a forest of Douglas fir. It was one of my most productive days in terms of distance. I had found my rhythm. I camped overnight and then went for pancakes in the small town the next morning.

I had heard there was a pancake restaurant that had a standing hikers' challenge: if you could eat three pancakes, your order was on the house. I was hungry and running low on money, so I decided to give it a try. When I entered the restaurant, I immediately spotted the cook. He was a huge man, bearing at least 450 pounds. He had his back turned to me. As he was tending to the pancakes, he shifted to the left, exposing a sign on the wall that read: "Never Trust a Skinny Cook."

"Sir, can I have three pancakes, please?" I asked.

The cook looked me up and down. A slim and fit 160 pounds when I set out, I was now about 145. "Why don't you start with one," he suggested.

And so I had my second sit-down meal of the journey. The pancake was bigger than the plate and three inches thick. I sat down and went to work. I managed to eat the entire pancake, which tasted more like birthday cake. Frankly, it was horrible. I felt so sick that I gave up the challenge. I paid $1.50 and left.

That night, I camped out and reminisced on what I had accomplished. I had been running for fifty-seven days. But my journey really was a series of moments, all different in length, strung together, and I had been present in every moment. My goal had been to finish the trail in a reasonable amount of time. I was curious about by how much I could beat the previous records, not for publicity but for myself. I had proven that a younger person could set an ultra-ultra-distance record and hold up physically. By any measure, I had bettered the fastest time by half.

I probably had the chance to create a marketing opportunity of my run and pocket a decent amount of money right from the get-go of my career, but that wasn't what I stood for. I was turned off by commercializing the experience and, for that matter, by the running-and-fitness market in general. By seeking publicity for the run and the record, I would have been participating in the very aspect of running that I despised. I had initially thought that would be part of why I was doing the run. But doing the run changed everything. The last thing I wanted was to diminish my experience by endorsing a power bar.

If anyone cared to check, there was proof I ran it in fifty-eight days, as I had signed and dated all the registers I passed. But I didn't want to find out what it meant to someone else, because I knew what it meant to me. And, inevitably, someone would say that my pace was impossible and that I must have scoped out the registers and driven the highway, hiking in at points and signing the registers.

I felt a sense of accomplishment, but not in the modern way. Some people actually feel a sense of accomplishment when they take a plane halfway around the world. They've traveled far to a foreign land. But what have they accomplished? When you can take a journey on your own body power—a mile, 10, or 1,727 in this case—that is an ultimate sense of accomplishment.

Some people want to learn what they are capable of, but they never push their limits to find out. I had done that. Though I had lost fifteen pounds, I felt no joint or muscle pain. I had gone through one pair of modified running shoes. The only problem I encountered was a queasy stomach the final week—which that hubcap of a pancake didn't help. The most interesting physical change was that my feet spread out as if they had been flattened by a truck. I later figured out that my shoe size went from 10.5 to 12.5. As my mileage decreased over the ensuing months, the size returned to normal.

On the final morning, I awoke with one question on my mind. "Am

I doing this run for me or for others?" The goal had essentially been accomplished: I had run it faster than anyone ever had. I didn't want to plant a flag with my name on it as a challenge to others to beat my time. I was actually pleased that I had concluded that the goal was somewhat hollow, for it was the trails of the mountains that were the experience, not the end point.

I had made my decision. I packed up and ran fifteen miles toward the Oregon border. Less than two miles from the state line—the finish line to the California section of the Pacific Crest Trail—I stopped.

I felt joy and humility, but most of all respect for the land and for myself. I took a deep breath, the best of the entire trip. And then rather than run the two miles to the border, I turned around and ran fifteen miles back to Seiad Valley. That afternoon, I caught a bus back to Seal Beach.

In making that decision, I eliminated any chance for commercialism. I decided not to formally cross the finish line. Anyone who bothered to check all the PCT logs from Campo to the Seiad Valley would see that I had run nearly all of the PCT in record time. But nobody would check, because I hadn't signed the final log. Partly, I was being defiant. The border was just an artificial line. But mostly, I was showing that I had done it for myself.

Chapter Eight

HEALTHY AS A HORSE

I 'm not Superman. I'm not even Tarzan, though I would like to believe we have a few things in common. Much of my initial approach to surviving in the wild was rooted in athleticism and my belief in my abilities. Being a runner and approaching survival from that place allowed me to see amazing locations. However, the better I was able to slow down and observe, the more I saw that I needed to become more patient and less reliant on my physical abilities to become a true survivalist.

The more I learned, the clearer it became that the books describing hunter-gatherers spending two hours a day gathering plants and cooking them up were fiction. It became apparent to me that surviving on any landscape takes all day. There is no way to run through an environment and understand it, no matter how fit you are.

Surviving in the wilderness takes an extraordinary amount of patience. But the body has to first adapt to extreme settings, to endure temperature swings, to cover great distances without proper hydration, and to summon deeply stored energy when near collapse in order to reach for that next stage.

The biggest obstacle for most people is the temperature swings. There are numerous tragic stories about raw mountain skiers being lost for days and freezing to death before being found, and there are equally as many stories of day hikers suffering from heatstroke or even heat-induced heart attacks.

From talking to medical professionals versed in survival in dire conditions, I have learned that the first part of surviving in extreme situations is oxygen delivery. The better the body can deliver and process oxygen, the greater the chances both for survival in life-threatening situations and for being able to operate when the body is pushed to its limits.

Dr. Sam Parnia, a noted critical-care physician and bestselling author, explains: "What has happened in Matt and others with his conditioning is that their lungs and heart have adapted to be able to maximize the use of the oxygen that they are taking in. Their breathing has become more efficient with time, and they are taking in more air for every breath than a normal person. Their heart is contracting so much more strongly that for every beat, they are pushing out more blood and more hemoglobin, which carries the oxygen to all parts of the body."

The ability to process oxygen more rapidly helps the body in numerous ways, from increasing endurance to preventing disorientation and cramping from dehydration at high altitudes. Altitude sickness is caused by low oxygen pressure at high altitudes. Elevations at which the effects are felt vary from person to person. Generally, people will feel some change in their breathing around four thousand feet. Above eight thousand feet, most people experience a shortness of breath and the inability to draw enough oxygen to compensate. Activities that cause the body to demand more oxygen, such as skiing or running, compound the problem and often leave people feeling dizzy, sick to their stomachs, or headachy.

The more fit a person is, the better they will be able to deliver and process oxygen at high altitudes. I have met several ultra-distance runners who are as fit as me and can run for long distances in the mountains.

The most amazing runner is Matt Carpenter. Ultra-distance runners often call him superhuman. On top of being genetically gifted, he has built his life around extreme running. In the wintertime, he relocates his family from Colorado to South America so he can live on a high peak and train year-round. Matt has won every high-altitude race numerous times and holds virtually every course record, including the fastest times recorded in marathons run at the unthinkable altitudes of fourteen thousand and seventeen thousand feet.

Carpenter's ability to process oxygen, known as VO2 max, is unparalleled. At the U.S. Olympic Training Center, it was registered at 90.2, the second highest on record behind a Norwegian cross-country skier. A reading of 60 is considered excellent for an athlete.

Though I have never had my VO2 max measured, I doubt it would match Carpenter's. However, once in a routine physical, a doctor was measuring my oxygen saturation level, which is the amount of oxygen in the blood. Anywhere from 95 to 100 percent is normal, depending on a variety of circumstances, such as altitude. The doctor explained that because we were at 5,800 feet, the highest possible reading would be 97. She turned to the machine and said, "And yours is . . . oh my . . . ninety-eight."

I once ran a high-altitude ten-mile race in Colorado against Matt Carpenter. The race had forty superstrong distance runners, including a world-renowned Kenyan runner, and was by invitation only. Entrants had to apply and send a résumé. Because the race was being held at fourteen thousand feet in snowy conditions, I had to trade out my sandals for running shoes. I knew that if I were sloshing through snow in my sandals it would slow me down. I found a pair of Nikes that were low cut and didn't have too high a heel and trained in them for a week before the race.

The race started at eight thousand feet. The Kenyan runner bolted off the starting line at a five-minute-mile pace. All of us, including Carpenter, let him go. He was sprinting at an out-of-this-world pace for an altitude that high, particularly considering we had ten miles to run.

After a couple miles, the group started spreading out. Matt eased past the Kenyan runner and took the lead. I was in fifth place. As we got higher up the mountain, my strength started to kick in. I passed the Kenyan runner, who was by now huffing and puffing in the altitude, and settled into third place.

I felt strong. When we did the turnaround to go down the first mountain, I was tearing it up. I actually had my sights on Carpenter. I thought I might not beat Matt that day, but I was certain I could pull off a second-place finish—which would have given me a big paycheck and a free flight to Italy.

As I charged down the hill, I saw a mountain biker who was preparing to ride up the hill toward the second peak. He was clicking pictures of me. For a split second, I lost my concentration. My running shoes had that small heel, making me less stable, and I rolled my ankle.

It wasn't bad, so I kept running. But after another mile, I was wincing. I thought, *I might be able to finish this race, but I will likely hurt myself.* So I stopped.

Carpenter ended up winning by ten minutes. I didn't have the fitness to catch him, but I felt like I could've beaten the other runners and finished second. Second place to Matt Carpenter is the best you could ever hope for.

While enhanced oxygen delivery can make you feel superhuman, there are other health factors that come into play in extreme situations. Toxins in the body often prevent it from working at peak level. The primary sources of toxins in our body are food, stress, lactic acid in muscles, and poor air, as well as electromagnetic toxins from computers and cell phones. Living in the wilderness strips away the toxins faster and more completely than a massage at a day spa. The natural climate reboots the body over time and leaves it free of toxins that handicap the body's functions and ability to quickly recuperate.

The lack of bacteria and germs in nature also means that I never get sick when I am out in the wild. In city atmospheres, people are exposed to germs and bacteria that cause viruses such as colds and flu. People are constantly washing their hands and using antibacterial products to kill germs. Those types of bacteria don't exist in the wild.

Diet plays an important role in the ability to endure extremes. The Paleo Diet is now in vogue. The basic philosophy of the diet is to eat the way hunter-gatherers ate hundreds and thousands of years ago in pre-agriculture days. The prescribed foods are things that are found in the wilderness such as meats, fish, leafy greens, fruits, nuts, vegetables, and seeds, rather than processed foods and grains or anything with refined sugar. This translates into high protein, low carbohydrate, high fiber, and a much higher intake of vitamins and antioxidants.

Balancing your diet is just as critical. Over the years, I found that eating only plants and grains did not provide me with enough energy to get through the day. I added fish to my diet, but I was still drained and tired. Eventually, I began hunting wild game and eating meat, and my sustained energy level was far higher.

I later consulted the nutritionist Ryan Koch about how I felt. "In a realistic primitive-living situation (the same situation that our ancient human ancestors found themselves in for more than two million years until the Neolithic era around ten thousand years ago), it has been my assertion that a person *must* procure animal foods to remain in good health," Ryan said. "Not only that, but anybody in a primitive setting will *crave* the essential nutrients found *only* in meat and especially fat if they are out there long enough (animal forms of fat-soluble vitamins A, D, and K). In other words, plants will not sustain any lengthy wilderness survival sojourn."

Perhaps the biggest asset to having more strength and endurance has been slowing down my metabolism. This means that I don't have to eat as often. Most Americans eat constantly, or at least three times a day. If they don't eat every five hours, their blood sugar drops and they start to feel woozy.

The fact that my metabolism is slower also helps me store energy in the event I can't find food for a day or more. A slower, more efficient metabolism means that even if I don't eat for a couple days, my blood sugar will remain stable and my energy level won't flatline. Like the Tarahumara, it also allows me to run great distances without needing to consume thousands of calories.

My lifestyle and my fitness constitution have merged. Though I didn't seek publicity, people in the running community took notice of how I lived and trained. *Trail Runner* magazine wrote an article about me.

"Matt Graham is what you might call a traditional runner, but not in the sense of old school legends of the sport like Bill Rodgers or Steve Prefontaine," the article said. "Graham trains like the ancient foot messengers of southwestern Native American tribes, running in homemade sandals and living off the land. He makes jerky out of coyote and raccoon, flint knaps spear points and arrowheads, rubs two sticks together and makes fire. He runs in sandals made of discarded tires or rawhide and yucca fibers. On long runs, he shuns manufactured energy bars for the natural energy from Pinole and chia seeds."

All of this helps explain why I would enter a twenty-five-mile horse race—without a horse.

It was a temporary job in Utah that resulted in a study of a biped versus a quadruped. A friend had asked me to help him and his wife do a traditional Dutch oven cookout for a group of trail riders in the mountains. When we got to the camp to set up, I realized that the group of riders was preparing for a horse race on Boulder Mountain. I cooked and talked with the riders.

The race was a three-day, fifty-five-mile event, with the middle day of twenty-five miles being the toughest. The trail headed southeast and then cut through the mountainous Utah terrain and picked up the old mail route. The trail crossed the Escalante River and headed upstream about five miles to an area called Big Flat, and then wound its way back into the town of Escalante. I was familiar with the route, as I had run it many times.

I started to think that it would be cool to run the race against the horses. I hadn't been on a long run in a few weeks, and the challenge was irresistible. The race had three legs over three days through canyons and over mountains, and the riders could switch horses.

I sought out the race director, a man named Krocadoomis, and asked him if I could enter the middle day—on foot.

Krocadoomis was a classic Southern Utah cowboy with a floppy mustache and ragged hat. He looked at me like I was crazy. He asked how in heaven's name I proposed to keep up against a group of horses.

"With my feet, in my sandals," I replied.

"You won't be able to keep up," he said.

I smiled. "I think I'll be all right."

I convinced him that I knew the land well. Even if the horses blew me away, I wouldn't get lost.

As he thought about it, a smile broke out across his face. He tugged at his mustache. No doubt he was thinking that the local papers might write about the sandaled runner against the thoroughbreds, which could help publicize his business. He agreed that I could run, but set down one condition. At the start, I had to run ahead, open the gate, and either stand aside or take off to avoid the charging horses, which would have a lot of built-up adrenaline in the beginning.

The next morning, I put on a pair of thin sandals with rubber soles made from old VW bus tires and lined up with thirty-seven horses and riders. A few of the riders had gotten to know me from the cookout. I had told them about running the Pacific Crest Trail, though I think only a few of them actually believed me. In any event, they were an eclectic bunch of cowboys and wealthy adventure seekers who weren't about to let a guy in sandals outrun their horses.

Krocadoomis signaled me. I ran about a quarter mile to the gate and opened it. Instead of waiting, I took off down the trail. I could hear the horses pounding. I reasoned the best way to avoid the stampede was to sprint the first couple miles and build a lead while the horses spread out. I kept running through the pine trees. After another mile, the pounding was becoming fainter and fainter. I was actually pulling away from the horses.

On the first leg, we ran down from the mountains into Boulder, about eighteen miles. The terrain favored the horses. There were sandy uphill climbs and some rocky patches that were relatively easy for horses to find their footing. Even the flats were mostly sandy. I assumed the horses could do twenty miles per hour in the flats. Nevertheless, I hadn't seen or heard a horse since the starting line.

When I got into town, I stopped at the rodeo grounds, the designated end point of the first leg. I waited and waited for the horses. I couldn't believe I was that far ahead. The race monitors weren't even there yet. I sat for about twenty minutes until people started showing up with trailers to move the horses to the starting point of the second leg. They were dumbfounded that I was already there. One asked how long I had been there. I told him. He marked my time for five minutes earlier than when he had asked.

Several minutes later, the first horses arrived. It took another fifteen minutes for all the horses to come in. The horses were loaded onto the trailers and driven to the other side of town.

At the second starting point, the race officials let me go first, but they released the horses almost immediately, stripping me of my lead. Once again, I heard the horses pounding behind. I picked up the pace and soon lost them.

I took off running, built a lead, and didn't see a horse until about mile fourteen. I was crossing a canyon and the horse was on the other side, where I had come down minutes earlier. At mile sixteen, I stopped at the vet check station.

The race officials told me that Krocadoomis, who was riding with the race contestants, had radioed in and told them to stop me. "We're not going to vet check you, but we have to hold you for the time it takes to vet check the horses," the guy told me. "To make it fair . . ."

Krocadoomis and several other riders arrived about fifteen minutes later. We talked for a while, and then they let me go. Again, twenty minutes or more had gone by, and they released the horses right after me. I was beginning to think Krocadoomis didn't want me to win.

It was just nine miles to the finish line. The challenge for me was that the final leg of trail was a flat, sandy road. The sand gave the horses a decided advantage. After a mile or so, Krocadoomis and one of the horses passed me.

At some point, I must have zoned out because I ended up taking a wrong turn. When I realized I was not on the race trail, I reversed course. By the time I had rejoined the trail, two more horses had passed me. They were about a hundred yards in front of me. Down the homestretch, it was three stallions, followed by me, and then the other thirty-four horses.

Most humans would have been fine with that performance, but I was determined to retake the lead. The trail came to a cliff with a steep drop-off—the one chance for me to make up some ground on the horses. I raced down the hill, foot over foot, bouncing off rocks like they were rubber, and I managed to pass one of the horses. At the bottom of the hill, there was a half-mile horse-racing track that led to the finish line. I sprinted my absolute hardest, but I wasn't able to catch the first two horses.

I ended up finishing third with a cumulative time of six hours, fifty-seven minutes.

My achievement was respectable, but certainly not unparalleled. Each year, there is a Man Against Horse Race in Prescott, Arizona. The fifty-mile race covers various terrains, including steep mountain climbs and flat plains. Some years a horse wins, but most years a man wins. In another, the Man versus Horse Marathon in Wales, UK, a man has won twice in thirty-five years.

So how can a man outrun a horse?

Stride and breathing are the two main factors. In experiments where a man runs alongside a horse at the same pace, it has been determined that the man's stride is actually longer than the horse's stride. This means that the man's legs are moving slower to cover the same amount of ground, therefore allowing him to cover more distance per stride and conserve energy.

Horses are quadrupeds. Once all four legs are working in unison in a gallop, the horse can't pant. Because panting is the only way for the horse to cool down, a horse in a gallop is constantly heating up. Eventually, the horse will have to slow down to a trot to cool off.

Humans, of course, are bipeds. Even when we reach our "galloping" speed, we can pant. Therefore, even if we are sprinting for a long period of time, we are able to continually cool down enough to keep running.

In the race I ran against the horses, I basically sprinted the entire way. This allowed me to take advantage of stride and breathing. The temperature was in the seventies, so I didn't overheat. To stay light, I didn't carry water. I stopped and drank whenever I found water and at the mandatory stops. I also plucked pine needles off the trees to provide me with a constant stream of energy. The horses ate only when they were at the vet check stops.

Terrain is also a critical factor in a man-versus-horse race. The horse will excel on flat, sandy road; whereas a runner will have an easier time on a narrow trail with several switchbacks mixed with long, flat terrain where the horse will need to rest. That accounts for the difference in the results between the hilly Prescott race and the flat-road Wales race.

The evening after my race, there was a dinner for all the riders. When Krocadoomis spoke, he singled me out. He presented me with a silver medallion.

"When I saw you charging down the canyon and coming up the other side, I finally understood how Geronimo could do it," he said.

EPIC SURVIVAL RULE #3:

KNOW YOURSELF

I am finding that learning survival is a complex process that is as much mental as it is physical. I realize that knowing myself is just as critical to surviving in dangerous situations as being able to endure nature's tests.

Though I know quite a bit about living in the wilderness, I am also just now discovering that survival is a full-time pursuit. The hunter-gatherers we read about in books or see on television look like they are working at it a few hours a day. They throw up a shelter, grab plants on the go, and cook up an occasional meal. But that's not realistic. It takes all day to survive in the wilderness. I no longer believe it is possible to scoop ice cream eight hours a day and be a hunter-gatherer after work and on the weekends.

My approach to the wild has been from an intensely athletic point of view, but I am altering that approach. When I was younger, being a runner allowed me to see many amazing places and work on my hunter-gatherer skills at the same time. That was my identity. But I am trying to find a balance where running great distances does not define me.

I am learning that the ability to slow down is equally as important as running sixty-five miles across a desert in one day. I am beginning to observe more. I don't think I have been missing the grandeur of nature or even the smallest flower petal, but I know that observation is going to be critical to shaping my life as a hunter-gatherer.

Now that I am fully engaged in the process, I realize that it takes an extreme amount of patience. I need to be able to call upon my patience so that I do not make a fatal move, such as trying to seek a shelter I cannot reach when a biblical storm is approaching.

The things I am discovering cannot be picked up from reading a book, and they are critical for me to learn. In everyday life, it is easy to get caught up in yourself. The land does not allow that. It demands that you look around and observe.

As I transition into the land and begin to more fully understand it, I am also learning about myself. I want to reach the point where I know myself so well that I can use that in my relationships. I want to be able to pause and ask myself, "Okay, what does my friend need today? What do I need to do to be better for this person?" *For me, being able to do that will mean that I know myself.*

Chapter Nine

FINDING MY PLACE

As a kid, I was always exploring or climbing. *(Personal collection)*

The pit house in the community of Salt Gulch, Utah, that I built and lived in for five years. *(Donna Simpson)*

In the Panama jungle, filming *Dual Survival*.
(Russell Fill)

Preparing to hunt with the atlatl and dart.
(Russell Fill)

On the road for a *Dual Survival* shoot.
(Russell Fill)

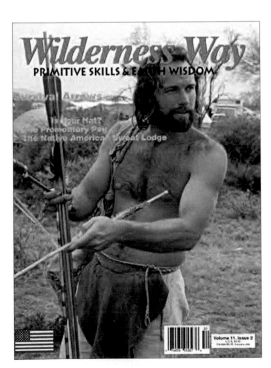

On the cover of *Wilderness Way*.

Addressing a group of students.
(Ace Kvale)

Starting a fire as the Fremont Indians did using the hand drill technique. *(Ace Kvale)*

A small coal is nurtured and brought to life with a gentle breath. *(Ace Kvale)*

The picture shows the focus in the instant the arrow is released. *(Ace Kvale)*

Three atlatls made by me and two darts made by A.J. Applying artistry and attention to detail creates an extension of the maker of the tool. *(Personal collection)*

Blistering sunset over the Kaiparowits Plateau. *(Personal collection)*

Rainwater filtering through and running on top of slick rock in the southern Utah canyons. *(Personal collection)*

This shot illustrates the powerful life-giving energy that storms can generate over southern Utah. *(Personal collection)*

The geological process sculpted by nature over thousands of years. *(Personal collection)*

In the summertime the canyons of Utah can reach over a hundred degrees, but in the winter-time they are blanketed with snow under freezing temperatures. *(Personal collection)*

This image illustrates how flash floods can carve out canyons over time. Despite the cotton-wood trees, there is no surface water. *(Personal collection)*

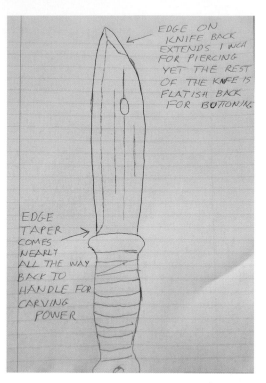

This is the ideal, all-purpose bush knife
that I have used for many years.
(Personal collection)

The finished product.
(Personal collection)

On a walkabout in the southwestern United States.
(Personal collection)

The inescapable grandeur of the Grand Staircase-Escalante National Monument hit me harder than any sight I had ever seen. Its beauty is absolute at every turn.

Situated between the small town of Escalante, Utah, and the even smaller town of Boulder, Utah, the Monument, as locals refer to it, is splayed out over 1.7 million acres, making it larger than Delaware. The Monument has three main areas: the Grand Staircase, the Kaiparowits Plateau, and the Canyons of the Escalante. The ever-changing geology is the triumphant result of ancient glaciers and flowing natural waters. A maze of canyons carved out over thousands of years is lined with streams and vegetation. The canyons are hemmed in by dark orange cliffs that transition to reddish-orange plateaus.

Distances are hard to gauge. From the peak of Boulder Mountain at eleven thousand feet, it's difficult to tell if you're looking out a mile or ten.

The mountains have horizontal stripes that look like they were drawn with hundreds of colored pencils. The flats are layered in shades of gray and green that appear to have been mixed together and applied with a paintbrush. The tones and hues of the landscape change constantly with the movement of the sun and placement of the clouds.

Highway 12 passes through the Monument and connects Boulder and Escalante. This section was dubbed the "Million Dollar Road" for the cost of building it in 1935—back when that was a lot of money for a blacktop road. Looking out in the distance from the higher ridges, there is certainly a million-dollar view. The winding road looks like a dark ribbon running along the silhouette of the plateau. In one section of S-curves with no guardrails, called the Hogsback, the ridge is about three feet wider than the roadway.

People often talk about the real world. To me, a place as natural as this is the real world, and the cities we live in are the artificial world. When I first saw the area, I knew I was home.

I moved to the Boulder area in March 1996. I first read about it in a magazine. I was living in the Sierra mountains studying primitive skills when I came across a tiny ad in *Backpacker* magazine for a survival school. The ad had a small arrow with a stone point and read "primitive skills and survival school." There were no pictures. I immediately thought, *That's the kind of place I am looking for.*

I dialed the number of the Boulder Outdoor Survival School (BOSS) and reached the owner, David Wescott. After we spoke for a while, I asked him if he would hire me. He chuckled and explained that I had to take a course before I could work for him. I told him about my lifestyle, and he encouraged me to come to Boulder.

Wescott was in his mid-forties and had worked as an instructor himself. In addition to owning BOSS and another survival school in Idaho, he also hosted the two largest annual gatherings for survival enthusiasts, Rabbitstick in Idaho and Winter Count in Arizona. He was also the cofounder of the *Bulletin of Primitive Technology*, a forum for people who want to share and learn primitive-living skills. I decided that working for him was ideal for me.

I didn't own much. I loaded everything I owned into saddlebags and strapped them on my mountain bike. My parents dropped me at the California-Nevada border on their way to a vacation, and I started pedaling for Boulder, Utah.

The journey was over five hundred miles, across the entire state of Nevada. I was so anxious to get to Boulder that I rode about 130 miles a day, which was the maximum I could do at that time with all the hills.

I rolled into Boulder in mid-March, before the hiking season began, and rode straight to the BOSS office. The only guy there was an instructor named Rob Withrow. He informed me that Wescott would be back in a few weeks. Rob was living in the office during the winter. He was a craftsman who built tools. He told me he was preparing to head up the mountain and live off the land for a while. I asked if I could join him, and he told me to be ready to go the following morning.

I set up camp near the survival school. When I arrived at the office the next morning to see if Rob was ready to leave, he told me he needed more time to get ready. He told me to check back in three days. In the interim, I decided I needed to orient myself with the land.

I climbed to the top of a nearby mesa to a large rock formation called Schoolhouse Ledge. I built a ten-foot circle of stone, where I planned to stay for four days and four nights. I went into the circle wearing nothing but a cotton sheet. I didn't bring any food or water. Like the Indians who went on vision quests, I was seeking clarity for my path forward.

Though I didn't experience any significant visions, I did succeed in readjusting my body to the land. I ended up staying only three days, but those days forced me to slow down, listen to the land, and allow it to open up to me. When I was sitting in the circle, I also found my new home.

In the distance, I could see a dark spot on an otherwise luminous rock formation. I was intrigued. Upon leaving the stone circle, I hiked across the canyon and up the mountain to examine the spot. It turned out to be a hole in the rock the size of a large door. I entered and crawled through a passageway. I went about ten yards and the passageway angled up slightly. Suddenly, it opened into a spacious cave. I decided to make the cave my home.

To this day, it was unlike any cave I've found in the area. It was very heat efficient. In fact, one time I was in the cave sleeping on a cotton sheet. I woke up and there was a bunch of snow on the ground surrounding me, but I was hot and toasty in there.

In just a few weeks' time, I knew that Monument offered me everything I wanted. There were areas where somebody without a lot of experience in survival, provided they knew a few basic things, could live well off the land for a period of time. That was what I hoped to teach others.

Becoming a survival school instructor was more difficult than I realized. When David Wescott finally returned, he explained to me that he had ten people applying to be instructors, but before any of us were hired, we had to pass a rigorous, fourteen-day survival training course.

I felt confident I would do fine. In addition to studying primitive skills on my own, in the month that I was waiting to meet Wescott, I met a local named David Holladay. He was an aficionado of the area, and he taught me about its riches and peculiarities. But I became a little concerned when everyone was packing for the trip.

The other guys were bringing far more supplies than me. I had my traditional five-foot piece of cloth rolled up and tied around my waist. I was wearing a breechcloth and a cotton shirt and sandals. The other guys were all in nylon pants, carrying fully loaded backpacks. They were also packing a second bundle that would be dropped off to us midway through the course. The idea was that after you endured the first part, the second bundle arrived to make you more comfortable. The second bundle contained food rations of nuts and lentils, a poncho, and a blanket.

Breck Crystal, an instructor I had met a few days earlier, asked me where all my gear was. I told him that my rolled-up cloth was all I had. He offered me his extra poncho and blanket.

The course, I soon learned, had phases. I hadn't read the brochure, so I didn't know how it worked. It started with the most challenging part, Impact. In this phase, we walked through the Monument for six days and ate very little. The only water we drank was what we found. We also only ate food we found, like spiderwort, wild onions, mustard greens, and yucca flowers. The idea behind Impact is to force the body into starvation mode. It also readjusts your metabolism. Because we consumed so few calories for those days, during the rest of the course, when we ate eight hundred or more calories a day, it felt like more than enough.

I handled Impact with relative ease. I felt a little light-headed at times. Some of the participants had mini-blackouts when they stood up. As the days progressed, everyone moved slower. Oftentimes when people sat down, it took them a while to get to their feet.

The second phase was Group Expedition. The primary focus of this phase was to teach students how to work with a map and compass, and how to cook. I told the instructors that these were all skills I knew inside and out already and that I was there mostly for the land connection. They only half believed me, but they did allow me to run off and do my own thing during that phase. They became somewhat convinced I knew what I was doing. I would glance at the map in the morning and then run off and meet up with them at the end of the day in a designated spot miles away. I was able to find them by reading the land features and gauging the twists and turns of the canyons.

In the next phase, called Sheep Processing, we processed a sheep that was brought to us. The premise was that students are removed from the meat they buy in the market and grill. A sheep was supposed to more closely represent what they might buy in a market, as opposed to wild animals like squirrels or raccoons. It also showed them how in the processing of a large animal you can use the skin for blankets and the bones to make knives, awls, or tools.

The instructors made an overly elaborate ceremony out of the slaughter. The sheep was placed on its side, and everyone was told to put their hands on the sheep. One person was designated to kill the sheep with a knife. Not only was this unnatural, the person with the knife was so nervous that he barely made the kill without the sheep suffering.

The process made me uncomfortable. I had been on a Navajo reservation and seen sheep sacrificed. The Navajo held a much different ceremony for the animal. One person would lead the sheep. As soon as everyone was distracted celebrating the life of the sheep, the guy leading the sheep would kill it in an instant.

Does the lesson have value for students? Yes, it does help them connect to what they eat. But at the time, I was 100 percent a hunter-gatherer. I never bought meat from a market. Any meat I ate, I hunted, so I had a hard time participating in the killing of a farm animal.

After the sheep was dismantled, no one wanted the fat so I kept it. The fat proved to be a great benefit for the next phase, the Solo. They dropped us off for a couple nights. We each were given a one-quarter-mile section of the canyon and told to stay in that vicinity and build a camp. For food, I picked cactus and fried it in the sheep fat, giving me far more tasty meals than the others had.

The final phase was called Student Expedition. We were given an eighteen-mile route to navigate. The lead instructor's confidence in my skills had grown so he designated me to lead to the starting point. After we all returned, there was one final challenge: a seven-mile run to finish off the course.

Of the ten people, Wescott hired two for the coming season, a woman and me. However, I soon learned that we were not considered full instructors, but rather interns. That summer, I ended up working on seven courses. Each course had seven to ten students and lasted from a week to a month.

At the end of that summer, Wescott sold the school to Josh Bernstein, an Ivy League–educated New Yorker who was well known in the survival skills world. Most of the instructors left town and moved on with their lives, but I stayed. While I hoped to continue working at BOSS under its new owner, the primary reason I stayed was because Boulder was my home.

I had settled on a mesa about six miles outside of town. I built a primitive-living structure called a pit house. I dug a hole nine feet in diameter and about three feet deep, making the pit house deep enough to stand in. The ground was caliche, sedimentary rock that resembles white clay. The caliche was hard enough to enable the sides to stay up without any reinforcement. Most of the soil in the country is sand, which would have needed to be reinforced or turned into cement to achieve the same result.

I built a crib-style roof that spiraled around to a center point and had an opening at the top. I covered the roof with sticks and bark, and packed caliche on top. The pit house was big enough to sleep in with a tiny fire burning alongside me.

I regularly went on walkabouts to explore the land. Shelter was easy to find in the numerous caves. Some of the caves were bigger than small buildings. They measured as much as seventy yards across and had high ceilings, making them big enough to throw a dart the entire length. These were particularly helpful in monsoon season, because even if I had a group of a dozen students with me, it didn't feel claustrophobic.

The caves were made by water. Every time water hits the bank, it carves deeper and deeper into the rock, and eventually you are left with a shelf that is protected. They all have sandy floors in the front near the overhang, but in the back they have non-broken-down oak leaves. These leaves are perfect for making a bed with very little work. They will also keep you warm, as the canyons do get cold at night, particularly in the winter.

Most of the caves are safe. But even if the caves have big boulders, then you can look up at the ceiling to gauge if there is a risk of falling rock. If you see cracks or loose rock, you avoid that spot. But even in caves with unstable parts, there is always a safe section. If there is a pile of rocks, typically the ceiling will have a fissure in that area, but if you proceed to a section where the ceiling is smooth sandstone with no fissure, that area will not fall apart.

The caves tend to sit twenty to fifty feet above the water line. There is a mixture of geology occurring. There are places where a river has cut through and created a cave, but the water is still flowing underneath. Even if a cave has a creek, it is not unusable. Over time, the water will change sides or shift, and a sandy floor will be created.

The land was full of surprises. I was constantly discovering something new. One day, I followed a creek up the mountainside to find the origin of the clear water. I hiked up the mountain about three miles and found a waterfall. The area was a world of its own. There were raccoons and squirrels roaming the land, an abundance of trout in streams, and a bounty of wild greens, such as watercress and mustard.

Because the area is so isolated and difficult to access, nobody ever goes up there. It became one of my retreats, and to this day, I have never seen another person there.

After spending the winter exploring the land, I was eager to begin teaching again. However, the following spring I wasn't rehired at BOSS. Josh Bernstein, the new owner, made changes to BOSS. He replaced many of the guides and didn't bring back any of the interns, myself included.

I ended up spending that summer guiding llama tours. The tours were luxurious by survival standards. Using llamas as pack animals, we would take people into the backcountry, set up camp, cook them meals, and basically wine and dine them. The company used llamas because they are not as temperamental as horses and they are more of a novelty. It wasn't very challenging work, but the tips were generous.

Truthfully, money was never an issue for me. I was earning about two thousand dollars a year, which sounds like a paltry sum but was more than I needed. I broke down my expenses and figured out that I could've easily lived on six hundred dollars a year—not including commercial travel. Half my income went toward going to California to visit family over the holidays.

Eventually, in the summer of 1999, I did go back to work at BOSS. Josh asked me to guide a course with David Wescott, the former owner, who was now guiding. Josh had asked me the previous summer, but I was booked with llama tours. I have to say it felt good to return to showing the land to people who could make their own camp.

At the end of that course, Josh apologized for not keeping me on the previous summer. He even gave me several presents. More important, he hired me as a full instructor. Though I ended up working there for the next eight summers, I soon realized that as well as I knew and loved the land, I was also becoming too rigid and single-minded in my pursuit of connecting fully with the land to become a teacher of its gifts.

Chapter Ten

RELATIONSHIPS

Tanning hides is a very meditative exercise. Hunters often discard animal hides, but using them properly fills a part of the cycle of life. When tanned properly, a hide becomes a durable piece of fabric that wears well, has no odor, and is something of a work of art. Wearing animal skins gives you an added connection to the land. There is no way to feel alienated from the natural world with a hide on.

However, if most people saw a fresh hide being skinned from an animal, they would be disgusted. There is a membrane layer like the one on a rack of baby back ribs, only it doesn't peel off quite as easily and it's an inch thick. The hide is covered in blood and pus, with pieces of fat stuck to it. The sight is so gruesome that not even the grittiest movies dare show it.

The smell makes it even less attractive to the uninitiated. In the early phases of the process when I'm stripping down a hide, I find that I breathe so much of the hide that my bowel movements smell like the detritus I peeled off the hide.

In the winter of 1999, I was working the hides of several large animals. I lived in a tepee on a piece of private land in the middle of the National Monument in Boulder. My days were occupied from morning to night with perfecting the hides, but my concentration was spotty at the time, as I felt the pull of a relationship.

I was dating a smart, multifaceted woman named Karen. She was a ranger and had moved to Flagstaff, Arizona, to continue her schooling. We both respected that we had commitments that kept us apart—her education and my work tanning hides.

The fall passed, and we hadn't seen each other for months. At Christmastime, she called me late one night and pressed me to come and live with her. "I need you here, not there," she said.

I told Karen to give me another three weeks. When she protested that that was too long, I told her I couldn't interrupt my project. "I've got to finish tanning these hides," I said in all seriousness. Then I surprised myself by adding, "Maybe we should call it quits."

Undoubtedly, she was thinking, *If Matt's priority is scraping flesh and washing blood off dead animals, then he doesn't love me.* She agreed, and we broke up.

I hung up the phone. A half hour passed. I felt sick to my stomach, far more so that I ever had smelling a hide. I knew what I had done wasn't right. I was being stubborn and selfish, and I quickly began to regret my rigidity. Even the most ancient hunter-gatherers had to factor human interactions into their lives. Maybe it was time for me to do some growing in that area.

So I made a split-second decision to win her back. I decided to leave the unfinished hides behind for my friend to finish. I hurriedly packed a bag, got in my car, and drove through the night to Flagstaff.

I arrived at 8 a.m. the following morning. With a bouquet of flowers in hand, I knocked on her apartment door. A guy answered. Karen heard us talking and rushed to the door. Defensively, she claimed it wasn't what it looked like. But I knew it was—and that I probably deserved it. After all, I had told her I was choosing animal hides over her.

Rather than returning to Boulder right away, I decided to stay on in Flag-staff. I had some money saved up from the guiding season. I had never rented my own place, so I thought, why not try the proverbial "normal life" and see what I could learn? Perhaps it would help me in my next relation-ship, or at the very least in my ability to communicate with my students.

I rented a room in a house and took a job working in a health food co-op. I bought a membership to a climbing gym, and I even signed up for three classes at the community college—theater, anthropology, and dance. There was no stewing about my ex-girlfriend. There was no mo-ment that wasn't full.

At the climbing gym, I became friends with one of the instructors, Jesse Perry, who was an expert climber. On his days off, we climbed routes in the mountains outside of town. Jesse also had an interest in primitive skills. I started teaching him how to make rabbit sticks, hunting boomer-angs, bows, and buckskin pouches and bags. At that time, my belongings were still very primitive. I didn't own any modern gear. Everything was made by hand. He commented that my room looked like a museum.

As spring 2000 drew closer, I decided to return home to Boulder, but I didn't like driving. I had owned a truck camper for four years and only put a total of twelve thousand miles on it, despite the fact that I lived in rural Utah. I decided to leave the truck with Jesse's parents and walk back home for the summer. Jesse asked if he could join me on the walk to see what he could learn about living in the wilderness.

We set out on May 17, 2000. I wore rawhide sandals and carried a wool cloth, tools for flint knapping, a few tablespoons of spirulina—dried algae so complete in vitamins it is called a "superfood"—and a water bottle. Jesse also wore sandals and rolled up all of his supplies, including a mechanical pencil and a blank journal to record our journey, in a cloth. He dubbed his traveling pack "my burrito." We also carried a bow and several reed arrows that I had made.

The route we chose was about 450 miles, roughly 100 miles longer than the driving route. I wanted to try to miss as much of modern civilization as possible. As expected, there were spectacular wilds, but also plenty of things I didn't want to see.

To reach the national forest trails without following the roads, we started out by crossing a vast forest. We were forced to climb over several barbed wire fences separating one plot of private land from the next. With each new area of land came a different pack of barking dogs. As we jogged, ran, and dodged our way through, I felt like exactly what I was—a trespasser. It was not a way to begin a teaching-and-learning experience.

Once we reached the federally protected land, there were more signs of man interacting recklessly with the land. The forestland was dotted with burned ponderosas. The trees served as a reminder of careless campers whose unextinguished fires were all too often sparked by windy, dry conditions and burned through the land.

On the second day, we experienced yet another contrast between the wilds and modern society. We awoke in a sea of aspen and Douglas fir banked by the distant glow of purplish mountains. After we ate thistle roots and cleaned up our campsite, we headed north and ran into Highway 180. We walked along the shoulder for several miles, as cars and trucks roared past us. The shoulder was littered with trash tossed out of car windows, such as Doritos bags, crushed soda bottles, and a torn T-shirt.

We shifted about four miles from the highway and paralleled it. At that point, Jesse felt like we were pushing our dehydration limits. It was early in the journey, but he was uncomfortable. He said he was considering turning around and walking back to the highway. I didn't respond. I pressed on, and he followed.

After another mile, we crossed a dirt road. The tracks were so old and so deep that it was clear that no car had driven on it for months. Sitting on the road right in the exact spot where we crossed was a bottle of Ocean Spray cranberry juice. It had been sitting there so long that the label was completely sun beaten.

I opened the bottle. We both smelled it. Undoubtedly, it was passed its expiration date. We each took a few sips. We realized that it was a little bit on the fermented side, but we decided to carry it with us and drink it later.

I could also see that Jesse's energy level was low. For the first couple days, we had eaten only chia seeds, thistle roots, steamed greens, and pine tree bark. The lack of food and the walking in the hot sun caused Jesse to half joke that he was feeling like he was tripping on psychedelic mushrooms.

His blood sugar was crashing from a lack of calories, and it was taking his spirits down with it. In the middle of the highway, I spotted a dead squirrel. I waited for a break in the traffic and then ran onto the highway and scooped up the squirrel. Not surprisingly, Jesse had never eaten roadkill, but we needed the nutrition.

I took the squirrel into the bushes to clean it. I asked Jesse to hold one end while I held the other. As I began to cut off the head with my knife, the look on Jesse's face told me he wasn't going to eat the animal, regardless of how desperate he was for the calories.

Jesse had grown up as a vegetarian. Eating meat was something he didn't want to take part in, but ultimately he knew that he would need to in a survival experience. But he was not ready to make the compromise, which I fully respected.

As our journey continued, Jesse continued to face the same dilemma. He was coming to the conclusion that he could not exist on chia seeds and greens. Food—particularly that which would be classified as vegetarian—was at a premium.

At one point, I managed to get a rabbit with the bow and arrow Jesse was carrying. I skinned the rabbit and cooked it up. Jesse was disgusted. Not only did he not eat any of the rabbit; he threatened to leave at that very moment.

"I don't feel like this trip is for me," he said.

I was sympathetic. I told Jesse that I understood how he felt. I asked him if he would at least continue with me for a few days to see how he felt. He agreed and settled for eating pine nuts that night.

As we continued hiking, I immediately felt stronger. Jesse, on the other hand, was feeling weak, and could see how energized I was. There was a stark difference in our energy levels. Later that day when we made camp, Jesse decided to eat the rabbit meat. He had concluded that there was more to learn.

Though Jesse's beliefs were being challenged, he was also going through the physiology of what happens to human beings when they don't have a lot of food. While their stomach is growling, that is not the real mental challenge. In fact, they are experiencing a flood of emotions they have never previously experienced.

The fact that I had killed and cooked a rabbit was not the issue. It was that he was in a deprived state of being. The feelings are so overwhelming and intense that they fool your mind to the point that you can become depressed because you are seeing and feeling too much. At the end of the range of emotions, when he ate the rabbit, his body equalized and gave him a clearer perspective on what it would take to survive.

That night, we climbed Humphreys Peak, up to some twelve thousand feet. We walked through a forest of aspens. As we headed down off the mountain toward the high desert, it was getting late in the day. Darkness was falling fast. The winds were kicking up. But I made the decision to continue hiking through the night, both to make miles and to drop some elevation.

We weren't dressed to travel in that kind of weather. I was wearing shorts, a buckskin top, and a felt hat. Jesse had on shorts and a cotton shirt. The temperature was in the forties but the wind chill dropped it to the twenties. Temperature-wise we weren't in the danger zone, but we were because of how we were dressed.

We continued walking to make those miles and to stay warm. In survival situations, when it is cold and you have enough visibility, you want to walk because it keeps you warm. However, all the while, it is scary because you know you will have to stop and deal with your survival needs. The dilemma is making sure that when you do stop, you have a clear head.

The light seemed to be coming from an otherworldly source. Every few minutes, places on the desert floor would light up, like sparks jumping off an electrical box. The flashes of light were coming from the ground and going up into the sky. One would be a hundred feet away from us, followed by another thirty feet away, and then another fifty feet in the other direction. Sometimes we both saw the flashes together; other times only one of us would see the flash.

After each flash, we would walk to the spot where it had originated.

When we reached that place, it was just dirt. Granted it was dark, but we never saw an explanation for why a flash of light would be rocketing up from the ground.

When we finally stopped that night out of fatigue, it was pitch-black. There was no moon. Clouds were covering the stars so we didn't have that light, either. It was nearly impossible to see. Feeling my way through our supplies, I pulled out our fire kit. I had packed a hard drill and piece of cottonwood root.

I took the stone knife out and then I started drilling into the board. I couldn't see where to cut the notch, so I had to feel the hole with my finger. I started carving the notch with my stone knife, hoping it was in the right place. I felt back and forth between the notch and the hole. It seemed to be right.

I kept drilling. In the day, when you are making a coal, you can watch for the smoke. But even if there was smoke, I couldn't see it in the dark. I had to wait for the red glow. That added an extra thirty seconds to a minute of drilling.

As I was drilling, I finally saw the salvation: a faint red glow. Jesse brought me some bark and I put it on the nest. I blew the spark into a flame. We then gathered a few sticks and ignited them. That light enabled us to see other wood in the area, which we collected and put on the fire.

After the fire was lit, I looked at the fire board. I realized that the notch wasn't touching the hole. I had been very lucky to get a coal. Normally, if the notch isn't directly in the hole, you cannot get a coal.

Both Jesse and I were grateful for the fire. We spoke only briefly about what a tough night it would have been without the fire. Jesse was on his first survival walkabout. He didn't understand the severity of the situation had I not gotten the fire going. For me, there was no point to take him down that path.

It was our second night out. Jesse was looking at the journey through fresh eyes, so he regarded it as mild drama. Could we have survived the night without a fire? Maybe. But without a doubt, we would have been in the early stage of hypothermia. I had been there before. Likely, I would have survived. I'm not sure about Jesse.

In a survival situation, there are a certain amount of priorities that must be met to stay alive. Sometimes those priorities come one at a time, but other times they hit you all at once. It is easy to get into that kind of predicament. When they hit you all at once, that's when you have to really know the flow of the land. That's when people get into trouble, when there are multiple survival issues that compound themselves. If you don't know how to prioritize those, you will die.

That night, likely because I knew the situation had been dangerous, I began processing thoughts on life issues I needed to address. Before leaving on the trip, I had briefly gotten back together with Karen. We had shared some wonderful moments. But during those first few days with Jesse, as I began sinking back into the land, I regretted that decision. Our relationship had not been a healthy one emotionally.

I knew I was seeing the land in a way she would never be able to comprehend. I was on my own path to become a teacher of the land. I hoped I would find students willing to learn, because I concluded that was more important to me than my own personal relationships.

Chapter Eleven

THE LONG WALK HOME

The major difference between walking from place to place and going on a walkabout is that you cannot avoid roads, people, and pollution. In the National Monument in Utah, I could easily go on a walkabout for a week without encountering any of those because I knew the land. But walking from Flagstaff, Arizona, to Boulder, Utah, despite picking the most remote, backcountry route, I knew that Jesse and I were going to cross roads and find disturbed areas, unregulated mines, and other things that would pain us to see when we were trying to have a very pure connection with the land.

From Flagstaff, Jesse and I walked ninety-five miles across the desert and into Grand Canyon National Park. At the south entrance to the Grand Canyon, we stopped to resupply in Tusayan, Arizona. The town is a bastion of low-end commercialism, and the embarkation point for every conceivable tour of the Grand Canyon—air, foot, bus, bike, mule.

The businesses lining the main drag told the story. Mixed with souvenir shops and "trading posts" were helicopter rental companies, motels, fast-food joints, and a place labeled "The Tourist Center." One motel featured a twenty-foot-tall Fred Flintstone pointing at a sign that read " 'Yabba-Dabba-Doo' Means Welcome to You." I felt anything but.

Tusayan is the home of Grand Canyon Airport, where helicopter and airplane tours of the Grand Canyon originate. At midday in high season, the skies are like LaGuardia Airport on a Friday afternoon. Seeing hundreds of helicopters in the sky eating up the environment and drowning out the natural sounds of the land frustrated me to no end. I felt like I had to do something.

I ran out into a clearing and frantically waved both arms at a low-flying helicopter. After seeing people wave back, I squatted, pointed my backside upward, pulled down my pants, and mooned them. Admittedly, it was not the most mature move, but it was the place I was in.

Jesse certainly didn't mind. In fact, he broke up laughing so hard that he doubled over and ended up rolling on the ground.

Jesse was as disgusted by the noise pollution and hordes of tourists as I was; yet he was also happy to be near civilization. As I set up camp, he jogged into town to buy some comfort food. He returned with two bottles of Guinness and a bag of Oreos.

That night, we were able to ease our frustrations somewhat. I taught Jesse how to make a fire with a small chert, a yucca stick, a few pieces of bark, and some elbow grease to create friction. He was extremely pleased when his efforts produced a wisp of smoke, followed by flames.

Nevertheless, the experiences of seeing trashed areas of the land and helicopters leaving a film of smoke on what would otherwise be one of the cleanest vistas in the country wore on me. I found myself growing bitter and cynical, as well as judgmental and preachy. Our talks focused on what civilization is doing wrong rather than what we can do right. It was bringing me down, and I'm sure it was bringing Jesse down, too.

After we left Tusayan, we hiked to the south rim of the Grand Canyon and stopped for the day. I began making a new pair of sandals. We were within sight of the tourist path, and I could see that many noticed me sitting there, hammering on the yucca fibers with a rock. Jesse joked that they probably thought I was a paid attraction, a frontiersman working away to show how rustic the area once was.

We spent much of the day staring off the rim of the Grand Canyon and enjoying the view. I had us wait until 7 p.m. I figured by that hour the rangers would be off duty and we could just slip into the park before dark, as we did not have a hiking permit.

It was about two hours before dark. We had plenty of light to get into the canyon, but we needed to leave. We started running down the trail. But just as we hit the first turn, we ran into three rangers coming up the trail. They stopped us and asked where we were going.

I told them we were going to Boulder, Utah. The rangers looked uncertain. The lead ranger studied our meager belongings and asked if we were planning to camp in the canyon. They held the fate of this part of our journey in their hands. I smiled at him. "We might take a nap," I said.

The lead ranger nodded. "Have a good trip, guys," he said.

Jesse and I headed into the canyon. We traveled several miles and then slept for the night.

After days of walking through the canyon, Jesse's knee began bothering him. It reached the point where he felt he was going to collapse under his own weight. Considering the dry and rugged crossing that lay ahead of us, we both decided he needed medical attention before continuing.

We made a plan. Jesse would hitchhike his way back to Flagstaff to have his knee checked out. When I reached Marble Canyon, which was about sixty miles away, I would call him and see if he was able to rejoin me.

After leaving Jesse, I spent two days with my friend Farlinger on the North Rim of the Grand Canyon, working on a flint-knapping project. Before leaving, I phoned Jesse. The diagnosis was an inflamed IT band along the outside of his knee. His doctor treated him with acupuncture and bee venom therapy and told him that after a few days' rest he could return.

I set out across the Grand Canyon. I stayed away from the highway that ran north of the canyon. The route was rugged. There were a lot of canyons coming in and out. It was by far the hardest part of the journey. I walked more than thirty miles without any water.

I dropped down off the plateau to a beautiful canyon called North Creek. It was lined with oak trees and full of trout, watercress, and edible greens. Once I reached the bottom, it turned to pure desert. For miles and miles there was only clay. The tallest plants were about six inches. The land was tranquil. The temperature was upward of a hundred degrees. There was no water at all.

In the morning, I would get up as soon as I could see and begin to walk. Even before the sun would crest the horizon, the mercury would start to climb. Once the sun rose, it felt like a fire was searing me. At one point, partway through the walk across the desert, I put on my yucca sandals to start breaking them in because my rawhide sandals were almost worn out.

The first life anywhere close to me was a cottontail rabbit. I thought I should try to get it for food. I picked up a flat stone and threw it with the sidearm motion used for skipping rocks in a stream. I hit the ear of the rabbit. It dropped its head into a crouch and started sprinting. As the rabbit raced away, I picked up another rock and took one more throw. The rock sliced through its throat, killing it instantly.

I didn't have any water, so I couldn't cook the rabbit. I cut it up into strips and hung the meat on a stick with string. It was so hot I figured I would have jerky within a day.

The following day, I entered the Navajo reservation and found a white water tank for cows. There was no surface water, so I climbed up onto the tank to drop my water bottle in and pull out the water. For purification, the only thing I had was grapefruit extract, so I mixed it with the water. The rabbit at that point was bone dry. I enjoyed a meal of jerky with my water.

I didn't have much farther to go. Fueled by the food and water, I hiked longer that day through the desert. As it started to get dark, I saw some lights in the distance. I could see it was the highway parallel-ing the cliff. I had no idea I was still at least eight miles from Marble Canyon.

Though I had a makeshift map, I was traveling mostly by memory. When I was on the South Rim of the Grand Canyon, I had traced a map of the region that stretched all the way to Marble Canyon. I had included the high points that would be easily identifiable, but there was no way to gauge distances.

I ran toward the lights, covering several miles fairly quickly. I came upon a hotel on the highway before Marble Canyon. As I approached, I saw two Native Americans leaning against a lit-up sign. When I reached them, I stopped and said hello to them.

One of the guys looked me up and down, turned to me, and said, "When I first saw you running out there, I thought you were Forrest Gump. But now that I see you, I know you're a prophet." It was a classic Indian line, mixed with a heavy dose of pop culture.

One of the guys introduced himself as Robert Mirabal. He was from Taos, New Mexico. I later learned that he was a world-renowned, Grammy Award–winning flutist whose flutes have been displayed in the

Smithsonian's National Museum of the American Indian. He explained that the following morning he was heading out on a raft trip on the Grand Canyon. A group was paying him to sit in their boat and play his flute.

We spoke for a while. I told him I had made a flute during the trip. It was the type of flute you blow into from the side. I couldn't play it very well. He looked at me and said, "I can play anything with a hole in it."

Sure enough, he took the flute and made it dance. The music was gorgeous. I was excited because he showed me that my flute had potential to make beautiful music. I vowed to keep practicing it until I was able to play it well.

Robert gave me an elk leather necklace. He said that he had worn it in Hawaii when he performed there. Because he had said he admired elk hide sandals, I took mine off and handed them to him. He told me to stay with him if I ever made it to Taos.

We shook hands. I walked a mile into the countryside to get away from the motel and camped for the night.

The following morning, I met Jesse in Marble Canyon. He was shocked at how tan I was since I had seen him. I told him that the sun had been hotter than I had ever encountered. He openly wondered if he could have made it.

"I have never met anyone who has so much faith that the wild will take care of him," Jesse said.

"If you're being respectful, the wild will take care of you, too," I replied. Then I added, "Let's go."

We continued on our journey to Boulder. There were times that we pushed the envelope by everyday standards, but I felt that because we did, magical things happened and the land opened up to the point where you can't deny that there is a large force taking care of everything.

Jesse was often concerned that we always seemed to be on the edge of dehydrating. At one point, we were down to a quart each. He was constantly concerned that we would run out. But just as the situation felt critical, water would appear. Once it was at a barely functioning desert spring. Another time we were dangerously low, and we ran into a man working for the government. He was making the area livable for bighorn sheep, and he directed us to the water source used for the sheep in drought conditions.

During one stretch when he became discouraged, I asked Jesse what he wanted. He responded that we were near Paria River and he had heard the fishing was good, so some fishing line would be nice. We walked a hundred yards. Sitting on the road was a brand-new spool of fishing line.

He was in shock.

"What else do you want?" I asked.

"Hooks," he said.

Literally in another hundred yards, I found a bag of hooks. He laughed at the absurdity.

"I prefer lures," I said.

Sure enough, we walked another fifty yards and found a bag of lures.

He was speechless. Had we not been in the middle of nowhere, he would have looked around for a hidden camera. Truthfully, I didn't know what to make of those instant gifts, either. That kind of absolute synchronicity is rare.

How did it happen? I'm not sure, but it did show that if you have a relationship with the land, it will respond in positive ways. For Jesse, it confirmed that mystical events can happen when you place absolute trust in the land.

The Paria Canyon-Vermilion Cliffs Wilderness had prairielike plateaus interrupted by deep canyons. Though the land was unusually dry, we found a narrow river to follow and it led us to an apricot orchard. We picked a bushel of apricots and made camp. I then built an oven from sandstone slabs and wet clay, and we baked a loaf of apricot bread.

But the area was also teeming with tourists. We passed groups of hikers wearing heavy boots and lugging overstuffed backpacks. The groups seemed to be divided into two schools of thought on us. Some seemed enchanted by our mountain-man appearance, while others cast a wary glance and kept their distance.

Jesse and I debated the experience these hikers were having. He pointed out that before coming out with me he had no idea that the outdoor retail industry had scared the public into believing it needed vast amounts of gear to camp out. He felt that the majority of people were being lulled into believing they were having a wilderness experience when, in fact, they were on something that better resembled a ride at Disneyland and only gave them the illusion of the experience.

Part of this was due to the fact that hikers were supposed to buy a permit to travel through the Paria wilderness and stay on the trails. While I understand this was primarily for safety, it also showed an assumption that the average person could not venture from the predetermined path and learn the wilds for themselves. Though I felt that Jesse's assessment was somewhat harsh, I wasn't sure I would've wanted to see some of the people we passed trying to survive in a remote area.

Once we entered Paria Canyon, we were alone. The canyon was filled with beautiful oak trees and knee-high brush. A narrow river with sandy banks ran through it. Eight-hundred-foot red rock walls towered above both sides of the river. The river was narrow enough that you could jump across it.

My yucca sandals were shot. I had given my backup pair of sandals to Robert Mirabal, so I had to go barefoot. I stuck the yucca sandals in a cave by the river. We joked that someone would find them and think they were archaeological artifacts because they were made in the ancient way. Even the straps were woven from yucca fibers.

I was barefoot. Like a kid at the beach on a summer day, I would run through the hot sand and then cross into the water to cool my feet. I did this for the thirteen miles it took us to reach Big Water.

We sustained ourselves on the apricots and bulrush shoots we gathered along the river. The apricots were a godsend. Without those, we would have both been in trouble because nature was not giving us much during that stretch. Jesse later wrote of our meager meals: "Dreams of pancakes and hash browns carried me through the blood sugar debt of the late afternoon."

After leaving the canyon, we crossed a stretch of desert. Parts were rocky and tough on my feet.

We moved at a pretty good clip because we had a goal in mind: a meal at the café in Big Water. When our reward came into sight, we sprinted to the door, only to be greeted by intense disappointment. The café had closed fifteen minutes earlier.

Jesse laughed and volunteered that we should press on rather than wait until it opened in the morning. The decision was made easier because we knew that we had a drop box waiting for us at the Big Water post office.

We walked across the highway to camp out in the desert for the night. The only food we had came from a ketchup packet we found on the highway. Luckily, it had not been run over. I tore it open and we shared the tomato paste for dinner that night.

The next morning, we went straight to the post office to pick up our shipment.

The town's postmistress was amused by us. "I've been waiting for you two," the small lady said. "For three weeks I've been wondering what's in these two boxes, but mostly, I've been curious who was going to pick them up." By her smile, I guessed that her two new customers did not disappoint her.

My box had a leather bag full of beans and rice, and my tire sandals. Jesse's had a nylon backpack, food, and a water filter. I strapped on my tire sandals. After walking in sandals that were falling apart and then going barefoot for thirteen miles, I was very appreciative to have new footwear.

Jesse and I traveled another fifty miles and reached Last Chance Canyon on June 5. We decided to make it our home for the next week to give us a reprieve from some of the ugliness we had encountered. The worst, which stuck with me, was when we had come upon a spot on an Indian reservation that was trashed. Inconsiderate campers had left bottles and plastic containers, as well as the coal remains from their campfires. We cleaned up the site as best we could.

Located just over the Arizona-Utah border, Last Chance Canyon is hot and dry and extremely remote. We managed to find a nice cottonwood tree to shelter us from the sun. Water was an issue. The streams contained visible cow dung, as ranchers in the area allow their cows to wander the public land. We made the water potable by filtering it through some fine mesh I had brought for such emergency situations.

During our time there, I busied myself with projects in hopes of evoking Jesse's interest. I carved rabbit sticks and made a cloak out of my blanket. Jesse, however, was content to do nothing. He went on a few walks to get a better view of the sunset. He did do some gathering of juniper berries and cooking, but I still felt as if he wasn't attempting to forge a close connection with the land.

After leaving Last Chance Canyon and hiking across the Kaiparowits Plateau, we reached the Canyons of the Escalante. A year earlier, I had buried a stash of food in a large bucket for my trips through the canyon. I located the rock formation. We made digging sticks from a juniper bush and dug for the food.

The journey was in its final leg, and our differing goals for the trip finally came to a head. The following night over a fire, we aired our grievances. I told Jesse I didn't feel he was making a true effort to learn the primitive skills I was teaching him.

He, in turn, explained that he didn't embark on this journey as a student. He simply wanted to live away from the workaday world for six weeks and take in the environment. It was a simple, honest, and straightforward conversation that not only reestablished our friendship but also ended up becoming a building block for me as a teacher.

With our hopes more clearly defined and better understood, the final week of our trip was the best. We spent two days in Choprock Canyon, which offered both of us the reprieve we were seeking. One wall of the canyon had a huge mural of petroglyphs—tiny, detailed pictures carved into the rock depicting an ancient society. The area was filled with cattails and bulrush. The spring water from the creek was the purest of the trip.

We also established a new level of understanding. While sitting in camp, I heard a bird chirp far up in the tangled vines of a tree. I went to investigate. As I climbed up, a bird flew out of a nest. I noticed there were eggs in the nest. As I tried to get my hand into the nest to grab the eggs, the momma bird dove at me and attempted to peck me with her beak. I didn't have any interest in killing the momma bird. I instructed Jesse to grab his rabbit stick to push the momma bird away so she didn't nail me with her beak.

The nest had three eggs. I was planning on us each taking an egg. I reached in and grabbed a small egg and handed it down to Jesse. He refused to take it.

Though I was frustrated, I was beginning to understand his perspective. That moment became a learning experience for me, too. Never again would I consider taking two eggs out of a three-egg nest because I know the momma would abandon the nest.

Three days later, Jesse and I arrived in Boulder on June 23, the eve of my twenty-eighth birthday. After I showed him around my space, we parted ways. Jesse hiked into town to plan his return to Flagstaff.

The trip was a turning point for me in several ways. I realized that I had a weakness as a teacher. I needed to figure out how to bring light to my students, not tear things down in front of them. I had to accentuate the positive of the wilderness in a way they could understand. It couldn't be a simple contrast where nature represented good and man-made represented evil. Even though I saw destructive things, I couldn't focus on the bitterness they evoked. That wouldn't do my life or theirs any good.

I ended up becoming close friends with Jesse. He sent me his diary. Reading his thoughts only underscored the learning curve I was on. The diary made it clear that I was more aggressive when it came to pushing survival concepts and less receptive to him taking them in on his own terms than I needed to be. I had to find a way to better convey the natural world to people willing to learn its gifts. I wasn't yet the teacher I hoped to be.

Chapter Twelve

AN ALL-PRIMITIVE WALK

Ötzi the Iceman was the first all-primitive man discovered by archaeologists. A natural mummy unearthed on the Austria-Italy border, Ötzi was nearly fully preserved and carbon-dated to around 3,300 BCE. He was wearing a coat, a loincloth, tights, and shoes, all made of leather, and a bearskin hat. Around his waist was an ancient fanny pack containing a primitive tool kit of a drill, a bone awl, and flint-knapping tools tipped with copper that were used to shape stones. He was also carrying a copper ax and several arrows made with dogwood shafts.

His shoes became the subject of much debate. Resembling sandals, they were much wider than his feet and appeared to be constructed for walking across snow. The soles were made of bearskin and the sides were animal hide. They were insulated with grass for warmth and had netting around the ankles. What stood out the most was the detail and thought that went into the footwear.

Ötzi was also carrying two species of mushrooms with leather strings running through them. One, the birch fungus, had antibacterial properties and was likely used for medicinal purposes. The second, a tinder fungus, appeared to have been used in conjunction with several plants and some pyrite for fire-starting purposes.

My friend Breck Crystal and I studied Ötzi's possessions in great detail as we prepared to embark on an all-primitive walk. We didn't necessarily want to emulate him, but we did want to show ourselves that we could create tools and supplies like our most ancient ancestors and walk into the wilderness and survive.

Our primitive tool search led us to Ishi, who was considered the last Stone Age American man. He was the last surviving member of the Yahi, a group that was part of the Native American Yana people in California. After living nearly all of his life with no connection to modern society, Ishi emerged in 1911. It turned out that after his entire people were massacred, he had lived completely alone for three years.

Ishi allowed anthropologists and archaeologists to document his way of life, and the methods he had used to survive. After his tribe died, he burned off all his hair in mourning. He fended for himself for years, but with no tribe to back him up, that lifestyle took a toll. He was actually caught stealing eggs, which forced him to come out of the wilderness.

Ishi immediately adapted to the modern world because it was so much easier. After living for a year in the urban world, he said that life became so comfortable for him that he did not want to ever return to the woods. He could eat whatever he wanted, whenever he pleased. He could work on his crafts without the fear of letting down his guard.

Many people I've talked to over the years have the initial impression that the hunter-gatherer lifestyle is a laid-back one. The fact that Ishi was starving to the point where he had to steal eggs shows that even somebody who grew up in that situation has to work so hard for food that they can become desperate. Nature is not cruel; it just cannot be slowed down.

Someone who lives fully in nature cannot say: "I don't think I'll do the hunter-gatherer thing this week. I'm gonna take some time off and get back to it next week." However, in our modern society, we can actually check out of our lives for a couple weeks and then resume them. But if a hunter-gatherer does that, he will die.

Closer to home, Breck and I also studied the Paiute Indians, one of the most fascinating and overlooked Indian tribes. The local Utah Paiute tribe was one of several in the Western states, and all of them had struggled to find and maintain a home. They lived a Stone Age lifestyle all the way through the 1900s.

Though information on the Paiute tribe was sparse, a friend of mine named Bill Latade, an archaeologist who lived in Boulder and was head curator of the Anasazi State Park Museum, had a private manual with photos and historical information. I was interested in their tools but also in their values and customs.

One of the most interesting bits of information I found about the Paiute tribe was their patience and dedication to craft. They made nets out of plants that were hundreds of yards long and stretched them across the plains to catch rabbits. It was the most impressive feat ever seen in a hunter-gatherer tribe.

It took a person roughly one year of constant work to make a hundred yards of the plant-fiber netting. Collecting a fistful of the fibers took several hours. They would then weave them together with their hands and feet, spending a year on one net. The long sections of net were woven just loose enough to trap a rabbit's head. When the nets were not in use for hunting, they were folded over and over to create a well-insulated six-by-six-foot blanket.

The big lesson for us was that a person could find the concentration and patience to twist these fibers with their hands and feet every day for a year.

I also read Father Silvestre Vélez de Escalante's journals. Father Escalante, for whom the nearby Utah town was named, and his companion, another priest, set out to travel from Santa Fe to Monterey, California, in 1776. They ended up in what would become Utah, the first two white men known to have set foot in the state. Father Escalante kept a record of their travels, detailing the survival tools and techniques they used, as well as what they encountered.

The goal that Breck and I had was to take a journey using primitive means. The BOSS season had ended in December 2000, and most of the staff had left Boulder for the winter. The final outing was something of a production. To promote the upcoming *Charlie's Angels* movie, Cameron Diaz, Drew Barrymore, and Lucy Liu had gone on a four-day course as part of a promotional shoot for *Marie Claire*. The three actresses proved they could handle themselves better than many who had taken BOSS courses.

Although it wasn't conscious, the contrast to the journey Breck and I were preparing for was somewhat comical.

Breck and I wanted to enter nature without any connections to the modern world. We both had a foundation for exploring the land, and we wanted to see if we could do it as the ancient explorers had, with only the materials they had used.

To build our primitive tool kit, we absorbed every story we could find, from Ötzi's shoes to Ishi's arrows. We looked at photos and decided what pouches, blankets, and bags to bring, and what foods we wanted to collect and prepare. We referred to Bill Latade's books about the tools the indigenous people of Utah had used to build the area's early living structures, as well as other illustrated books showing how Native Americans made their tools.

For food, we dried jerky and made pemmican, a jerkylike food consisting of dried berries, meat, and fat. The process took two months. We started by drying rose hips and berries. Next, we took deer that had been killed by motor vehicles and left roadside. The dried deer meat was pounded down along with the rose hips and berries. We then heated that up and mixed it with rendered raccoon fat.

The purpose of the fat was to preserve the meat. Rendering animal fat to its purest state keeps it from turning rancid. The fat is rendered in water and then the layers are scraped off. After several processes, only the superwhite fat is left. The white fat is the best. It keeps the meat from going bad and can last hundreds of years. Pure dried meat jerky, in contrast, lasts only about six months before it begins to lose flavor and feel like cardboard.

After rendering the fat and cooking the meat, we rolled the pemmican into small balls that we stored in parfleche. Parfleche is a method of storage for dried food used by Native American tribes. It is made by taking rawhide from an animal, stretching it out tight, and then staking it until it dries. After it dries, the rawhide is pounded with a round cobblestone until it starts to loosen up enough to be folded without breaking.

Our pemmican had just the right mix of meat and sweet berries. It was high in nutrients and tasted far better than it should have. That was actually a mistake. The pemmican was so good that it was hard not to eat it right away. Before we even started, our supply began to dwindle.

Along with the deerskins from the roadkill, we also collected deerskins from hunters (as they discard the skins). We processed the skins and made loincloths, vests, shirts, shorts, and moccasins. I tanned a full-size elk and made a blanket. We also wove together yucca and agave fibers to create the nets for our iceman packs. I made a pack very similar to Ötzi's, while Breck made a more modern design in the shape of an X.

To cook our food, Breck made a clay pot and I hollowed out a gourd and oiled it really well so I could use a hot rock in it. We made our own knives out of stone and put shiny, wooden handles on them. We used the tanned skins for sheaths and sewed them with yucca fibers.

For hunting tools, I made a bow out of hickory with sinew backing for extra strength. We made our own arrows with stone tips. The bow-string was made out of the tendons from the deer. Those same tendons were used to tie the fletching—the guiding feathers—onto the arrows. The shaft was built with cane I had collected in Arizona and some tamarisk, a heavy wood that resembles a willow.

We went overboard on the footwear. We figured that if Ötzi had made snowshoes more than five thousand years ago, we as modern hunter-gatherers should have footwear for all different conditions. We each made heavy winter moccasins, plus three pairs of sandals. We also tanned hides.

Most of the craftwork was done outside of the BOSS survival school office, which happened to be adjacent to the Anasazi State Park Museum. This provided a nice sideshow for museum visitors, who often sat and watched us after touring the exhibits.

The process of building our primitive-living gear took nearly three months. Though it was enlightening, we became so carried away that rather than building the tools that made the most sense for our location, we built the ones we liked from each different story. Pretty soon, our backpacks weighed fifty pounds each.

Aside from wanting to travel and live like the Native Americans, we were both hoping to reconnect with some of the traditions of the past. We hoped to re-create a trade walk. We brought no cash. Instead, we made trinkets and extra hides to trade for whatever supplies or food we needed.

Our initial destination was Chaco Canyon, which wasn't then but is now a national monument. Historically, Chaco Canyon had been the epicenter for trading in our area. Though we couldn't trade at Chaco, we wanted to see the area, and we knew that there were trading posts run by the Navajo nearby, where we hoped to trade for food and dry goods.

After venturing through Chaco Canyon, we planned to continue across the mountains and canyons. One possibility was ultimately ending up at Winter Count, the annual primitive-skills gathering on the Ak-Chin reservation in Arizona, some four hundred miles and two months away. But any destination was secondary to the feeling we wanted to achieve. We wanted the journey to be something that two men could have embarked on a thousand years ago.

The first steps of the walk felt magical. We were two travelers dressed in long buckskin shirts, breechcloths flapping in the wind, stone knives dangling on the waist, walking down a spring-fed canyon. On our backs were agave fiber packs holding elk skin blankets and buckskin pouches filled with pemmican. Clutched in our hands were bows backed with sinew for added strength, and arrows made of cane tipped with finely honed, razor-sharp chert, a sedimentary rock. Nothing we carried was modern. Everything was handmade from the land and could have been hundreds or even thousands of years old.

I was elated. I felt like we were headed to an ancient city, where people grew corn, beans, squash, and cotton. There, we would trade for what we needed by offering up one of our intricately tanned hides.

Leaving town, we headed straight into the backcountry, through a lush river valley. Within an hour, we were in a place with no people in sight and no hiking trails. But after a couple hours, the practical aspects took hold.

We stopped to take a break to adjust our packs because they were digging into our shoulders. From then on, we had to stop constantly to fiddle with the straps on our backpacks, which seemed to become heavier with each step.

By the end of the day, we were completely worn out from the loads we were carrying. We hadn't done any test walks with the fifty pounds of gear, so we had no idea how taxing it would be. We had forgotten the most basic lesson: be familiar with the load you are carrying.

It was the same thing as the backpacker who buys out the REI store versus the seasoned backpacker who pares down his gear to the bare minimum. Despite the fact that our gear was handmade and paid homage to primitive cultures, carrying so much of it—regardless of how spiritual it might be—was no different from overloading on modern gear. If you haven't used the gear or carried it, there is no way to know how it will perform or how you can perform with it. In short, we had a load of ancient tricks that we didn't fully understand, and we were as uncomfortable as we had ever been in the wilderness.

We stopped for the night to make camp. We were miserable. We built a fire and crawled under our hide skins. The temperature dropped to near zero. In order to preserve our rations, we hadn't eaten the first day. Because we weren't conditioned to go without food, the night was even more uncomfortable.

Intuitively, we knew that food can be hard to come by in the winter months, but at some level we were denying that. We wanted to believe we could walk out onto the landscape and survive off the land at any time. But the reality is that on this landscape everything hunkers down in the winter. This fact was not lost on the native people, who did the same in the winter. The difference was that they spent the fall preparing by stocking up on food.

After the first day, for many reasons, it became evident that the journey would be tougher than we imagined. By the end of the second day, before the sunset, we realized we could not make the trip with the amount of gear we had.

Breck and I discussed what to do. Clearly, we needed to rethink our primitive walk. We decided that in good conscience we couldn't ditch our gear, because we had put so much love into making it. So on the second day, we hiked back to town to drop off our gear and restock with our more technical—and infinitely lighter—gear that we used for teaching outings.

Neither of us felt like we had failed; rather we felt like we had not properly prepared. Back at the BOSS office, we regrouped and repacked. Our packs now held the essentials, such as a knife, a few clay pots, a wool blanket, sandals, wool socks, nylon shorts, and a wool shirt. We exchanged our hide footwear for sandals made from tires.

We also brought a bag of knives and arrowheads we had made, as well as the pemmican. Still, it wasn't like we had gone to the North Face store. Even with a fire burning, sleeping under only a wool blanket on December nights where the temperature hovers around zero still requires a tough constitution physically and mentally to ward off feelings of freezing.

We set back out with our packs at roughly one-third their original weight on what was now a somewhat-primitive walk. We dipped down into the canyon. The changing clouds elevated our spirits. Just before sunset, we discovered a patch of savory oyster mushrooms. It felt like we were on the right path.

That night, a rainstorm rolled in. We found shelter under an over-hang in the cliff that gave us protection and lined it with cottonwood leaves. We set up a stone slab just outside the cave and dug a pit to make a fire. We gathered sagebrush bark. Using a fire made from yucca, I spun a coal.

With the fire started, we made a stew of oyster mushrooms, thistle greens, and pemmican. After dinner, we sat in the cave in silence, writing in our journals. The temperature hovered around twenty-five degrees, but the cave was so toasty warm that we were both in shorts.

But as we began to look ahead, it became apparent that we were seeing different things. We were camped in an area where we had guided, and therefore knew the terrain well. Soon we would be miles away, in a new land. I had the mind of an explorer. I was anxious to experience the land like a newborn baby, seeing, feeling, hearing, and smelling life for the first time.

Breck, however, was in a different mental space. As hard as we had prepared for the trip, we had never addressed the emotional aspect. He was thinking of his girlfriend, whom he had been dating for a year. Ahead for him was the time they would be apart.

Over the next few days, it was clear that his focus was wandering. We talked on and off about what he was going through. His emotions were pulling him back to his normal life, even as he needed them to be wholly invested in the journey.

Tensions between us began to rise. Admittedly, I wasn't offering any comforting words. I didn't have the personality or experience to empathize with what he was going through, because I hadn't put any energy into a long-term relationship at that point in my life. There was no way I could motivate him to finish the journey and then deal with his emotions. As much promise as the journey held, it was also going to be very tough physically and mentally. If one person was slightly off, it would eventually break the other.

We both realized we were processing different thoughts and needed time with those thoughts. So at the end of the first week, without any acrimony, we split up. Breck turned back. I pressed on.

I watched him split off in the distance, as we both climbed out of the canyon. He traveled in a northwest direction to get up on the rim of the canyon, and I headed straight north.

It began to snow the day we parted. The first major winter storm was approaching, and the animals were preparing their homes and storing food. The wind was cold, but the scenery was spectacular. I was surrounded by snowcapped mesas topped with clouds that looked like whipped cream. Above the clouds was a searing blue sky. The snow was not only beautiful but also gave me moisture for the hike.

Once Breck and I had shed our all-primitive gear, the journey had lost its purity. We had started out with a very specific goal in mind. We had painted a clear picture in our minds and for others. We had told everyone we were going out in primitive gear, that we were going to re-create a trade route, and that we were going to collect salt along the way. We were going to trade tools and goods like our primitive ancestors did. We had set a certain expectation, rather than keeping our goal vague and doing it for the sake of exploration and for ourselves.

But now that we had parted, I was still on a journey. Even though I no longer felt a calling to go to Chaco Canyon or to the Navajo trading post, I needed to consider my circumstances in no less a way than the indigenous people of the area would have done. I was entering a potentially dangerous survival situation.

The fact is, the Boulder area is a pretty hard environment to live in during the wintertime without a house or regular access to a grocery store. I knew that I hadn't gathered enough food to make it through the winter. I also didn't have that much money at the time. It wasn't like I could check out of my lifestyle, rent a place, and buy some food. But aside from not having the financial means, I didn't have the desire to do that.

I figured that as a follower of primitive-living skills, my best alternative was to migrate south. That meant walking until I reached lower elevations, where there would be more natural resources. At the same time, I established a second goal: to walk the four hundred miles to Winter Count, the annual primitive-skills gathering in Arizona. In reality, because of all the zigzags, my walk would be closer to six hundred miles.

As I thought about what had been lost, I began to feel that more could be gained by walking to Winter Count. The goal that Breck and I had set was to go through Chaco Canyon to *feel* what it had been like back when twenty thousand Native Americans traded goods there. But the fact is, Chaco Canyon was now a national monument where people came to see the ruins, not trade goods.

Winter Count, however, was a real gathering of 450 like-minded people. All my friends would be there. In essence, it was our Chaco Canyon. We met there to trade goods, as well as lessons and stories. It would be a similar experience, only now I was actually chasing something real rather than an ideal that had disappeared long ago.

After Breck and I parted, I walked through an area called Little Egypt, which is known for its resemblance to the rolling sand dunes of the Middle Eastern country. It was surreal to be walking across sand dunes that were being carpeted with a thin layer of snow.

After I topped out of the canyon, I ended up on the Kaiparowits Plateau. Snow began falling more heavily, accumulating to about six inches. It was a peaceful snow, not the kind that drifts. There was very little wind. I was wearing sandals with no socks, but the snow was so powdery that as long as I kept moving, I was able to stay warm. The canyon had numerous caves in its walls, which provided a sense of security should the snow pick up.

As the sun began to set, I looked for shelter for the night. I found a rock overhang and I hunkered down. I built a reflector wall and then made a fire. The fire was built outside the cave, up against the reflector wall rock, facing toward me in the cave. This allows the smoke to escape and the heat to be contained, but also does not tarnish the cave wall as making a fire inside the cave would.

Once the fire was lit, I crawled into that space. I felt a sense of peace and joy. I sat in the rock ledge, watching the snowfall. I was warm and cozy in a cocoon of rock, with the heat from the fire, and I had put it up in a matter of minutes.

The truth is, that shelter can be put up faster than most backpackers could set up their tent and camp. Even in a tent, most people would be uncomfortable with such low temperatures and snow. The reflector rock made all the difference. It is set up outside the cave, and the fire is then built on the inside of the wall. The wall contains the heat in the cave and can easily add twenty to thirty degrees of warmth to the space.

The evening was made even more blissful now that I had restored purpose to my journey.

EPIC SURVIVAL RULE #4:

DEAL WITH DANGER

In the wild, the demands are immediate and constant, and I am the only one who can guard myself against the danger.

Sure, cities are full of dangers, but much of modern society is consumed by guarding against these dangers. People protect themselves with everything from proper shoes to safer cars to door locks for their homes. Houses are insulated against cold, and air-conditioning and heat provide comfort and safety.

When these people venture out into the wild, they strive to reach the point where they feel at one with nature. They go backpacking or hiking because they want to leave their houses. Yet most go on a shopping spree at the REI store and stuff their backpacks full of gear because they are so afraid that if they don't have these things, they will not be able to survive.

I also often hear about people in the wild being scared of the animals. Dealing with wild animals is rooted in respecting them and their environment. They sense someone who respects their land.

Once I was in the backcountry and a mountain lion spotted me. It just stopped and stared. Then it began moving its head rhythmically from side to side, like a Michael Jackson maneuver in slow motion. It kept staring intently into my eyes. At that point, I felt like there was a soul exchange, that we were both reading each other. That mountain lion was like a brother, if you will, and it felt as though we had a mutual understanding that he wouldn't hurt me if I didn't hurt his land.

The demands of the wild can be unpredictable and sometimes life threatening, but I am finding that dangerous situations can be handled in different ways. I try to avoid putting myself in danger, but that is not always possible.

Chapter Thirteen

GOING IT ALONE

The Kaiparowits Plateau is massive and intimidating. Once home to the ancient Paiute people, the area stretches for forty miles and has varying terrain. There are extensive flat clay surfaces, narrow canyons, and several undulating ranges. I was entering it in what would be my first winter with no home other than the untainted landscape, which made my spirit soar and allowed me to feel like I was living in a long-lost culture. I had four hundred miles to go to reach the Winter Count gathering in Page, Arizona, and much to discover.

After walking for two days along the Kaiparowits Plateau, I got down to the flats. There were no trees. The bushes grew only about six inches tall. There was nothing for me to keep a fire going. I was tired so I just sat down on the barren flatwash, scanning the horizon for shelter. The temperatures were dropping severely. It was going to be one of my coldest nights ever in the wilderness—and I had no shelter.

I started to have doubts about my skills. I felt like I was in the wrong place at the wrong time—a potentially deadly mixture when you are alone in the wilderness. I was frustrated that I wasn't farther south. I was even more upset that I hadn't made it to a proper shelter for the night. I also realized there was no way I could use the fire-starting techniques I had used in the cave, because there was no fuel to burn.

I had a sense that I should get up and keep going. I knew I was near a highway. I thought, *Maybe I should just call this journey good, run out to the highway tonight, put my thumb out, and end in Page, where I could have a hot meal and a warm bed.*

As it was getting dark, with less than a half hour of daylight remaining, I sat back and scanned the wash. Rather than continuing down a path of self-doubt, I began to use my powers of observation.

Not far, maybe a quarter mile away, I noticed that there was a little bit of a pocket in a rock. It wasn't big enough to be a cave. What was it?

I went over to investigate. The pocket was about five feet long—not long enough to stretch out in, but long enough to curl up for the night. Shelter.

I couldn't start a fire, so I began to explore options the land could provide for warmth. Nearby, I spotted six-inch-tall rabbit brush growing wild. A big wash cut through the brush. I looked more closely. The roots were being eroded by the wash, and the rabbit brush was actually falling.

I walked through the wash and started pulling the loose rabbit brush. The base was twiggy, and the ends were feathery. It would make ideal bedding. I gathered the rabbit brush in bundles and stuffed the rock pocket with it. Then I burrowed out an area and crawled inside. The shell of rock was completely insulated by rabbit brush. I slept very well that night, as the land had granted me just enough to give me the courage to continue.

The next morning, I woke up fairly early. I continued walking toward the dam that stretches across the Colorado River that would take me toward Page, Arizona. From there, I walked another ten miles to Page.

Just as I was coming off the Kaiparowits, a ranger stopped me. He asked where I was going. I gave him the best description of my route that I could. I told him I was going to follow the Colorado River and then try to find a way into the Grand Canyon, which was a hundred miles away from where we were. When he heard the words *Grand Canyon*, it sparked his authority. He told me I would need a permit to enter the Grand Canyon. I explained to him that I was doing a walkabout on a whim, and that I hadn't planned to stop at a ranger station. The ranger insisted I needed a permit. I appealed to him that I had once done trail work in the Grand Canyon for five months, and added that I didn't have the financial means to buy a permit.

But he couldn't be swayed. He told me that I would have to go to the Paria ranger station. Because Jesse Perry and I had been to the area on our walkabout, I knew the station was at least ten miles out of the way, meaning twenty miles round-trip. I nodded and gave him the "sure, whatever" treatment.

I continued into town to the Navajo trading post. I had a few things I didn't need, and I wanted to add one more layer of warmth. I ended up trading a pair of moccasins and a stone knife for two thin wool blankets. I now had those two blankets, plus the wool one I had brought. The idea was that I could ditch my pack basket and roll my gear in a blanket. I would have to rely on my tire sandals for the rest of the journey. Though I knew I would have to walk through snow, my feet hadn't been cold. I was more concerned about warmth at night, though as it turns out, the human obstacles were going to give me greater troubles. The trade was the type of calculated risk I often have to make.

After I left the trading post, I headed off the highway, away from the roads. I ended up in a maze of slot canyons. A slot canyon is a tight, narrow canyon where you can often touch both sides by extending your arms. The walls range from ten to three hundred feet in height and can prevent you from seeing what lies ahead. It was very confusing terrain, but I knew that if I could navigate the canyons, then the route would take me over Echo Cliffs.

Moving from the southeast over the top and going to the northwest side, I continued another thirty-plus miles and walked over Echo Cliffs and into Marble Canyon. A hole-in-the-wall place, Marble Canyon has a hotel and a gift shop but not much else. (The Marble Canyon Lodge is where I had met Jesse when he returned to our journey.)

Just as I was about to cross Highway 89 into the Navajo reservation, the same ranger drove up in his truck and motioned me over.

The country was wide open. With a pair of powerful binoculars, you can see for miles. Clearly, the ranger had been tracking me.

The ranger asked me if I had gotten my permit. I told him there was no place on my route where I could get one. He then informed me that I would have to go to the Lees Ferry ranger station, which was eight miles out of the way, or sixteen miles round-trip.

I tried to explain my situation. I told him that I was exhausted. I was crossing some 120 miles alone with mostly primitive gear. My respect for the land was absolute. After all, I had only a small roll of supplies, on which I was traveling hundreds of miles. I appealed to him to let me continue my journey without any bureaucratic interference.

But he wouldn't budge. And he didn't offer me a ride, either—not that I would have taken it.

In nature, I teach my students to think for themselves. Rules, I tell them, are a guideline. To me, that incident showed that this ranger was not thinking for himself. He could plainly see that I had no money and that I respected the land. He knew my story was legitimate; yet he was not willing to let me go.

There was another dynamic at work. I walked everywhere in my life.

The result was that some days I looked good, while other days I was really dirty and looked tired and hungry. Authority figures often assumed I was a homeless bum, and they would harass me in different ways. It was ironic. People who are supposed to be protectors harassed me. It left me with a bitter feeling toward authority. It was also frustrating because I was trying to lead a humble existence and be a steward of the earth, yet people who likely were not living as respectfully tried to defile my existence.

I had no choice. I walked in the direction of the ranger station. I decided that the fastest way was to run it both ways. I stashed my pack in the woods and ran the eight miles to the ranger station.

When I arrived, the situation turned into a fiasco. First, the ranger on duty insisted that I call the ranger station in the Grand Canyon, where I was headed. The ranger on the phone asked me where I planned to camp every night. Being honest, I said I didn't have a clue. The Grand Canyon ranger asked my route. I told him I had seen a place to enter the Grand Canyon near the Little Colorado, and I asked if I could get in there. He wasn't sure; no one had ever been over there.

I explained that I used to work in the park and asked him to waive the fee given my history with the area. The ranger seemed sympathetic, but said he could not go against "regulations." The permit was $55.

To put things into perspective, I had about $350 to my name. For most people, paying $55 would not be a big deal, but it literally cost me one-seventh of my money. Federal park officials don't realize that there are people who live in the bush full-time, and when they are forced to cough up those fees, it can be a significant chunk of their wealth. That ranger had no comprehension of what that meant. That was why I was upset with the process.

When I asked the ranger what the money was used for, he told me it was for "the knowledge and resources we supply to the public." So I turned that around on him and told him that if he would waive the fee, I would let the ranger station know if the route was doable. He gave me the whole "dude, man" speech.

"Believe me, I totally get it, man," he said. "I understand what you are doing. If it were up to me, dude, I'd pay you for that. But I've got my boss listening over my shoulder."

I stopped him. "I'll just do the route and let you know if it's viable; then you will have information to give up to others," I said. "It's on me."

I felt I was due a favor, particularly if I discovered a route they had never traveled—not possibly couldn't travel.

I paid the money that I really could not afford to spend, and ran back to retrieve my sack. I then trekked thirty miles across the Navajo reservation and came to the entry point for the route I had asked him about, which dropped into the Little Colorado valley. I studied the gazetteer map that I had torn from an old book, but it wasn't very helpful. It was not like a regular topo map, where you can see the trails and details of the land features.

Just as I was about to head down, I encountered a Navajo man. I was standing at a fork in the trail. I asked him if I could get down into the canyon.

The man smiled. "Oh sure," he said, gesturing in no particular direction.

"Sorry . . . right or left?" I asked.

"Just dat-a-way," he said. Then he turned and walked away.

So I was left to my own devices to choose. I guessed right. Soon I ran into a sheer cliff. Just below it was a trail the Navajo had built to get down to the bottom of the canyon.

It was a steep, switchbacky trail, but it seemed purposefully carved out. When I reached the bottom, I discovered the purpose for the trail. The Navajo had built a fishing camp at the very bottom, some two thousand–plus feet below the cliff. As there were nets strung across the river, I knew the fishing camp was still active.

No one was there. I looked around. There was no evidence of anyone having been there recently, probably for at least a month. But the camp was still intact.

I didn't invest the energy into fishing there, because I couldn't physically see any fish. The water was turquoise, but I could see only a foot down into it. I didn't have any fishing supplies. Even though I was skilled at catching fish with my hands, without being able to see them, that wasn't an option.

In fact, the only food I was getting was trapped mice and rats. I would stop every night and make traps. That was my skill level at that point. I was carrying a bow and arrow, but I wasn't a skilled hunter by any means. Part of the journey for me was to develop those skills. So I was living off that little bit of meat and a few rations, such as pinole, that I had brought.

The country was very dry and barren. From the vague map I had, I determined that I would have about forty miles of dangerously dry travel. I was walking through a very shallow canyon that was about thirty feet deep and cut through the clay beds, which made me conscious that I needed to be aware of my water sources.

After several miles, I found a pocket of water containing a few gallons. Water had gathered in a depression in the rock from rain and also from meltwater that had flowed down from higher elevations. There was a cave next to it. I sat in the cave and hydrated.

I began to carefully consider how much farther I had to go without the ability to carry water. Given the heat and the lack of any consistent water, the journey would be a rough one. Instead of moving on, I decided to use my ability to be patient. Though I was avoiding a survival situation, I was also putting myself in one. But my experience was telling me that staying was less risky than moving.

I needed to listen to the land. I had to have the patience to stop and wait for the land to tell me when it was time to move. That was a hugely valuable—and intuitively difficult to grasp—tool. Generally, only a hunter-gatherer has the patience and ability to execute this strategy. Basically, I was risking starving to death, because I felt my heightened relationship with the land would deliver me what I needed to survive.

This was a major step for me. The impatience of youth was being conquered. The inaction I was taking was the polar opposite of what the kid who ran everywhere would have done. He would not have been able to sit in a cave for an undetermined number of days and wait for rain. But I could.

There was an inherent risk. Every hour I sat in the cave, the water supply in front of me dwindled, as I drank it and watched it seep into the ground and evaporate into the dry air. When I first hunkered down in the cave, there was no sign of weather, no clouds or noticeable dew in the air.

I had water for the time being, but not much food. I had cornmeal to make ash cakes, which are cooked directly over the heat of a fire. With cornmeal, I cook it on a rock at a forty-five-degree angle until it hardens, and then I finish cooking it directly on the hot ash. I used the leftover corn niblets to trap mice. I ate mice every day. Though I was consuming less than five hundred calories a day, it was enough to sustain me. Despite the lack of food, I was much more concerned about getting enough hydration to make the forty-mile crossing through the barren flats of the Navajo reservation.

The landscape was relatively flat. The reddish clay surface was intermixed with shallow washes of sandstone. Those washes were where I would get water after it rained. Actually, the moisture that fell in the area was more a slushy half-rain, half-snow mixture. The temperatures were reasonable, fifty degrees in the day and around thirty degrees at night, but once the sun rose, any of that slushy mixture that fell into the wash would melt fairly quickly.

The issue was that the deepest washes were no more than a couple inches. In harder stone, there would be slick rock pools as deep as ten feet. These can hold water for two months, whereas the small pools dry up in a matter of hours. Immediately after a slushy rain, the small pools would hold water just long enough for me to move across the land.

While waiting for rain, I stayed in the cave all day. I worked on a few crafts and condensed my load. But for the most part, I sat and meditated. After a day and a half of being in the cave, I started seeing a little cloud buildup. It wasn't dense enough to release any moisture, but I felt that it would in another day or so.

I had been living with the land so much that at home in Utah I was actually able to predict the weather. As incredible as that sounds, I could tell people weeks in advance when it would rain, how many days it was going to rain, and how hard. I hoped I had connected with the Arizona land closely enough that my premonitions of rain would be correct.

On my third night in the cave, the clouds opened up and dumped a significant amount of rain on the land. It rained hard, filling the pockets with water. I set out the next morning. The rain allowed me to stay hydrated as I traversed the thirty-plus miles across the flats.

I followed the Little Colorado River down into the Grand Canyon. The river is magical in the wintertime because the water is still a comfortable seventy-five degrees. I passed an old salt mine and grabbed some salt for the journey. I knew that a white man wasn't supposed to take salt, but in many ways I felt closer to the Navajo and Hopi so I took a small amount and continued on my journey.

By the time I reached the Grand Canyon, I was pretty spent. I had no reserves. I was cold. I didn't know it, but I would soon be in grave danger.

Chapter Fourteen

A CLOSE ENCOUNTER
WITH DEATH

The winter solstice in the Grand Canyon changes everything. The grandeur of nature's remarkable creation dims. When the sun is at its lowest point on the horizon and shines the least amount of hours, you cannot see it at the bottom of the canyon. The day is in constant shade. When the light fades, it drags the temperature down with it.

Because I had lived in the area years earlier, by the time I was several thousand feet down in the canyon, I had an idea where I was. My original plan was to stay in the bottom and hike along the Colorado River and then pop out. However, with the low temperatures and lack of sun, I knew that wasn't a viable option.

My entire body was beaten. I was cold and hungry. I was wearing a very thin wool shirt with no pockets and a pair of nylon shorts. I had to get into the sunlight. Rather than cross the canyon, I decided to climb out.

I climbed up toward the rim, thinking that at the very least I would be exposed to some sunshine. My idea was to eventually hike to the South Rim and resupply some food staples. That would have been twenty-five miles. I had traversed more than two hundred miles, weaving my way through deserts and canyons, from Utah into Northern Arizona, so this seemed very doable.

As I was hiking up, it started snowing. Within an hour, the snow was coming down in sheets. I was wearing sandals, as I had traded away my lined moccasins for an extra blanket. I had known the trade was a risk, but I thought I would be farther south when the heavy snow came. I had no socks. I was generating heat by moving, but my body temperature had dropped slightly, so I did not have much heat to give. The snow was falling so rapidly that it rose somewhere between my mid-calf and knee. The wind was also picking up, limiting my vision. The visibility was so bad that I couldn't tell I was in the Grand Canyon.

It was getting later in the day and colder with each step. I needed to warm up my feet. I left the trail and undid my bundle. I unrolled one of the thin blankets, cut off the bottom, and then cut it into two big squares. I removed my sandals and wrapped my feet in those two squares of wool, and I then tied my sandals back on.

The realization hit me that I was going to die dressed in a thin wool shirt and nylon shorts. I took the saddle blanket I had brought and cut a slit in the middle. I threw it over my head and tied it around my waist like a poncho. I was in such a dire survival situation that I was cutting up my blankets. This would render them useless later, but there wouldn't be a later if I didn't make it out of here.

I continued pistoling toward the top. I reached the top right at dusk. The sun had gone down, and it was getting dark fast. With every step, I sank into the snow. My legs were numb from the cold.

I finally stopped. I couldn't go any farther. My thoughts turned to surviving. If I could make it through the night, I could reach sunlight in the morning. I looked around to see if I could find some insulation for shelter. It was no use; everything was blanketed in two feet of snow.

I needed to get a fire going. Without a fire starter, I would have to spin a coal. I looked at my hands and started to move them. I could feel the onset of hypothermia. I tested my hands by rapidly opening and closing them. My fingers barely moved. I was telling my fingers to move, but they wouldn't. They were too cold to spin up the fire.

I had reached my outer limit of exhaustion. I was out of food and energy. Many times I had pushed my body in terms of lack of calories and against the cold, but I was aware that I had gone too far.

I stood there in the blowing snow. There was no dry place to lie down. I knew that if I lay down in the snow, I would freeze to death and die in the night. The nearest village was still at least fifteen miles away.

Though it was the middle of the night and too overcast for the moon to light my way, I decided my best chance to live was to try to walk the fifteen miles. With my physical gifts and my mind's ability to focus, I believed I could plow through the snowdrifts and make it to safety.

I could feel frostbite on my nose and cheeks. I tucked my face into my makeshift coat to try and warm my skin and ward off permanent damage. My body was completely defeated because I didn't have the caloric resources to maintain warmth. My core temperature was barely on the functional side. I could feel hypothermia setting in, as my joints tightened and my pulse slowed.

I continued walking directly into the wind. A hard, sharp, icy snow was slashing my face. I was freezing. I didn't have the energy to shiver. I was beyond that stage and I was not really conscious of what was going on. The functions of my body felt like they were slowing down.

Moving was my last-ditch effort to stay warm. I don't know how I was moving. My body had never been that far gone. Several times I collapsed in the snow. On one fall, I hit hard and ended up on my back. I lay there for a while, not fully conscious.

I summoned the energy to get to my feet. I walked another few steps. But eventually, no more. I fell down. The lights in my brain went out.

Time passed. I woke up and came to the realization that I was lying in the snow, unable to move. I looked up and saw lily-white flakes falling from nature's ceiling into my eyes. Even in my depleted state, I appreciated their elegance. I said to myself, "Thank you, God, I had a great life. I really appreciate it."

I was certain I was going to die; yet I was okay with it. It was time to go do something else. I closed my eyes.

Then I heard a voice. "Your mom's gonna be pissed if you die in the snow," it said. "She'll be forever cursing your primitive skills and earth-living life if you die here."

I couldn't place the voice, but it didn't matter. I wouldn't call it motivation because I certainly did not want to die, but upon hearing the voice, I slowly rose to my feet. Putting what mattered over mind, not trying to muscle my way but just attempting to move, I continued both my trek to town and my life, which, at that point, were intertwined.

I later found out that it was minus eight degrees with a wind chill of at least twenty degrees below zero. Even for the Grand Canyon on the shortest day of the year, the conditions were particularly dire.

How did I find the energy to get up and walk when I seemingly had none?

I learned that it was a classic boost of adrenaline triggered by fear that can occur when the body is challenged to its limits. Likely, my body temperature had dropped below the danger level of ninety and was nearing eighty-two, the trip line for loss of consciousness. In a study, Vladimir Zatsiorsky, a professor of kinesiology at Penn State and an expert in the biomechanics of weight lifting, found that an ordinary person can summon 65 percent of their strength in normal circumstances, whereas a weight lifter can summon 85 percent. In competition, a trained athlete can go 12 percent above those figures. But with an adrenaline surge, the short-term push can be double those figures.

As Dr. Sam Parnia explains, under extreme stress, the adrenal gland dumps adrenaline into the bloodstream, blood pressure surges, and the heart pushes oxygen to the muscles. "Performance is multiplied, and every fiber and muscle in your body functions a hundred times better," Dr. Parnia says. "You become superhuman for a short period of time. For Matt, because he is so fit and can deliver oxygen at a much higher level, his adrenaline surge would be even more extreme."

I also attributed surviving to my ability to deal with fear. I had reached the brink. I knew what it would feel like to die. Fear was present, but it did not consume me; therefore I did not succumb to it. Fear can lock people up and prevent them from thinking clearly. It can debilitate the mind and the body. However, people who have the ability to relax have faster reflexes; whereas more fear translates to slower reflexes, as a doctor once told me.

Coping with fear requires calm. In my experience in the wild, anytime fear creeps up, I try to relax into it rather than allow it to consume me. If you tense up, everything about you starts to fall apart. Climbing initially taught me how to deal with fear. It was get up the rock or die. That type of fear showed in my clunky technique. But once I learned to respect the rock and be in touch with my movements, awareness trumped fear.

Lack of fear also brings clarity. I was lying in the snow, hours from freezing to death. My skills and judgment were being questioned. The result of not answering would be final. But I inherently knew I had a connection to the environment that had taken me down.

I did not feel that I was being reckless. I was not attempting to challenge the land out of hubris. I was not trying to prove anything to anyone, or find anything. I am not a pioneer, nor am I an explorer. I was a person living the way millions of people over hundreds of thousands of year have lived. The wall I found myself up against was a circumstance of my life.

I have always believed that in the wild if you're open and honest with your environment, there will be a connection—you take care of it and it takes care of you. If everything you do has an honesty and integrity to it, even if you are in a difficult situation, you won't feel a high degree of fear. If you look at the big picture—I'm alone in the wilderness, stuck in subfreezing conditions with no communication—you will be uncomfortable. But if you can step back and isolate that one spot where you are and focus on it, you won't be afraid of the unseen and unknown. For me, those were the skills I had devoted my life to pursuing.

The adrenaline surge likely gave me energy to push on, and my ability to focus on where I was allowed me to use that productively—not that the rest of the trip was easy by any means.

I eventually reached the south village. Unfortunately, I was on the east side, which I soon found out was closed for the winter. My immediate thought was that I would die if I didn't get inside a warm building. I planned to use what little bit of cash I had and check into a hotel. But when I reached the hotel, it had a sign that informed me it was closed for the season.

I looked around. There were two feet of snow on the ground and no signs of any life at all.

I said to myself out loud, "I can't make it another two miles to the next village." That actually made me smile, because in all the time I spent alone, I had probably not uttered more than two or three sentences out loud.

I was spent. I had barely made it to the doorstep of that hotel. My first thought was to find shelter somewhere, anywhere. I was freezing, so I needed more than just a dry spot. Perhaps there was a doorway, or even a garbage Dumpster. I looked around. The buildings were square, brick box structures. I didn't see any doorways or awnings.

I started walking back out into the road area. It was covered with snow. But I noticed that there was one set of tire tracks. Though they were filling with snow, they made walking a little easier. It also gave me hope that someone might drive through.

I felt like I wasn't going to make it. I didn't know what to do. I walked a few hundred yards, and then I saw my salvation. A white van with heavy chains clanking hard on the road was coming right at me.

Given how I looked, I thought there was no way they were going to give some crazy-looking mountain man a ride. I thought about lying down in the middle of the road, but quickly reconsidered. I put out my thumb.

The van stopped. The driver, a man bundled up, pushed the door open from the inside and offered me a ride. I pulled myself in and thanked him.

It turned out that he worked for the town and was picking up employees who were out drinking late at night. Where, I had no idea, because the place looked deserted.

He dropped me at the nearest hotel. The desk clerk didn't seem all that shocked at my appearance. I checked in and went to my room.

I unrolled my pack and spread out the contents to allow the warm air to dry them. I went into the bathroom and ran my hands under lukewarm water. When I looked in the mirror, I realized my face had some minor frostbite on my cheeks and nose, but I would be okay. I took a warm bath, dried myself, and then crawled into bed. I fell asleep immediately. What happened next was majestic.

Chapter Fifteen

VISIONS AND
VISION QUESTS

In life, time carries youth forward, and young men and women become adults. While there are traditional and nontraditional paths to maturity, the first rite of passage in life is often when people leave home. Leaving home, be it for college, military service, or any other reason is the first step into adulthood. That separation carries with it a sense of heading off alone to stake your own claim. Even in the cocoon of a university, surrounded by friends and supported by parents, young adults experience independence, and begin to search for their path forward. In the Native American cultures, they would escalate this experience by going on a vision quest, which I have found can be a powerful survival tool.

Native Americans have a connection with the land that is absolute. Many of their traditions and ceremonies have become precedents for the modern customs that people use to better their spiritual lives. Some of the Native American tools for growth, such as the vision quest, have not translated to modern society, though in this case many people would greatly benefit if it did. Vision quests are a way for the Native Americans to advance their lives. As a survivalist, they have moved me forward both spiritually and emotionally.

A typical Native American vision quest lasts four days and four nights. (Going out into the wilderness for two full nights and one full day can also yield benefits.) It takes place in a location with only the most minimal of possessions. Typically, a stone circle is constructed and the person does not leave the area. The vision quest includes four stages: solitude, immersion in nature, fasting, and community. Being alone forces one to look inside his own soul. The return to nature creates connection. Fasting opens the body up to absorb possibilities. The return to community with this newfound elixir helps others.

Most vision quests are purposeful and forced. Native Americans go on them to find their spiritual center, to reinvent themselves, and to chart their life's direction. They use them to guide themselves in many pursuits, to open cathartic doors, and to recover from life-altering events. Once the body has opened itself up to nature, the mind has an ability to see it in an unconfined way. In other words, you can't sit in your apartment, turn off the TV, and have this experience.

Vision quests are part of the fabric of nature and owe their power to the natural world. Being alone in nature causes the body to go into a more relaxed state. When the body reaches that state, it becomes more receptive. On these quests, the person often experiences visions because their mind becomes comatose from lack of food and sleep.

Starving yourself depletes your energy. When you have a normal level of energy, you are an "outputter." The body constantly wants to push that energy somewhere. The idea on a vision quest is that you reach a point where you have no energy to push out anymore. So instead of pushing out energy, the body receives energy from other aspects around it, primarily the land and other spiritual sources.

I have talked to people who have studied aspects of people's energy. Some people are born with more of their own energy, while others are born as empty vessels. People born as empty vessels usually have greater intuition and psychic abilities, and they are able to receive things. Some people have energy but lack a highly developed knowing center (the center of wisdom people are born with).

To break it down, there are energy channels that run through your mind, body, and spirit. Some people are always receiving that energy; they are not putting any out. These would typically be shamans. They would also be people who would be able to sit on a vision quest and find a profound spiritual revelation.

According to spiritual readers who have read me, I am less of a receiver than most. When I sit on a vision quest, I don't see the proverbial white buffalo. One aspect about me that is unique is that all my energy centers are fully developed, along with my knowing center. That means I have the ability to be a teacher and to share the knowledge I was born with.

The land will share with anyone who puts themselves into the state of receiver. Most people who go through the first three stages will have some form of vision. Native Americans believe that people who need it the most receive it far more frequently. Therefore, if somebody who is already well grounded walks into the vision circle for four days and four nights and doesn't have some powerful vision, the shaman's explanation would be that you were grounded enough in that moment that you did not need to see anything big.

I have used vision quests whenever I need clarity. Upon moving to Boulder, before I applied for the job at BOSS, I went to the top of a mesa to see what the land had to tell me. I spent three days and three nights (the typical time for a vision quest) in a circle there to connect with the land where I would work and I found my home. But I have also experienced powerful visions while out in nature that have helped me survive.

My journey on the "revised" all-primitive walk that led me into the Grand Canyon was similar to a vision quest. Like the Native American men on a vision quest, I was experiencing extreme solitude. I was purely connected to the land. My body was beaten down. I was nearly starving. When I arrived at the motel, I was not consciously seeking anything. Even though I was hungry, food wasn't on my mind. All I wanted was a hot bath to raise my core body temperature to a reasonable level and a warm bed for a night's sleep.

I crawled under the sheets and pulled up the covers. I was focused on nothing. There were flashes of my near-death experience. I wasn't planning for tomorrow. I didn't even say a prayer of thanks for my life being spared. I went right to sleep.

I have always been a lucid dreamer and a light sleeper. I owe that to living in the wilderness. Even during sleep, there has to be a level of awareness because there are so many variables. As I developed the ability to detect danger in sleep, I also found that I could easily recall my dreams and separate them from reality.

That night, I began dreaming before I fell into a deep sleep. I was aware of the precise moment when my dream started to take shape. I liken it to the feeling when you start dozing off and fuzzily enter a dream state before fully checking out. This dream started the instant I pressed my head into the pillow—the first pillow that had been under my head in a couple years.

I saw my spot, my place in life, my oasis. It was something of a wilderness area that didn't have a lot of distinctive features to it. The ground was covered with sand and a few trees dotted the land, but it was a very nondescript place. As I prepared to walk toward that spot, a man riding on what appeared to be a hovercraft approached me from the distance. The vehicle had no wheels and was elevated two to three feet off the ground like something out of the movie *Back to the Future Part II*. By the time the vehicle reached me, my sleep was getting deeper and deeper.

The hovercraftlike vehicle stopped in front of me. The man was of a tan Caucasian ancestry and held his face with a certain amount of compassion and wisdom, much like the actor Morgan Freeman would if he were playing the role.

"Hop in," the man said.

Puzzled, I replied, "Where are we going?"

"I'm taking you to your house," he said casually, as if he were giving a friend a lift home. "Your wife is there waiting for you."

"House? Wife? I'm not married," I said.

His demeanor turned stern. "Matt," he said. "You have dedicated your whole life to teaching primitive skills and teaching people how to live with the earth. This is what's been waiting for you." He punched the phrase *whole life*, though I wasn't sure what that meant.

I hopped on, and off we went. I was an adventurer on his strangest adventure ever. A mysterious excitement took hold.

The drive wasn't long. We pulled up to a house. It was the most beautiful structure I could conceptualize. The house had a white-picket-fence feel without actually having a white picket fence. The awning and entrance to the door were formed from a log that created a natural upside-down U shape. All the materials were of the earth, but the house had been built with a human mind for detail.

Just as I stepped off the hover car, the front door opened. An absolutely gorgeous woman walked out. She was dressed simply, which accentuated her natural beauty. There seemed to be a celestial light radiating from her. She walked up to me.

"Don't you think it's kind of strange that we are married and we don't really know each other?" I managed.

The corners of her mouth turned up as she spoke. "But I do know you and you know me," she said reassuringly.

Instantly, I understood what she meant. My fuzzy thoughts became clear. We walked into the house and back into our lives together. From that point forward, there were no words in the dream.

We spent an entire week together getting to know each other again and making the house our own. I worked on the physical tasks, while she took care of the more graceful pursuits. For instance, I would build a table out of wood, and she would make a flower centerpiece. Everything we did had a synchronicity to it that didn't require us to ask the other what to do. Our roles seemed perfectly defined. We were blissfully intertwined.

Six hours after the man picked me up, I awoke. I felt enlightened. The dream had taken up the entire six hours. Many experts claim that the longest dreams only last up to twenty minutes and occur only during REM sleep, but this was not the case. I know this because I recalled the precise moment the dream began, just as I crossed from consciousness to sleep.

The dream I had was unique, and I knew it. It was born of the harrowing experience I had in the Grand Canyon. Unlike a vision quest, I had not been seeking answers to my path forward. I have been on full vision quests for four days and four nights and not had such clear visions as I did in that dream. Still, I could not ascribe any larger meaning to it at that time.

My most complete visions have come on hunter-gatherer-type walkabouts in which I am living off the land for long periods of time. Most of the time I'm alone, but sometimes I'm guiding groups. Sometimes when I guide on long courses for a couple weeks, students will have visions that make me jealous. In part, this is because the previously uninitiated body is more receptive once it is put in a state where it can be an "inputter," rather than "outputter."

These visions tend to be highly stylized versions of something missing from their lives. Perhaps their late father said something so profound that it wouldn't be in an ordinary dream. Recalling such visions, the people are often in tears because they feel they have been touched by something real and lasting.

When students go through the starvation phase, they often start dreaming that they were with friends feasting at a banquet. Some people wake up and brush it off, saying it was merely a dream. But if the dream endures, others like myself enjoy the experience so much that when we wake up, even though we know it was a dream or a vision, we feel like we actually ate the feast and hung out with friends.

Like many people, in my early days I may have discounted such an experience, dismissing it as an ordinary dream or a hallucination, from being too hot and too dehydrated. But I have had visions that show these accounts are more than mere dreams or the side effects of a depleted body, and that things are connected more deeply.

I began to understand this on a long walkabout. I ended up staying out in the canyons alone for forty-two days. After about eight days, I found a cave in which to rest. I didn't sit in the cave as a substitute for a vision quest circle. Instead, I decided to purify my body by drinking mint tea and not eating any other food.

I stayed in the cave for a week. I was wearing a breechcloth and had another piece of cloth tied around my waist. Those were my only belongings. My primary daily activities consisted of leaving to collect mint and water from a spring that was deeply cut into the ravine below, and building a fire at night. I did little else.

On the fourth night, I was lying on the cloth by the fire preparing to go to sleep. My skin was exposed to the night sky and the fire. Seconds after I fell asleep, I heard a noise coming from the spring down the hill. I woke up.

I sat up and looked at the fire. The smoldering flames looked real, but I realized I hadn't woken up my real body because I could still feel the fire on my shoulder. It was a variation of my body, my consciousness, in a place that was identical to where I had fallen asleep.

I was puzzled. I didn't know how to react to this feeling. I was listening to the spring and looking down the hill, all the while wondering what I might be looking for. The sound was real. What if there was something dangerous down there, a potential threat?

For a moment, I lay back down and thought about waking up my real body. I've always been able to control dreams before they veer to an uncomfortable place. But I stopped myself. I stood back up. This feeling of being in two worlds was too bizarre not to explore. Because I had somehow created this state, I wondered if I could go further. I wondered if I could fly.

I lifted my body up. Sure enough, it elevated. *Yes, I can*, I thought. *I can fly.* I flew around the comforts of the majestic alcove. The cave was as large as a football field. The feeling was amazing, if not totally believable to me.

When I started to lie back down, I could still feel the fire. But before I lost this ability, whatever it was, wherever it was coming from, I decided I should explore.

I looked up at the sky. There was a three-quarter moon, so I knew the desert would be bright. I lifted my body again and began to fly.

I flew up over the canyon rim, across the desert, and through the countryside. I kept flying, over the San Bernardino Mountains, all the way to California. By the time I reached California, the sun was starting to rise.

Typically, there is a layer of smog over the San Bernardino Mountains. It was there but different. The smog had risen to the tops of the mountains and created black swirls in the sky. From above, it was the most polluted air I had ever seen. It looked like carbon exhaust that was magnified forty times to the point where I didn't think anyone could live or breathe. In the city below I didn't see any activity, not even any cars driving around. I decided to explore.

I descended. As I approached the foothills and closed in on the city, I spotted an apartment complex. The two-story building had a courtyard in the back. I landed on the walkway of the second floor. Through an open door, I heard a mother and child talking. They sounded fine. But the entire situation felt weird. The cars in the streets were all abandoned. As I walked toward the door, I heard somebody walking up the stairs.

I reached the door and leaned into the apartment. I saw that the mother and child were in the kitchen. They were occupied with something and turned away from me, so I decided to enter and slip by them. Despite knowing I was dreaming, I didn't want to get caught because I was unsure what the repercussions would be.

As I made my way through the apartment, I saw there was a screen open, so I went out the back to the courtyard. Once in the courtyard, I decided to depart. But the courtyard had wooden latticework around the top. I flew around like a trapped bird looking for a break in the lattice. All the while, I still felt the fire on my left shoulder.

The fire kept me focused. I knew I had to return to my real body. I found a gap in the lattice, squeezed through it, and returned to my cave faster than I had arrived. Back in my real body on terra firma, I opened up my eyes. The fire was crackling in front of me, real as could be.

I sat up, dumbfounded and bewildered by the whole experience. I didn't want to analyze it, so I lay back down and fell asleep.

I woke up the next morning, the fifth day after the fourth night, when visions typically occur. The vision was over; my awareness was heightened. I didn't know what to make of the experience, but I did know it portended well for me. It was time to return to my tasks of the journey.

I had collected a couple of squirrel skins that I wanted to sew into pouches. I decided to go out and hunt some more squirrels. I also wanted to make a bone awl. The area was pretty void of bones, but I remembered one cave where I thought I might find an old cow bone.

I walked to the cave where I thought bones might be. I entered and walked the length of the cave but didn't see a single bone. Right as I was on my way out, I saw one little white speck in the sand. I reached down, flicked the speck, and saw there was more buried there. I kept digging. Two feet underground, I felt something and spread out the sand to have a look.

I uncovered what was certainly an ancient, thousand-year-old Anasazi bone awl, perfectly made. Six inches in length, the awl was carefully made from the femur of a deer. The tool's maker had split it from a section of the femur and ground it down. It was still sharp. The craftsmanship was exceptional. The person had done a far better job than I could do.

I ended up using that bone awl for the next two weeks until a cave took it away. It disappeared into the sand again. Small things such as awls often get lost in the sand and disappear easily. The bone awl was left to be discovered again.

To this day, I still don't understand the significance of the vision I had of the mother and child in their apartment with a virtual wasteland surrounding them. There are possibilities. I don't know if it is part of my life that is yet to be lived, or something that I will discover affecting someone else's life. Perhaps it is as a simple as the fact that a night gave me something intangible that I had never seen (the flying scene), and the following morning gave me something tangible that I had never seen (the bone awl).

The awl was like finding a needle in a haystack. To this day, none of my friends has ever found an Anasazi awl. I just happened to walk out that morning needing to find a bone to make an awl. Those awls exist in museums and ancient burial grounds, but to stumble on one randomly in a cave is a one-in-a-billion chance. That led me to tie the discovery to the vision I had.

I do know that visions make it possible for me to be alone for long periods of time. They make me feel that there is a greater being watching over. They make me feel that I am not alone. As such, they have become an essential part of survival.

Visions have allowed me to discover that I can confidently live alone for extended periods of time with no human contact. But rather than feeling alone during those times, I actually feel more connected. Many survivalists who spend time in nature feel as though they have to read to keep themselves from losing all touch with others. Absorbing the stories from the pages keeps them from feeling alone. But I found that if you don't read, then you can open yourself up to visions that ultimately make you feel even more connected to life.

Chapter Sixteen

THE SOLSTICE JOURNEY

Though I had been teaching survival skills courses and leading hunter-gatherer trips for ten years, I felt that I was lacking a bigger piece to the puzzle. Despite the fact that I had spent weeks and even months at a time in the wilderness, I couldn't impart to students what truly living in the wilderness meant. I had to experience it for myself.

As an instructor who regularly took people out on monthlong hunter-gatherer trips, I was seeing in both myself and my students that we were making headway toward a place on a physical, spiritual, and mental spot, but then it would just end. I wanted to break those barriers and see what it was like to be out for an extended period of time.

At the time, I was thirty-three. I was feeling a lack of discipline in my life in terms of setting goals and sticking with them. At times, I felt like a wanderer in need of grounding. In the wilderness, one of the most critical things you must learn is how to be spontaneous and turn on a dime. Even if you make a plan, you have to be able to adjust that plan in seconds. Plans can be dangerous. You don't know when it's going to storm, when you are going to come across the next animal, or when natural obstacles will alter where you camp.

As I tried to carry that philosophy through my life, I was always adjusting and changing on a dime, sometimes to the point where, honestly, there were days when I needed to make a decision, but I couldn't. I was feeling indecisive with my life, so I sat down in the spring and wrote out a few goals on a piece of paper.

The biggest goal was to take a journey. I wrote down that I would go out for three months. Then I thought that was too short. I put it down for six months—from winter solstice to summer solstice. I had been out in the wilderness for extended periods, such as on the revised all-primitive walk. But on all those journeys, I would occasionally come across a town or some other aspects of modern civilization. The goal of the Solstice Journey would be not only to live in the wilds but also not to venture into a town at all, to never spend a dollar, and not to use any modern resources.

Questions swirled about my capabilities as a survivalist. How long could I live alone in the wilderness with no connection to society? What would I learn? How would it change me? What would it do to me physically, to my skills, to the whole collective picture of my core as a survivalist?

As I wrestled with these questions, I came to the conclusion that I needed to find answers even if it cost me my life—though my time in the wilderness had made me confident enough in my abilities that I could survive and even thrive pretty much anywhere.

Throughout the summer teaching season, I became more and more excited about the prospect of the journey. But at the same time, I was also nervous. I had been very comfortable going out in the fall and staying out for a couple weeks or a month. But the thought of going out in the middle of winter, the absolute hardest time of the year, and committing to stay out for six months with no human interaction whatsoever was daunting.

Setting the journey's starting point at the winter solstice gave me time after the teaching season ended to mentally prepare, and also to gather food and make any clothes and supplies. I was living in a cone-shaped wickiup on BOSS land that was only large enough for my body and one other person if need be. I had lived there for two years. It was a small brush hut with a fire area in front of it, but it served as a nice base to prepare for the journey.

I packed as sparingly as possible for the journey. I had a canvas back-pack that I loaded up with everything I could, but I tried not to take too many gadgets. I took one set of regular clothes, moccasins, a coyote-skin blanket, and a down blanket. I took a saw and a machete, along with feathers and fletching to make arrows and extra hides to work on. On my feet were running shoes.

For food, I loaded up basic rations of rice, lentils, olive oil, nuts, seeds, as well as dried fruit and dried greens. I also brought dried nee-dles, mint, and acorns. I packed three pounds of fresh meat and about fifteen pounds of dried elk meat. It was enough to get me started for the winter, and about all I could carry.

As I contemplated the journey, my emotions were running high. I was scared because I knew how important this goal was to my life as an instructor. If, for some reason, I didn't complete the journey, I felt that it would strip my life of validity.

I had reached a crossroads. I was at a point where I said to myself, I can't teach long-term survival skills if I don't have the ability to do it on my own—which meant the ability to complete the six-month Solstice Journey. But I was afraid. I was afraid that if I returned short of my goal, my spirit to teach would be crushed for years. If that were the case, I would be forced to justify the path of my life.

I had already gained a reputation among many in my community as the best hunter-gatherer and primitive-skills teacher in the Western

Hemisphere. For me to not be able to accomplish a goal that on the surface for someone like me seemed fairly simple would have been unacceptable. I would be devastated. I knew that if I did not make it, my life would go in a different direction. Maybe I would have to quit teaching and do something else—like go into television.

On the winter solstice, the shortest day of the year, I walked into the wilderness in Southern Utah with the goal of living there for six months. The day I left it was snowing. The land seemed to be reaching out to me. It wasn't very cold, and the snow added a beautiful white carpet, too. I hiked through mounting snowdrifts out into the backcountry.

About eight miles from town, I set up a base camp. I built a wickiup that was larger than the one I lived in. The wickiup is basically a low tepee built out of bark and debris with walls that are thick and insulated naturally. The one on BOSS land wasn't large enough to have a fire inside, but I knew that a fire would be key to surviving the subfreezing nights.

The traditional brush structure would serve as home base. It took several days to gather the materials and rough in the frame and another week to layer the bark and debris for the walls. Knowing that the temperature would drop to below zero at night, I built walls that were four feet thick. The insulation consisted of a mishmash of sticks, debris, bark, and pine needles. The end result was perfect. Every night as I lay down to sleep I would giggle because I had such a nice house.

I had previously built wickiups large enough to make a fire inside. They draft really well because they naturally breathe throughout, unlike a tepee. In a tepee, the venting comes up the side under the liner; whereas the entire wickiup is porous. When I built this wickiup, I completely closed the top to hold in the heat. At first, it appeared to draft fine, but after about three weeks, I developed a cough.

The issue was that the air was not circulating fast enough. I did two things to remedy my cough. First, I crawled up to the top and pulled debris aside to open a hole in the top. Though the structure immediately became cooler, the hole eliminated the smoke. However, on some nights, it was so cold—I later learned that temperatures dipped into the minus-ten-degree range—that I sacrificed my lungs for a night. Over time, I adjusted the size of the hole to regulate the temperature.

I also became more careful with the type of wood I burned. I started using more sagebrush and cottonwood versus oak, which was a bit harder on my lungs. In experimenting, I learned that pine wasn't great on my lungs, either, but the oak was the worst.

Every other morning, I would wake up and jump in the partially frozen creek and then come back to dry off in front of a warm fire. On other mornings, to keep me warm enough I would go on a long run and use the heat generated from my body to jump into the creek. After that daily ritual, I would meditate.

The journey had phases. The initial shock was jolting. Realizing it was going to be such a long period of time, I began experiencing feelings of loneliness. I had just enough food that I was not starving, but I was extremely hungry every day in the beginning. I spent a lot of time conserving energy by meditating on the hilltop, doing yoga, and writing and reflecting. Time began to lose its relevance.

During the first month, to stretch out my provisions I ate one meal a day. I spent my waking hours working on arts and crafts. I built several hunting tools and also worked on turning some skins that I had brought into extra clothing. Twice a week, I would fish. Every other day, I would hike ten miles to explore the land and return at night.

The only person I saw in the first month was Dave Neesha, a close friend of mine who had a camp a couple miles away. He was out for forty days. Oddly, knowing that Dave was camped a half mile away with his girlfriend made those feelings worse because I seldom saw him. It would have been easier not knowing someone was near me.

After Dave left, I didn't see anyone for two and a half months.

As I started settling into the land, the feelings of loneliness dissipated. All my senses became heightened. I was flowing with the land. I was hearing more, seeing more, feeling more, and the land was communicating with me. Once that happened, I lost the pangs of loneliness.

At night, I read benign nature textbooks. Normally on a walkabout, I wouldn't bring books, so as to keep the world out. But on this journey, I wanted to have something to keep me company. I didn't want to end up drawing a face on a rock and pouring out my heart to it. I read nature books and guides on tracking and plants. Though the books were not engaging in the least, I ended up reading each one six or seven times. To this day, I still remember nearly every word of text.

I also had a tiny AM/FM radio. I turned it on only every now and then. For some reason, the radio only worked at night and even then not very well. Over fuzzy reception, I would hear callers pouring out their hearts about being lonely. If I had a phone, I might have called in and given them a real earful on being alone.

The entire time I probably said five sentences aloud. I would say to myself things like, "I have to get up. I need to gather food. The leak in the shelter has to be mended." Every once in a while, I would hum or sing just to have the soothing feeling of my voice coming through my throat. But carrying on a dialogue with myself seemed weird. I also never had any desire to talk to a rock, a tree, or a pinecone.

It was wintertime. By then in a native culture, people would have procured and stored their meat. Winter is not an ideal time to gather meat, or anything else, for that matter. I managed to find some onions and spiderwort, which is still available that time of year. Every morning, I would make pine-needle-and-yucca tea. But for the first month, I mostly lived on the rations I had brought.

My body felt mostly normal during this time. Though I lost about ten pounds over the first two months, I still felt strong. I had no noticeable physical issues or changes.

And then wham!

Just beyond the two-month mark, there was a cataclysmic shift. I believe it was a mental trigger than took me down physically. Because I had originally set the goal at three months and then changed it to six, I kept thinking that if I had stuck to three months, I'd be on my way back to enjoy a big meal, see my friends, and listen to music. Instead, somewhat frighteningly, I was three months from returning home and my body felt like it was shutting down.

I had zero strength. My whole body felt like it was full of lead. Because I was pretty lean to begin with, I had become emaciated. If someone saw me, they would've thought I was starving to death. Maybe I was. In the mornings, it took every ounce of energy just to get up and gather food.

The following two to three weeks going into the third month were the lowest point. I was trapped in an out-and-out survival situation. I reached the point where I didn't know if I would pull through. During that time, I limited my movements. I forced myself to conserve every bit of energy I had.

I barely had the strength to file a dart or work on any type of craft. In the afternoon, I would try to hunt. Every once in a while I would spear a rabbit, though only maybe once a week in my depleted state. My energy was so low that I wasn't able to cover the distance I needed to come across animals.

My savior was that my fishing skills were still tuned to the point where I could sense where fish were and hit them with a spear blindfolded. I knew that if I pushed myself to walk down to the creek, I could toss my spear and land a few fish.

On the days that I fished, I would also collect greens for salad. I would return to camp and cook the fish. I would then wilt the salad greens, put the cooked fish on top, and add a little olive oil. That became my second meal of the day.

I was sleeping fifteen to sixteen hours a day. When I woke up in the morning, my body—despite the fact that I was probably down to 140 pounds at that point—felt like it weighed 1,400 pounds. I felt like I had lost at least thirty pounds, which sounds like a lot, but I had seen students lose that much weight in a month. On survival treks, people can lose a pound a day. My arms and legs felt like lead. I have never felt them so heavy. My eyes would open in the morning, but I couldn't even lift my head off the ground. I would roll over, muster every ounce of strength, and push my body up. Once I was in a sitting position, my head would begin to spin.

After an hour, I would finally stand. I would walk down to the creek and collect the food I needed, as virtually all my initial rations were gone. Bending over was draining, but somehow I managed to pick greens. I would steam a large portion—more than most people could eat in one sitting—in a clay pot. Then I would make pine needle tea. At the end of the meal I was so tired, I had to lie down and rest.

For several days, even weeks, I didn't know what was going on with my body. I felt certain that my body was shutting down and I was slowly dying. If I did die, I reasoned that I would not die a fraud. I was either going to complete these six months alone and return to my life as a teacher of primitive-living skills, or die. There was no third option.

After three weeks in that heavily depleted state, I slowly started to feel stronger. As fit as I was entering the wilderness, I now realize that there were leftover toxins that I needed to expunge. Though I ate healthy when I was in society, my diet contained carbs and other sources of energy that hunter-gatherers don't consume on a regular basis.

My metabolism had also slowed down because I was eating less. This made it far more efficient and meant that I didn't have to eat as often. Americans eat constantly, so if we don't eat every four hours, our blood sugar drops and we feel woozy. When you reach the point where I was, even if you don't eat for a couple days, your blood sugar remains stable.

Coming out of the energy crash, I felt like I was in a living situation rather than a survival situation. I began to notice intriguing, positive physical changes.

I could smell things I'd never smelled before and hear things I'd never heard before. Walking through the canyon, I could pick up the scent of an old track or an animal that was out of sight. When I was fishing, I could hear fish cutting through the water. Even though I couldn't see the fish, I could actually tell how big it was and which direction it was heading from the sound.

I felt as if I had finally achieved a hunter-gatherer body. I had very little body fat, and concentrated muscle tissue. I could go for a couple days without eating much at all. I felt super-light, and I could move like a wild animal.

But perhaps the most inexplicable change was that the tartar literally started peeling off my teeth in chunks. Initially, it left a sour taste in my mouth. However, I noticed that when all the tartar had flaked off, I had a sweet, calming taste in my mouth.

I also had a small black hole that looked like a cavity in one of my rear teeth. For months, it had caused occasional pain. But the pain subsided as the tartar flaked off. One day, I looked at the tooth in my compass mirror. There was a stain on the enamel where the hole had been, but the hole was gone. It somehow seemed to have repaired itself. I have no idea how, and the spot never returned.

Later, I told a dentist what had happened. The dentist poked at it and said there was a spot under the enamel but no cavity. I was told that it takes seven years of a dedicated diet of high greens and meat and zero sugar for enamel to repair itself. The natural world had done that in two months.

In April, the fourth month of my Solstice Journey, I crossed over the final threshold of survival. The three-month point—the original goal—had come and gone. My strength had returned, and I knew my six-month goal was feasible. Mentally, physically, and emotionally, I was all in.

I have found that when I first put myself in a new situation, the loneliness is adjusted to the time frame. For example, if someone were to go out for four days in a similar situation, they would probably feel very lonely the first day. On the second day, they would begin to get into a groove. Then on the third day, they would feel great because the journey was nearly over. But because I was going to be out six months, I put that issue on the back burner.

I had never been in a survival situation for that length of time. I also was well aware of the challenge. When I started, I had only a month and a half's supply of food. That meant for several months I would have to live solely off the land. Prior to the Solstice Journey, the longest I had lived exclusively off the land was about forty days.

Admittedly, I was nervous about it, not because I didn't think I could succeed, but because I didn't want to have to return to town and answer questions about why I was back so soon. I had told everyone I knew that I was going out for six months, both to let them know where I was so they wouldn't worry and to mentally align myself to the challenge.

The final three months were almost otherworldly on many levels. Spring had arrived. The weather became noticeably warmer and the earth was coming alive. Part of the feeling emanated from the fact that I had made it through winter. The animals were coming out of their hiding places and returning to their daily routines. The plants were starting to push through the soil. Mustard greens were opening up, and the watercress was thickening.

I started to wander the land, traveling light. I would leave my base for weeks at a time. Some days I would cover thirty to forty miles. Using both the sun and the land features, such as the unique rock formations or large bunches of nettles, I would find my way back.

I could also roam with ease. I didn't have to stay in my shelter. My strength started to return, a welcome feeling after the scare I initially encountered. I felt an enjoyment I hadn't felt in some time. I began to make daily circles through the canyons, and ultimately covered hundreds of miles. I spent time in the Kaiparowits hunting and foraging for food. I'd gather greens, hunt rabbits, and trap squirrels.

The first time I left my home base, I was away for two weeks. I returned for a few days, and then was gone again for six or so weeks.

I roamed the Kaiparowits, which is about 110 square miles. Though roads intersect it, there are ways to parallel the roads for long distances if you know the terrain. The sad truth is that anywhere you are in the United States, you will be within thirty miles of a dirt or paved road. However, if you know the land as well as I know Southern Utah, you can parallel the road. Because of that, I crossed a highway only twice the entire six months.

The stress of keeping time was completely removed. I didn't measure time in any conventional way. It was almost easier to lose track of time and settle into the experience. Counting the days by nicking a log sounds like survival. It's akin to the guy who gets lost counting the days until he is rescued. For me, that would have driven me nuts.

I was able to keep track of a month passing by the moon, as I knew that from one full moon to the next was a month, but I was a week or two off in either direction until spring grew near and told me the time of year. Even in the spring, I would wake up and think, *It's probably late March, possibly even April.* When I ran into a hiker toward the end of the six months—only the second human I had seen in four months—I asked him what day it was. When he told me it was June 6, I started keeping track of the days because I was so close to the end.

By the end of June, I was mostly ready to return. From a practical standpoint, I was scheduled to teach that summer and needed to prepare. I was ready to see friends and return to my life. Yet I was also wondering what would happen to me on a physical, mental, and emotional level if I stayed out for a year.

For one, it would be easier. June 21 to December 21 is the posh time of the year to live off the land. Summer reaps the rewards of spring's work, and the fall is especially productive because that time of year wants to give you a lot before winter comes.

But it was time to return. I had achieved my goal. I felt like I could continue with my life's pursuits honestly.

On June 20, two days before the summer solstice, I walked toward Boulder. I had my supplies in a buckskin bag I had made from skins I brought with me and tanned. When I reached the top of a sandstone peak, I could see Boulder. It was but an hour's walk. But I stopped because it wasn't the twenty-first, and cheating my goal at this point seemed rather ridiculous.

That night I looked out at the lights of the houses and farms dotting the landscape. Civilization was below me, but in no way did I feel above it. I felt humbled. I lay down under my blanket and fell into a gentle sleep.

I awoke just as the sun was coming up and the lights were going off. I packed up and walked to the BOSS grounds, where my wickiup home was. I had a couple handfuls of nonbiodegradable trash in a small bag. It contained a couple cans, some plastic, and dead batteries from the reading lamp I had taken. My six-month footprint on the earth was not even a pinkie print.

When I discarded the trash, I remember thinking I had created more trash than that in a day, and many people create more in an hour. That was very rewarding. For six months, I had lived with the land and improved the land. I harvested from plants in a way that improved their growth. I hunted animals in a way that preserved plants for other animals and sustained the cycle of life. I had emerged with only a small bag of trash.

When I reached the BOSS grounds, I walked into the office. Several staff members were mulling around, typing on their laptops. It was a weird sight. When I had left six months earlier, most people at the school were decidedly antitechnology. Though I had been gone only six months, I was able to immediately see what I considered a radical change that others would later describe as an unnoticeable transition.

Steve Dessinger, a staff member and friend of mine, was sitting at the desk closest to the door. He looked up from his computer and said, "Oh, hi, Matt." And then he looked back down and continued working.

I paused for a moment, swept my eyes around the room. No one else even noticed me. I stepped back outside and went to my wickiup to unpack. As I unloaded, I couldn't help but think that was a rather perfunctory hello to give someone after they had been gone for six months in the wild.

Later that night, everyone threw a party for me. There were hugs all around. Even Steve acted like he was seeing me for the first time. Each person approached me differently about the journey. Some people noted that I looked very skinny, like I had barely survived. There were plenty of jokes as well, particularly about my mental state. One guy said, "Wow, you don't even seem crazy." My friend Mojo pushed me for spiritual answers, which I could not properly convey.

I ate plentifully and healthfully. The only thing I had missed was ice cream. After being back for a couple days, I had some ice cream—which I dearly regretted. It made me not feel so good. Ice cream had always been my one weakness, my one craving when I was on the trail. So my Solstice Journey had another unintended consequence: it got me over ice cream.

Aside from that, along with the incredible physical and mental journey, there were spiritual aspects to living alone for so long, and they have informed the person that I am and become part of my teachings. Few people, if any, spend that much time alone. I had no idea what I would do if I had a heart attack. During that time, I actually felt clearer, and in odd ways even more articulate despite the fact that there was no one to talk to. When I came out, I was much clearer and more concise with everyone I spoke to.

There were benefits small and large to living in the wild as I did. I never got sick. It's impossible to catch a cold or a fever in the wild because there are no people to transmit them. Injuries were also not a concern. Most people get injured when they sit around at a desk, or in a car or airplane, and then explode out of them into a flurry of activity. But in the wilderness, even on the least productive days, I am constantly moving to the point where my body is always prepared for the next task. The muscles stay looser and more nimble.

Once in a while I would do yoga, but I never stretched. I didn't need to. People always say they feel old after they exert themselves. But the fact is that they merely *feel* old because they sit around, then they are active, and then they sit around again. There is a reason why Australian aboriginal hunter-gatherers, old men who are eighty, are chucking a spear the day before they die. They are physically active and remain uninjured their whole lives because they maintain a steadiness and an evenness to the physical tasks in their lives.

Those six months were the happiest time of my life—despite the fact that I almost died after the first three months. I felt so tuned in to and in touch with everything. Not having a backpack of modern supplies makes you inherently more dependent on nature. This is something we can apply to our lives. When we find ourselves in an unfamiliar situation or in a new city, we tend to go with what is most familiar, rather than reaching out and touching that new situation in a physical or even psychological way.

I found that I tapped into parts of me that I didn't know were there, and lessons that could be applied to my life sprang from them. For example, one day I would notice the colors of the leaves on the cottonwood trees. The next day I noticed the colors were duller, even though the sun and clouds were the same. This puzzled me. What had happened to my perceptions? The next day, the color returned.

What that told me was that nobody perceives things the same way. The way you see red may be different from the way I see red. This helped me to reserve judgments against people, because I came to understand that they might have different perceptions than me.

Looking back, the Solstice Journey was the single most important event in my life. For those six months I had a goal that was simple and to the point: to not come into a town or civilization. I knew that returning even for a day would be a huge distraction.

People are always interested in what I packed and if I had any human contact. Some are looking for inspiration through my journey. Others want to poke holes in my experience. But that does not bother me. The journey was about maintaining as much of a nature connection as humanly possible.

That time showed me, in some way, how nature thinks. I got into her mind in a way that I think very few humans can.

EPIC SURVIVAL RULE #5:

LEARN FROM OTHERS

I am no longer afraid to fail, because failure no longer means death. If I fail, I can learn from it and use that in my teaching. I am chasing survival and catching it.

The things I am learning from devoting my life to the wilderness cannot be picked up from reading a book. The only way to learn them is to live fully in the wilderness. As I do that, I feel like I will be better able to show this real world to others who live in the man-made, artificial world, and that will make their lives fuller.

After experiencing the wilderness through climbing and running, I have over time learned the necessary skills to live in the wild: tracking animals, crafting stone and bone tools, building shelters, and making fire with friction. I am now applying those skills to the most extreme survival situations and turning them into living situations.

I hope I'm far enough along that I can translate what I know to others who don't live in the wilderness. In everyday life, it is easy to get caught up in yourself. The land does not allow that. It demands that you look around and observe. Now that I am able to completely observe the physical aspects of the land itself, I am doing that in my own relationships. When I am with people, I find myself thinking, Okay, what does my friend need today? What do I need to do to be better for this person?

The lessons of the wilderness are feeling more and more applicable to everyday life outside the wilderness. Like other people who become proficient in their fields and help others, I feel like I am entering a new phase of my life from which others can benefit.

Chapter Seventeen

SURVIVAL TOOLS

H unting tools are essential to the survivalist. They are necessary to gather food to keep you nourished while living in the wilderness. But they also become a part of you and one of the defining aspects of your respect for the land.

The first thing to understand about any survival tool—whether it is a hunting tool, blanket, piece of clothing, or fire to keep you warm—is that you must establish a connection to the earth to have a pure experience with it.

At first, it may seem foolish to trade out your lighter for a friction fire kit or your sleeping bag for a wool blanket, or take your knife out of its sheath and make a leather one—or better yet, leave your knife behind and learn how to use tools made from stone. Maybe you can even create your own clothing.

In the beginning, you may suffer, but you will never experience the layers of survival if you don't try. The layers have always existed. A hunter-gatherer leaving the village to go on a vision quest could be the equivalent of a backpacker leaving all the last gadgets behind to establish a deeper connection in the woods. This is not to say that people should not go out on the land with their backpacks full of gear, if that is what they love to do. Natural tools may be harder, but they also may be more fulfilling.

There is a complex simplicity involved in the making of all hunting tools. Every region of the globe has had many generations of tool crafts-men to evolve the hunting tools that work best in that landscape. The list of primitive tools is extensive. They include the spear, the blowgun, the bolo, the sling, the throwing club, the rabbit stick, the bow, the boomer-ang, and most historically, the atlatl.

The atlatl and the bow are similar. Both project a "missile" toward the target. The atlatl is generally smaller and the rope attached to the dart makes it ideal for fishing, while the bow (and arrow) is better suited for shooting game over distance. The boomerang is most effective on open terrain.

The most versatile of all hunting tools is the atlatl. An atlatl is a de-vice for throwing a spear that gives it greater velocity. The atlatl creates an extra joint to the arm. It throws like a cross between a javelin and an elongated arrow.

Much of the information I have learned about tools and hunting has come from trial and error—an atlatl that backfires can maim you if you're not careful—and from studying what others have done. All teach-ers should be students, because there are no absolutes.

Having the right balance of tools allows you to draw energy from the land and gather food in extreme circumstances. Developing a sense of the land is crucial to being able to build and ultimately use the tools properly. These hunting tools require a clear sense of mind and an awareness of animals to be effective, and they must become an extension of the hunter. Experiencing that connection is powerful. It can also determine whether you live or die in the wilderness.

The first thing to understand about primitive hunting tools is that you can't just step into a survival situation and make a primitive bow, atlatl, or boomerang or any other tool and expect to hunt effectively with them if you have never used them before.

It takes a week to dial in to that tool. You can't speed up the process and be successful. The atlatl is the only tool I have been able to make and use on the fly to catch small game. But making a bow is a process. I've never known anybody who has made a bow and caught something in the same week. The bow doesn't have that quality.

To make an effective bow, you must cure the bow. You can speed up that process by doing it over a fire for a couple days, but it still must be cured. To make the arrows, you essentially have to follow the same process. However, if they are rapid cured, the spines will not be what you are used to. So even in a survival situation, someone who has a great deal of skill and experience will need four days minimum to get something dialed in—to say nothing of learning about the land where you need to use it.

While the boomerang works far better when it is cured, it is possible to make one that is "green" and use it. The boomerang is inherently easier because it can be used to go after a flock of animals, thereby lessening the skill needed to make a kill.

For the average person in a survival situation who has never used any hunting tools, the boomerang, club, and spear are the best options. Those would be followed by traps. In a pure survival situation, the simplest tools are the more effective for the average person.

The fact is that it takes years of dedication to know how to create and use a tool properly. The defining quality to hunting tools created from the land is that they are very specific to that landscape. For instance, a long bow works well in one environment while it doesn't work great in another. Same with a short bow. Hunting boomerangs are the same, depending on the type of game you are hunting.

As the archaeological records of mankind show, the atlatl has been used longer than any other hunting tool. It is responsible for us being alive. Ancient European cultures used it for ten thousand years before they decided to start using a bow.

The atlatl has gained a reputation for its ability to kill woolly mammoths. But the fact is that the atlatl was not used just to kill woolly mammoths and then put away once the mammoths were extinct. It was used as the primary and sole hunting tool up to a thousand years ago when this continent had the exact same game as it does now.

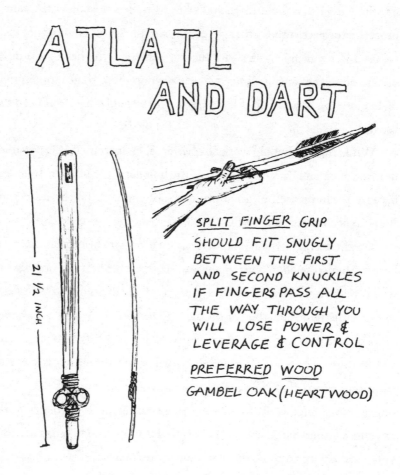

ATLATL AND DART

21 1/2 INCH

SPLIT FINGER GRIP
SHOULD FIT SNUGLY
BETWEEN THE FIRST
AND SECOND KNUCKLES
IF FINGERS PASS ALL
THE WAY THROUGH YOU
WILL LOSE POWER &
LEVERAGE & CONTROL

PREFERRED WOOD
GAMBEL OAK (HEARTWOOD)

Many people think that once the bow came into existence, it was the superior tool and people simply stopped using the atlatl. That's far from the case. Once the bow was employed, there continued to be an overlap. Even in our records of when the bow was first starting to be used and understood, that same tribe often preferred the atlatl. The bow for a long time was considered a tinker toy or a very specialized shooting weapon.

Though the bow and atlatl appear to be similar, there are differences. The advantage to a bow is that there is very little movement involved in its use. The lack of movement from the hunter to flag an animal when he takes a shot allows him to slowly draw back the bow. When he releases the arrow, he is not waving his hand toward the target, as he would be while throwing an atlatl. Actually, in my opinion that's the only modern advantage to a bow over an atlatl. In primitive times, using a bow also allowed people to carry a quiver of arrows for going to war, as they obviously could not retrieve a dart in battle.

A D bow is about sixty inches long. In Southern Utah, the best materials for making the bow are Gambel oak, Russian olive, and serviceberry. However, I prefer other materials, such as juniper and ash. Those make a better bow with sinew. I have since learned that maple also makes an effective D bow.

The bow's string is a two-ply twist sinew that is loosened with saliva, finger rolled on a log, and then stretched between two trees. The tip of the arrow varies. A long tip is used for a quick kill, whereas a stone tip is used for a stun.

Though bows are very much in favor, I prefer the atlatl to the bow because it is very Zen. People talk about archery as being Zen, but once you throw an atlatl accurately there is nothing that will create that kind of focus and meditation. Using a bow and arrow has a Zen feeling because you are controlling your muscles to stabilize the shot. Once you reach that point, you aim and let go. But with the atlatl, you put your entire body into it. That teaches harmony and helps you reach it.

BOW AND ARROWS

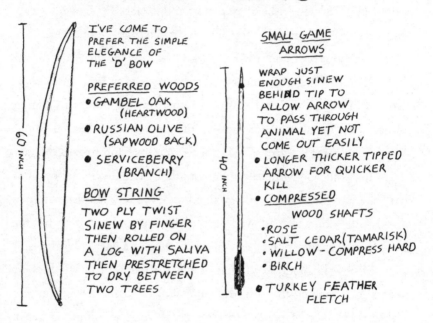

I'VE COME TO PREFER THE SIMPLE ELEGANCE OF THE 'D' BOW

PREFERRED WOODS
• GAMBEL OAK (HEARTWOOD)
• RUSSIAN OLIVE (SAPWOOD BACK)
• SERVICEBERRY (BRANCH)

BOW STRING
TWO PLY TWIST SINEW BY FINGER THEN ROLLED ON A LOG WITH SALIVA THEN PRESTRETCHED TO DRY BETWEEN TWO TREES

SMALL GAME ARROWS

WRAP JUST ENOUGH SINEW BEHIND TIP TO ALLOW ARROW TO PASS THROUGH ANIMAL YET NOT COME OUT EASILY
• LONGER THICKER TIPPED ARROW FOR QUICKER KILL
• COMPRESSED WOOD SHAFTS
• ROSE
• SALT CEDAR (TAMARISK)
• WILLOW - COMPRESS HARD
• BIRCH
• TURKEY FEATHER FLETCH

60 INCH

40 INCH

Archaeological records and studies show that nothing uses more explosive brainpower than throwing an object, even in our frenetic technologically driven society. For that split second, throwing an object requires the greatest amount of concentration a human can muster, without exception.

The atlatl becomes an extension of your arm, the tool an extra joint in your body. The dart itself becomes spiritually and physically connected to you in some way. As you practice throwing it, you start to "know" the dart. Throwing a rock would be very Zen if you kept the same rock all the time. But people don't do that. They go down to a creek and start chucking rocks. The first rock is no different from the sixth. However, if you retrieved the first rock every time you threw it, you would become attached to that rock, and it would become an extension of your hand and your arm.

The dart becomes a part of you because after you throw it, you collect it. Eventually, you start developing a relationship with that particular dart. You understand the way it flies, how it reacts to natural obstacles, such as wind and rain. If you stick with that same dart long enough, you will know it so well that you will be hitting dimes out of the air—or at least a pinecone from ten yards.

Darts are also more durable, so they last longer than arrows, which is helpful while practicing and hunting. The dart can be crafted from any type of stiff, thin piece of wood that has enough flex to give it the proper flight. Cane is a great source for that, as are willows and birch.

Depending on the continent, the atlatl itself has many different shapes and designs. The Eskimos used a one-finger design that had a very neutral hand position. They primarily used it for hunting on the ocean shores, meaning that they just needed a straight down flick to hit a fish or a seal. Unlike the bow, which is inevitably bent to a slight imperfection, the atlatl can be made to be ergonomically perfect.

I made my first atlatl shortly after I moved to Boulder. It was based on a drawing I saw in Larry Dean Olsen's book *Outdoor Survival Skills*. I worked very hard to perfect the design and spent hours practicing with it, but I could never get that atlatl to fly correctly. Based on my limited trial and error and the impatience of youth, I dismissed the tool altogether. For years, I considered the atlatl to be outdated, impractical, and even silly.

Several years passed before I tried again. I read more books about the use of the atlatl in primitive times, and I studied different models that had evolved over centuries. I came to the conclusion that I had made a crappy model. It wasn't worthy of flight, and certainly not of leading a hunt. I decided to try again.

Using a picture of an atlatl from the Southern Utah area, where I felt most connected to the land, I constructed a different atlatl. I reasoned that what worked thousands of years ago in my area should work now. Again, I was met with frustration. Though this atlatl worked better, I was not satisfied with its flight, nor did I feel any connection to the tool.

I decided to do some homework. I gathered every book and photo of all models of atlatl from aboriginal times—models from my area to the Great Lakes region to Arizona. I built several replicas of each one. In all, I ended up making about thirty atlatls. All of them were beautifully finished and delicately constructed. I then began throwing those atlatls until I decided which one fit me the best. The winner ended up being a slight hybrid to one that had been used in my area a century earlier, built in what is called the Basketmaker style.

The Basketmaker atlatl uses a six-to-eight-foot dart with a four-pronged tip. The dart has what appears to be a traditional arrow fletching on it, though in fact it is more streamlined. The grip on the Basketmaker is designed for either a baseball grip or a split-finger grip. I feel like I have more control with the split-finger grip, which fits like rings between the first and second knuckles. With the proper split-finger grip on a twenty-one-inch atlatl, I can throw a seven-foot dart farther than a hundred and fifty yards.

The models most commonly used in this area have the exact same type of arm that I use, but they have stiff loops made of sinew lines that make them rigid. The lines are covered in buckskin. Since experimenting with those, I have found a few ancient models that have handles carved from a sheep's horn. I have also used the sheep's horn specimen as a model for one I made out of hardwood.

The atlatl tip has an advantage over an arrow tip because you can place a fair bit of weight on it without making it top heavy. Hunters have used all kinds of material from splayed-out tips for fishing to barbed tips for bigger fish and small game to stone tips for big game. There were even a couple archaeological specimens with fist-size bludgeoning balls for stunning larger animals or knocking out smaller animals.

The challenge in hunting with the atlatl is how to conceal your body movement, or how to project where the animal will be by the time the dart reaches it. As an atlatl hunter who wants to put meat on the table, you are always positioning the shot in such a way that you are simultaneously evaluating the present and several seconds in the future.

Here's an example. If a rabbit ran out in front of me and stopped out in the middle of an open area, I would not take a shot at it because the rabbit would jump the shot. I would pause and follow one of two options. I would first try to figure out a way to position myself so the shot is concealed. If that were not possible, I would slowly push the rabbit into the brush so it felt more protected. Then I would take the shot into the brush.

Every hunt demands a different throw. If you were hunting game that required extra power or distance, you would develop a throw that had more length. If you were only fishing with the atlatl where you needed more control at short range, you would make a shorter throw.

There are other differences, notably in how the darts are retrieved. When fishing with a bow and arrow, I need to tie a string to the arrow or dive into the water to retrieve it after each shot. With the atlatl, I can fish in the winter without getting my feet wet. I can throw the dart from the bank, hit the fish, grab the dart, and if I have a barb on there, pull the dart back out.

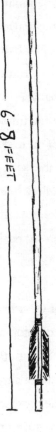

6 – 8 FEET

FISHING DART FOR SMALL TO MEDIUM SIZE FISH

CANE SECTION FORESHAFT WITH FOUR COMPRESSED FIREHARDENED HARDWOOD SPIKES WEDGED IN & SPLAYED OUT

I'VE CAUGHT FISH UP TO 15 INCH. WITH THIS BUT CARE NEEDS TO BE TAKEN WITH RETRIEVAL OF LARGER FISH – BETTER TO USE HARPOON TIP

PREFERRED WOODS
- SERVICEBERRY
- RUSSIAN OLIVE
- WILLOW
- BIRCH
- CANE

Another reason the atlatl is better is because in deep water sometimes I will hit fish that are eight feet down. That is possible because the dart is so long that it travels fairly straight. But when an arrow shoots off a bow, it inevitably has the archer's paradox, which is flex. To a degree, both the atlatl and the arrow experience the same thing.

The archer's paradox says that the arrow is streamlined in flight, so in order to travel well, it actually has to flex around the handle of the bow and then straighten out. When an arrow hits the water, it is still bending. Depending on the direction the arrow bends, it will kick in that same direction in the water. But a dart off the atlatl, even though it somewhat does the same thing, is so long that it stabilizes and keeps driving straight through the water.

With both the arrow and the dart, there is also refraction that takes a skilled fisherman a while to learn. When the arrow (or the dart, to a lesser degree) is in the water, it's also going to appear to bend, so you need to aim below your target. When light passes from the water to the air, it bends, and that causes the fish to look like it is in a slightly different place than it actually is.

There are other distinctions between the bow and the atlatl. With the bow, even when hunting small game, you rarely make the kill shot with one arrow. This means you need at least two arrows to be an ethical hunter and likely three or four unless you have uncanny precision. With the first arrow you are hoping to hit the game in the head, but if you can't because of timing or distance, you try to hit it in the gut or the leg to disable it. Nevertheless, the animal is going to keep running. You have to grab that other arrow and make another shot. You want a dead animal; you don't want a wounded animal that gets away. With one arrow, that could easily happen; whereas, if you use an atlatl and put a dart in a small animal, the animal is not going anywhere.

There is also a distance difference between an arrow and a dart.

When using a bow, the lighter the arrow, the farther it flies. With an atlatl, what makes the dart fly farther is the thrower. Modern carbon arrows are especially lightweight, as are aluminum ones. But a proper atlatl thrower can rocket a dart past an average user of the bow who flings a lightweight arrow.

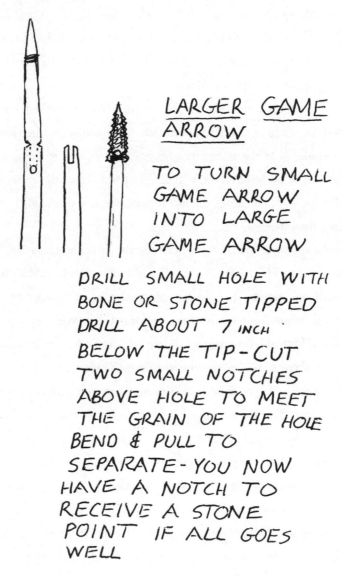

LARGER GAME
ARROW

TO TURN SMALL
GAME ARROW
INTO LARGE
GAME ARROW

DRILL SMALL HOLE WITH
BONE OR STONE TIPPED
DRILL ABOUT 7 INCH
BELOW THE TIP - CUT
TWO SMALL NOTCHES
ABOVE HOLE TO MEET
THE GRAIN OF THE HOLE
BEND & PULL TO
SEPARATE - YOU NOW
HAVE A NOTCH TO
RECEIVE A STONE
POINT IF ALL GOES
WELL

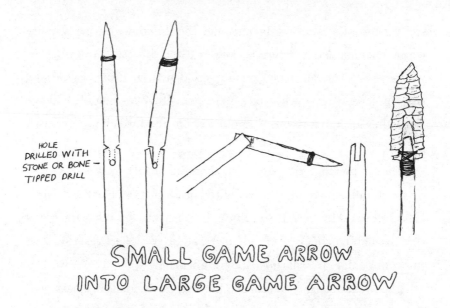

HOLE DRILLED WITH STONE OR BONE-TIPPED DRILL

SMALL GAME ARROW INTO LARGE GAME ARROW

Years ago, I was chucking darts way out across the BOSS land. Steve Dessinger, who was a program director at BOSS and eventually became its owner, was excited about how far I could throw. He asked me to try one of his lightweight aluminum darts. So I pulled back and let it go. The dart sailed way farther than either one of us anticipated. We stood there and watched as it just kept going and going. It must have traveled well over two hundred yards before it dropped into a clump of trees. We searched for an hour, but we never did find it. (Sorry, Steve.)

Watching a boomerang soar through the air and return to its thrower brings out the kid on the beach in all of us. But a hunting boomerang is far more than a toy. It is the heaviest of the three primitive tools and can inflict the most harm per throw.

A hunting boomerang is basically what it sounds like: an oversize version of a returning boomerang. The difference is that a hunting boomerang doesn't return because it is too thick and heavy. Today, specialized boomerangs are generally molded from inexpensive fibers and polycarbonate, which makes them virtually indestructible. On average, they are about twenty-eight inches long and weigh just over two pounds.

The boomerang originated with aboriginal tribes in Australia. It is best used in open terrain with very little brush or trees like the Australian Outback. However, if you throw it overhand, the boomerang can slice through tall, thick grass and brush and be very effective. A sidearm throw through grass will choke it off immediately.

The highest-quality man-made hunting boomerang is a heat-bent boomerang. An explorer named Paul Campbell taught me about these. Campbell learned the perfect construction by observing the indigenous people of California.

The construction of this boomerang is very deliberate. I take a fresh-cut sapling and heat it gently over a fire for a half hour. This allows the wood to be bent into a boomerang shape. I bend it by placing it between two trees and creating a vise. I allow the resulting boomerang to dry for two weeks before using it.

Boomerangs are best for hunting birds, particularly large ground birds. The bigger the flocks, the better. Ideally, you can catch flocks out in the open and then let the boomerang go. You hope to catch them on the ground, but even if they cluster and fly away, you can throw at that cluster. The boomerang has such a wide circumference that you will likely clip one or two of them. That was how the aborigines used boomerangs. The Hopi and Anasazi Indians in Arizona hunted quail in a similar fashion. I've also seen hunters make a clean harvest of a wild turkey with a single shot to the neck.

HUNTING
BOOMERANG

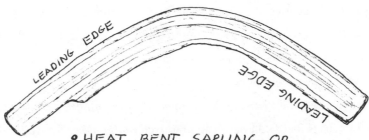

LEADING EDGE

LEADING EDGE

- HEAT BENT SAPLING OR STRAIGHT GRAINED BRANCH LET DRY THEN SHAPE
- BLUNT EDGES FOR DURABILITY NO KNOTS
- UNCUT LEADING EDGES ALSO FOR DURABILITY

PREFERRED WOODS

- OAK
- CATCLAW
- MESQUITE

Boomerangs were also used when the rabbit population swelled in Australia—hence the nickname "rabbit stick." The men would start chasing rabbits until they got a close shot and then whip the boomerang into the pack. It is also possible to chase down a single rabbit until it tires and take it with one shot.

Because of the complexity and mastery required for each tool, I stopped using the boomerang. I was feeling too spread out among the three and needed to maintain my focus on the atlatl. The closeness to the hunting tool is paramount to its success as a protein gatherer.

Whatever hunting tool a survivalist chooses must be durable enough to deliver on a hunt. But more important, it must connect with your spirit and be an extension of you. You must enjoy constructing the tool and practicing with it to develop the necessary accuracy to make you not just a hunter but also a harvester.

For me, the atlatl has also led to meeting like-minded people who want to master its use. While doing research on the atlatl, I discovered that an organization called the World Atlatl Association hosts competitions across the United States and in Europe. There are people so serious about this skill that they drive around the United States from one competition to the next trying to increase their scores. This is a pure labor of love, as there is no prize money. When I became proficient at the atlatl, I decided to test my skills in competition.

Every competition runs by the same rules. A participant throws five darts from a distance of fifteen meters, followed by another five darts from twenty meters, at a standard archery-style bull's-eye. Scoring is straightforward. The rings are worth one to ten points, a bull's-eye being ten. If you miss the target's rings, you get no points. The highest point total wins.

I have competed against and beaten the world champion on several occasions. We will usually throw five bull's-eyes at fifteen meters. At twenty meters, things are more challenging, especially if there is wind. From there, hitting a few nines is not uncommon. The trouble comes when a dart drops on you and gives you a six.

My highest score in competition was a ninety-three out of a hundred, which put me fifth in the world overall. I had actually scored a ninety-six in practice. The highest-ever score in competition is a ninety-eight. All four people ahead of me compete regularly, giving them a chance to raise their scores. But for me, the competitions were not about recognition but rather about a chance to bring me closer to my primary hunting tool.

Chapter Eighteen

HUNTING FOR A CONNECTION TO OUR FOOD

Hunting is a very sacred act that cannot be undertaken lightly. The ego must always stay in check, because the hunter is taking a life. Hunting tools should feel like an extension of the hunter and connect with his spirit. A person must also know the territory to be an effective and respectful hunter. He owes it to the land and to those creatures that have inhabited it longer than him.

To most people in our society, the act of hunting conjures up images of a guy wearing camouflage and using a rifle to shoot a wild animal. Very few modern-day hunters use primitive hunting tools. I am not opposed to people who hunt with guns. Unfortunately, when you hunt only with a gun, you have not earned it. If someone has never hunted with a primitive tool or never gone out and spent a lot of time with the animal he is hunting, then he has not developed the proper respect for the animal or for his power to take that animal's life. If a hunter picks up a rifle and pops a deer, that person has skipped a bunch of steps.

I haven't necessarily taken the easy road, but I have taken what I believe is an ethical road. I have never just picked up a hunting tool, be it a stone ax or an atlatl, and killed an animal. I have learned to make that tool, made multiple versions, and worked to understand how to use them. I have then lived in the animal's environment.

I have killed only two animals with a firearm. Friends of mine who owned expansive alfalfa fields were given five depredation tags. The authorities hand out depredation tags to people who have fields in an effort to control animal population. They asked me to help.

I, too, lived on land that contained alfalfa fields. I was given a gun during that late fall. I practiced with the gun for a week and could hit a baseball-size target at forty yards without fail.

On the day of the hunt, I stalked within twenty yards of a deer. I lined up a head shot, which would allow me a clean hide when I skinned it. That day, I shot two deer under the depredation tag guidelines.

My girlfriend Kirsten was living with me for the winter. The real upside to killing the two deer was that I would be able to preserve and store the meat for the winter for us. Had I not killed the deer, we would have gone through a long period of eating rice and beans. Together, the two of us had only three hundred dollars to make it through the winter until we started teaching again.

That was the first time I truly understood that when you are looking after a loved one, you excuse yourself for cheating the natural process to provide for them. There are many people who need to supply food, and for them the gun offers a quick and sometimes the only answer.

I don't feel that I am better than the person who goes deer hunting with a gun once or twice a year. But I believe that understanding the process of what I have created opens up my mind to different ways and techniques. Today, when I pick up a knife, it's not the same as most people picking up a knife. I appreciate the tool thoroughly—probably more so than somebody picking it up for the first or second time—but it has taken me time to go through the process.

When I hunt with an atlatl, bow, or boomerang, I feel more responsible than the time I hunted with a gun. That is purely a personal feeling, and in no way do I want to sound sanctimonious. However, hunting with my hands continues to unearth conflicted feelings.

It is hard to describe the feeling of killing something with your hands. The killing process feels horrible, yet is has a positive side. I have just killed something so I therefore feel bad about ending the life of a living thing. But I know that I will feel good in a moment when I get the nutrition I need. So I am thankful for the blessing of food from the land, but at the same time, I cannot help but show a certain amount of sorrow that I had to kill for it. That conflict will never abate.

Hunting animals respectfully is an experience in self-reliance and connections. Once in Hawaii this meant building an atlatl and hunting for a wild goat. I had observed and taken in the land for ten days. I watched the goats make their way through a streambed in the canyon. While studying the goats, I noticed that they could effortlessly navigate the hillside that was almost ninety-degree angles, but in the cobble-stones of the creek bed, they had trouble finding their footing. I decided the easiest method was to run down a goat.

Grasping the atlatl in one hand, I charged the goat over the cobble-stones. I chased it for a half mile, at a full sprint. I could see it was struggling. It kept slipping and sliding off the rocks. Eventually, I caught up to the goat, straddled it, grabbed its horns and pulled them back, and slit its throat.

It was traumatic. It's always emotionally tough to hold an animal in your arms and kill it. Hunters are removed when they trap an animal or shoot it—shoppers in supermarkets far more so. Despite how much I needed that meat, I literally cried my eyes out. I thanked that animal for leaving its body behind and prayed that its spirit would live on. It never gets any easier to kill something that is like me. I always accept an animal's meat as a great gift.

Every time I kill an animal—be it with a weapon, with a trap, or by hand—I say a prayer.

It's usually along the lines of "wherever your spirit may go, I hope it is a great place." But when hunting with students, I never push or enforce any spiritual concepts with them. Each person must find their own spiritual balance.

Trapping is important to surviving in the wilderness. It takes far less energy and effort than hunting, so if you are depleted, it is an easy way to get meat. In survival situations, I have trapped many different types of animals with many different traps.

I primarily teach students how to trap squirrels, because they are easy to lure. If a student has trouble, I step them down one more notch and teach them to trap mice. The easiest way to trap squirrels is to make what is called a Paiute-style deadfall. The trap is basically two slabs of rock. One of them is lifted upward with two upright sticks. Another stick is used to set a trigger mechanism that is usually baited. If you have no bait, the trap can be set over a hole. When the squirrel trips the trigger, the rock falls on its head and kills it instantly.

If students want to learn how to trap bigger animals and do it in a legal manner, I will teach them how to build more challenging traps, such as snares. I teach them to use the snares on squirrels so they can later cross over to larger animals. Snares are looped cords or rope that tighten when an animal walks through them. Generally, they are draped over some brush close to the ground. Sometimes sticks are placed in the ground to guide the animal into the snare. The snare is attached to a bent branch that snaps when the animal walks into the snare.

I also teach students to skin and dress the squirrel. A squirrel's anatomy is the same as a deer's, which is similar to ours except for the stomachs. When students and I trap a squirrel, I will demonstrate the cleaning and preparation process once very slowly. I expect them to be able to do it all the way through the next time.

I walk them through the anatomy, and I show in detail how to skin and clean the animal, showing them what to leave in and what to remove. You want to eat a certain amount of organs for the nutrients and fat content. The best for this are the heart, lungs, and kidneys. Others, like the gallbladder, stomach, and spleen, can be removed.

Because the animal has given his life for my benefit, I use all parts of the animal for something, except for the stomach. I discard the stomach because as bait it won't draw the type of animals I want to trap. Any leftover bits of flesh or meat after the cleaning process can be used for bait. Animal fat is ideal for cooking greens. Fat also supplements the lean meat to keep our digestive systems functioning properly. The bones can be filed into small tools, such as needles for sewing. The hides can be tanned and made into skins for a variety of uses. Even the skin of a squirrel can be made into a pouch.

To cook the squirrel meat, I lay it out flat and grill it. I try to carry rock salt to sprinkle on the meat.

In any hunt, I show my students how to read the health of the animal by looking at the fur. If the fur is dull or tattered, then they should probably leave the animal alone. Likewise if the animal is stumbling to the left then it probably has plague or is sick. Then when we catch the animal, I teach them how to look at the health of the liver and the glands. That will tell the overall health of the animal, and that also determines how much you should cook it. If the animal is extremely healthy, you could theoretically eat it raw. But if it has any signs of bad health or disease, then you want to make sure it is thoroughly cooked. In the worst case, if its liver is half gone, it has pussy glands, or it has worms in its meat, you might consider not eating it at all.

Students are less squeamish about the process than would seem to be the case. With small animals, they focus on what needs to be done. However, I find that with large animals, such as a deer we find dead by the roadside, students have trouble. The smell of skinning and preparing a deer is overpowering and weakens me even after all these years. The process itself is also physically taxing, as it takes over a day just to remove the skin.

Many people don't realize that we digest smell. It doesn't just go into our nose; it goes into our stomach and through our digestive system. When you have a bowel movement, that smell you have been inhaling comes out of you. Interestingly, this happens in cities as well—which should tell us something about how important our air is.

In our daily lives, we are often far removed from how much we destroy and kill animals and their environments. But we all do, maybe not with our own hands, but with our actions. As humans, we have to figure out what it takes to live, but also how to create balance and harmony in our environment to ensure long-term survival for all species. All of this makes hunting a major topic of discussion on the survival courses I teach.

Many of the people who come out on courses do so because they are looking for a gentler way of living with the earth. Some of these—I'd say about one in ten—extend that gentleness to veganism. Everybody on the course is respectful of one another's position. But what is interesting is that almost without exception, the vegans usually end up becoming re-solved during the course to the fact that it is better for them to eat meat, both for their health and for the ecological balance of the land.

I've illustrated this to students in many different areas and climates. When we enter an area, we can easily identify greens, roots, and berries that will make it possible for all of us to get enough nutrition and live healthfully for a couple months. But those are the prized foods for all mammals in the area, because they contain starches and sugars that sustain living things.

Once a student and I debated what respect for animals meant. He was a vegetarian, which is a very respectful way to live. He was on a course with me, living off the land in one very remote spot for ten days. The three-mile canyon was very rich in berries, which he would collect and eat each day. There was a huge squirrel population feeding off the same bushes. After a few days, I noticed that the berries were nearly depleted. I explained to him that we were starving the community of squirrels because we were not taking a balanced approach to the land. My reasoning was that it was better if we trapped a few squirrels, and then they would have less competition for the berries, thus creating a more balanced ecosystem. This is part of striking a natural balance in the environment to ensure survival for all.

Every environment is different. But the fact is that if a person finds the magic berry bush, there might be only one magic berry bush in a few-miles radius. A lot of animals are relying on that berry bush. If you just say, "I'm a vegan so I'm only going to eat berries," then you have just messed with a lot of animals.

The bottom line is that we should always strive to find the balance to live within a particular region. This is incumbent upon us, because as humans we have the intellect to look at the land and say: "There are only x number of wild onions, x number of berries, x number of cattails. I have to take into account a certain balance that will help everything out here. That means I will have to kill some animals, which will also help the cycle of all the animals competing for that bush."

That is how nature's balance is not only maintained but also improved.

Many vegans on courses examine the balance of the land and decide its natural order makes sense. While they end up eating an animal, both for the nutrition and the need to keep the land in balance, they almost always do not know what they will do when they return home. They are then forced to take that experience back home and figure out what to do with it. Some return to veganism, while others do start eating meat. They realize their body felt better, or they feel like they can buy free-range meat and create a better harmony that way. But no vegan goes home and starts eating hamburgers.

This becomes a topic of discussion. A mammal is very much like our flesh and blood. It has a family, raises its young, and acts a little bit like us instinctively. Killing something similar to us is difficult. Inevitably, the discussion turns to where the line is. Chickens are okay, and so are pigs—unless the person had a pet pig as a child. Dogs are not.

But when my students see hunting in the wild firsthand the talk always returns to the same place. It shows how far removed we are from our food supply. Most everyone has seen cattle grazing in a pasture, but few people think of that when they buy a Saran-wrapped tenderloin at the supermarket. The fact is, it is very hard to see the true cost of our food unless we watch it being separated from its environment.

Hunter-gatherers see this up close. It is very simple. They look at the land. They see what nature is telling them, and they look for a balance to make it better while still surviving.

Every time a hunter-gatherer walks into an area, he must have an awareness of the food supply. He must find the ability to live off the land. If he doesn't, he will die. Having the ability to live off the land makes you a seer.

Probably the deepest I delved into these issues with a student was on my journey with Jesse Perry from Flagstaff to Boulder. Jesse had a visceral reaction to hunting animals and preparing them to eat. When I killed the first rabbit for food, Jesse broke down in tears. Later, when I wanted to remove bird eggs from a nest, he adamantly refused to help. But what he saw caused him to question how we live in modern society.

After our journey, Jesse wrote a diary. He talked about being disgusted with the hypocrisies in himself that surfaced, the same hypocrisies that even the most caring consumer encounters every day. He crystallized the issue by writing that he wondered why he had trouble taking eggs from a bird's nest when he was famished in the wild, yet he had no problem eating an omelet at Denny's.

The fact is, we are so far removed from these issues in our culture that we must question ourselves at some point. Farm-to-table is a large movement, as is the humane treatment of animals we eat—all of which go through slaughterhouses. We want prime-grade, humanely raised animals. So we wrestle with this dilemma. We write books about these questions. We try to figure it out. But we ultimately cannot figure it out fully for the simple fact that most people are too far removed from the land.

Chapter Nineteen

TEACHING "THRIVAL"

The life-or-death element of being a true survivalist is not the daring. It's not being dropped by a helicopter into a jungle and trying to find your way out, like a TV show. It's not about the risk of doing something extreme like climbing the face of an icy glacier. It is more about the unknown, like walking through a vast desert without a map or compass and without knowing where your next water source is. Most consider that risk, but I consider it trust.

That separates me from other teachers. There is no trust from many teachers. For them, it's all about numbers. A lot of teachers must have a calculated formula that they can apply some kind of mathematical equation to. They need to know in terms of what the typical body goes through to get from point A to point B: "Am I going to compromise my health by not eating for a day?" They want it to be a scientific formula. And if the conclusion of science is that it is going to somehow be harmful, they won't do it.

I regard myself as a scientist, too. Science doesn't have all the answers, and neither do I. I'm seeking to find out if the facts are true. The point is that when someone thinks he is the final word on something, that's when you get yourself in trouble. This applies even more so in nature.

For most of human evolution, survival has been a way of life. Hunter-gatherers grew up learning nothing but survival. They had no choice. Nature is constantly changing and adapting. The tides may recede only to rise again, but no two tides rise alike. Survival skills must be applied in a fluid way that dovetails with the part of the earth you are on, and you must learn that area's rhythm, timing, and changes to be able to live there.

One truism I have learned from teaching survival courses and living in countless survival situations is that survival manuals offer only a vague outline of the necessary skills needed to live in the wilderness. Even worse, they can often give people a false sense of security that they have the means and tools to live off the land when, in fact, they do not. Survival is not a certificate you can pick up at a conference.

The fact of the matter is that if someone is relying solely on a survival manual, I guarantee they will not be able to live off the land for very long. A survival manual will force them to build things they don't need. Living off the land requires an intellect and understanding to ask questions. Where should a shelter go? What time of day should it be built? What time of day should a root be dug, or should that root even be dug up? Should an animal be collected in this spot because they are plentiful, or is there a reason why it should be collected somewhere else?

In short, the critical question is, where and how should a person invest their energy? In snow and cold, how and where someone invests their energy and the type of relationship they establish with the land will determine their success and maybe even their fate.

The lesson I underscore over and over is that there is no one answer for a survival scenario. I am the last person to give an answer. I want to know every single detail because I know there are a million different ways to answer the question.

One of the differences between what is understood about survival and the reality of survival is that many people believe that they have to use their energy while they have it. Survival books often teach that right from the get-go, the moment you are out the door, you must build tools quickly because you are going to start losing energy. But the land will resist that thinking.

When someone goes out full bore, he or she ends up making the wrong decisions. They haven't established any connection with the land. Consequently, they build things hastily, and those things don't work for them. The land wants to see somebody sit for a moment, contemplate, and ask themselves, "What should I do?" That's the space that many people don't utilize. A lot of people might think that is the lazy or the passive approach but it's really not.

Sometimes a student will see me running around at ninja-pace speeds. That is also a part of timing and rhythm. However, I have already built a deeper land connection so even if I am moving at a fast pace, I may actually be seeing more than the student moving slowly. Rhythm on the land is like a river that sometimes flows calmly and other times has rapids.

I teach my students that when they find themselves in a new environment, the best thing they can do is take a slow walk around and just observe everything. I tell them, don't try to collect anything or be productive. If there is a spot where it feels like they need to spend some time, I tell them to sit down or lie down, whatever their body feels like doing, and be open and receptive. Listen to the wind, the birds, and the animals. Then look around. Perhaps they will notice a nest in the distance, maybe berries over in nearby brush. But what I do know is that they will actually start seeing things that they wouldn't have seen if they charged out like a bull. This is a beginning level message, but it is critical.

On a more advanced level, there have been times that I have lain down and not fallen asleep. I have gone into a lucid stage and actually seen things about the surrounding area and had visions of them without having physically been to that place yet.

Most people go through life at one pace. In the wild, I have found that it is best to take in nature's immersions painfully slow and then at warp speed. After moving at the extremes, I then try to fill in the middle once I understand nature's timing. I think of nature as a song. The most beautiful note never feels old, even if another note does or does not follow it.

I tell students that if they reach that deeper level, if they sit down and start looking around and noticing more on either a physical or sub-conscious level, they will start absorbing the things around them. This will relax them and allow them to ask, "Based on my knowledge, what can I do with this to make this journey easier, more respectable, and more comfortable?" Not only will that cause them to make clearer decisions, it also will give them strength because they will have received energy from the land.

I emphasize that another important aspect is to take different paths. Instructors regularly fall into the rut of traveling the same path, rather than experimenting with a new one. For example, if the first time someone collects water and follows the main trail because it was easier, then the next time they should take a brushy path back, and start investigating the land to see what they can come to discover. Many, many times I have found food because I deviated from the main trail.

This thinking can tie into anyone's everyday life. If a person stays on the same path with their head pointed down, they are not going to be exposed to new opportunities. When you walk somewhere, try a different route. If you happen to be moving to a new place, go out, sit down, take a look around, and feel it. Maybe if you do that, you might find the place is not for you. The basic lesson is that you want to receive information. In nature, that is heightened and you can end up in a survival situation if you don't have that awareness.

Many people I've encountered know what works in their everyday lives, but somehow, they think that when they get out in nature, they have to change the rules. But they don't. If they start taking charge of things in their work life or in any situation when nobody knows who they are, they aren't going to get any respect. It's the same thing in nature; it will push you out. Even though those are two different worlds, the principle is very much the same. There is a clear overlap. People can act in nature the same way they do in a community. The difference is that if they try to impose themselves too much in the wilderness, they might annoy an African lion.

I can't lay down a formula for survival in a day or even a week because that would be called a survival handbook. Even if I did, the manual would be virtually useless. When I spend a couple weeks (or more) with students, they learn the patterns of nature that teach them how to look, where to look, and how to identify the characteristics of survival.

How do you not use excess energy? How do you receive energy without food? How do you make a shelter and a bed? How do you enjoy your experience in your shelter and your bed? How do you identify an edible plant? How does it feel? How does it smell? How does it taste? And most important, how do you knock out the word *survival* and turn it into a living situation?

By addressing these questions myself, I have become a better teacher. Students come out legitimately feeling comfortable they can survive, that it was a rewarding experience, and that they would do it again. They feel like they can even jump right back into it the next day if they had the time. The biggest lesson I have passed down to students is how to legitimately live off the land if they so desire. I've accomplished that by teaching them how to learn the natural flow and how to think for themselves.

Some people who enter my classes are thinking, *I have to learn how to survive.* But the word *survive* brings fear to people. That is a kind of fearmongering that comes from the exploitative TV survival shows. But I disabuse my students of that notion. I tell them that what I teach is the land. My ideal candidate is someone who wants to learn the land. I teach people how to live with the land in a fluid way that doesn't feel like a survival situation.

They watch as I apply the surroundings to our needs. I help them break things down to see what we need and what we don't. This makes it easier to navigate through complicated situations, such as tough terrain or a difficulty finding food. Interestingly, I've found that when people return to their normal lives, they use this to eliminate their baggage, to employ a pop psychology phrase. Baggage comes from holding on to things that made you comfortable when you were younger. However, these security boxes can end up blocking you from growth experiences, even though at one point they provided security.

Everybody has a different reaction to the wild. When city slickers set that first foot on the trail, they feel like they are in an alien environment. They walk for a while, get a few miles in, collect a few plants, and start to sense something magical unfolding. They become more focused and less distracted, as their daily worries fade into the landscape. On day two, they begin learning what it takes to live in the wild. Generally around day three, they wake up and feel like they have been hit upside the head because their diets are so different that detox kicks in. It is at that point that they realize there's no Chinese takeout available in the wilderness and the cuisine will be exotic plants and whatever we can hunt down together.

Even very healthy people feel like their strength level is cut in half. Everyone deals with this differently. Some feel like they are dying and moving only out of reflex; others wonder what it means. Jesse Perry likened the feeling to "an altered reality reminiscent of past experiences with psilocybin mushrooms," which, come to think of it, might be an inadvertent sales pitch to some prospects.

The positive people soon start to experience heightened sensations. All the rhythms in their body slow down, and they start seeing a lot more detail. For instance, they will see a bumblebee flying. It's no longer just a bee. They see every stripe and the fuzz on its abdomen. They see how its wings are moving to the point that they might even count the flaps per minute. They study the tips of its feet. This is the beginning of understanding their relationship with nature.

One thing I consider myself effective at is foreseeing problems that students may have. I pay such close attention to my students that it almost gives me headaches. I look at everyone individually and assess their skill level. I read over their medical histories to determine where the dangers may lie, see what medications they take, and find out specifics, such as if they have ever had an allergic reaction to a bee sting. Going into anaphylactic shock when you can rush to the emergency room is trying; going into anaphylactic shock in the wilderness is deadly.

Based on what I learn about each group of students, I am mindful of the footing and pacing. Many instructors push their students to the edge, to the point where they are going to tweak something, break an ankle, or have a bad fall.

The potential to foresee accidents is critical. Because students are off their normal diets and their blood sugar levels are reduced, extra care is required. I believe that everybody in America is partially diabetic, and it shows when they go out on a survival course. I haven't had anybody reach the point where they pass out for minutes at a time from low blood sugar, though many people have become dizzy and seen stars.

An instructor can't prevent accidents simply by safe protocol alone. In fact, the rules are less important than being able to properly assess all students' skill levels and coordinate that based on the environment.

I try to stay one step ahead and keep my students in balance. Once they dip too far, there is nothing in the wild that I can pick off a tree and immediately give them for an energy boost. There is no Snickers bar solution. When we first go out, I have students eat continuously to maintain their blood sugar.

I have had some students who are over seventy, so inevitably there are going to be some issues. The only serious problem I had occurred when I was guiding llama pack trips. A lady sat down on a log to tie her shoe and her artificial hip popped out of place. She was in excruciating pain.

I was a trained WFR, or wilderness first responder, meaning at that time I was certified to put a hip back into place. A WFR is a step up from being an EMT, but it's not as trained as a Wilderness EMT. As a woofer, I had learned to put shoulders, hips, and joints back into place, dress wounds, and give shots of adrenaline. We had also learned interesting techniques about sucking open chest wounds and how to build valves in the event of a chest puncture.

Hard as I tried, I could not get her hip joint back into place. I gave her Advil for the pain. She was a real trooper, but we needed medical help. Unfortunately, the company I worked for did not believe in guides carrying cell phones. We had no form of communication, and we were eight miles from the nearest trailhead. I took off sprinting and ran the eight miles in forty minutes to find a phone to call for a medevac.

The helicopter arrived twenty minutes later. The paramedics couldn't get her hip back into place, either, so they gave her relaxants and took her to the hospital. Fortunately, she made a full recovery.

The most practical aspect of survival in the wild is making a fire. I learned the process by reading a Tom Brown manual that contained poetic information about that magical tool. I took that manual, along with fifteen others books, to the Sierras and studied the skills, for myself and so I could teach students.

At first, I struggled mightily, rubbing two sticks together for hours and ending up with nothing but blisters. But over time I became very proficient and learned that there are many natural ways to bring dead wood to life. One thing I learned from Tom Brown's teachings in his books and from my students who have taken his classes (as I have not met him) is that he often leaves out important information to push students to gain more out of the process. I liken it to the way a Shaolin monk makes his pupil sweep a floor for two years before teaching him a basic karate kick.

Brown favored what is known as coyote teaching. The term comes from Native American lore. Many Native Americans believe that animals contain different aspects of our personalities. Coyotes are considered to be slightly mischievous with traits of a child. The coyote still needs to grow up, but at the same time, it has that childlike ability to draw people in.

Brown had a different way of doing things. I always had an innate ability to figure out things for myself. I looked at books as very rough guidelines. For me, it was still about looking at all the pieces and trying to figure out the solution. But as a teacher, I prefer a more direct approach.

The only coyote teaching I will do is to figure out a way to trick somebody to sit in a particular spot for a long period of time. Or if I realize that someone has citified personality traits, I might use those in a coyote way to spark their interest in the natural world.

As a teacher, I pride myself on giving very clear and fully detailed lessons so the students understand the information without being frustrated. What I have learned is that as a species we are so incompetent in the wild that if I started trying to trick people and make it harder, they would give up. It is critical for me to maintain their passion so that they feel successful. I prefer looking at methods that are rooted in ancient ways because they are just as good in the present day as they were a thousand years ago.

There are many secrets to making fire without any man-made aids. The main thing I teach my students to focus on are the properties of the wood. If you are creating a primitive fire, most are variations on what is called friction fire. This involves figuring out a way to create friction between two pieces of wood. The main issue is that the grains of wood must be short. They cannot be long and splintery; otherwise they will not create a coal that is fine and dark. A fine, dark coal holds together.

All friction woods have an infinite amount of combinations that work for different reasons. If the wood is too hard, then I find that I have to apply too much muscle to create the bond or friction. However, if a root is too soft, then one piece of wood drills through the other. To check for hardness, I push my thumb into the root. If it dents easily, the wood is too soft.

The wood should be bone dry, meaning it's dead, though in nature there are always exceptions to the rule. For example, sagebrush almost works better when it's completely alive. There are short pores through the structure that create the right type of char. Sagebrush works well for a bow drill (where you wrap the drill stick in the strands of a bow) but can be very difficult with a hand drill (where you rub your palms on the drill stick). In snow or heavy rain, I look for places where trees have grown over and created pockets where I can dig and find dry material. Sometimes I have to carve into the wood several inches to find out if the wood is dry enough.

The bow drill has a mechanical advantage. But without any technology, it requires more work to build. In North America, the hand drill was used almost exclusively by all Native American tribes. The way the user kneels and places his hands together resembles someone praying. With the proper knowledge, skill, and wood, it is extremely efficient—like a big, primitive lighter.

The environment often dictates what I can do. If I am in a snowy area at a higher elevation, it is unlikely that I will find usable firewood. Knowing this in advance, I will collect wood at a lower elevation before climbing. If I am in a pure survival situation, I will hike down or wait for it to stop snowing. The word for that is *patience*. I build a shelter, hunker down, and go hungry for a couple days. The advantage of fasting and staying put is that I slow down my metabolism. I don't fight the land. I let it teach me and then when it opens back up, I find my rhythm.

Safety is paramount. I always clear a large ten-to-fifteen-foot-diameter circle of all the debris, grass, leaves, and everything around that area. I kick it out with my feet so I am left with only soil, dirt, or sand. Then I build a fire pit in the very center of that area. More important than building a solid ring is clearing the area.

The reason most forest fires start is that people don't remove surrounding grass and debris. They come upon a place where there is dead grass, put some rocks down for a border, and build a small ring for the fire. Problem is, the fire is almost always going to jump the ring and catch the grass.

Another issue I have seen is roots. If someone digs a fire pit and hits a dead root and builds a fire on it, there is a chance that root will smolder back into the tree. If the ground is not damp and there is even the smallest bit of an air pocket allowing oxygen to travel along that burning root, it will reach the tree and the tree will catch fire.

The following morning, I teach students how to deal with the remains of fires. Some people just leave them. I insist my students go overboard to instill a certain order of respect with the fire. I have them take the leftover coals and grind them up into a powder. I then have them mix the ground coals with leftover dung from whatever animal is around and spread that mixture into the brush so it creates a fertile soil and provides nitrogen to the plants. This leaves the areas better off than when we arrived.

Sometimes when I'm trying to avoid any impact at all and I just want to stop and cook up a lunch meal, I will find an island in the river or build a fire right on the edge of the water line, where it is free of all debris. When the fire is done, I will scoop the ashes back into the water, knowing that area will eventually get flooded over. The ashes don't harm the water because it disperses the coals over miles and miles. The island is also a safe place for inexperienced people to build a fire, because the fire is unlikely to jump the water and start a forest fire.

In my mind, human roots lie deeper in fire than in any other force of nature. If we go way back in time to when man first harnessed fire, most likely they saw lightning strike. They probably kept their fires going by adding sticks for kindling. They sat around those fires in the cold and were warmed. The fire sculpted the people. Instead of using their furs to keep them warm, they hung out around the fire, and it created community time.

Fire also moved us away from being animals, as most animals run

from fire. That was a huge shift. Archaeologists always say what separates man from animal is his ability to use tools, but we're now finding
out that animals can use tools, such as a monkey using a stick to plunge
into an ant hole and pull out the ants. But likely it was fire that separated
man from animal, as undomesticated animals do not sit around a fire.

The common word used in the survival community for converting a survival situation into a living situation is *thrival*.

Thrival is a state that occurs when the layer of desperation and the
feeling of fighting to live (or be rescued, if that's the case) is replaced by
the joys of nature. At that point, a person achieves a feeling of place and
belonging. It is a place where nature is no longer their enemy numbering
their days on the planet. In thrival, they become a big part of nature and
would be able to live there indefinitely, if they so chose, because they
have succumbed to nature's turns and are beginning to thrive, instead of
just survive.

There are two primary aspects to thrival: starting the journey being
open to possibilities, and being able to stay with the journey knowing
you will never completely arrive at a destination or conquer it. The process of thrival begins simply enough, by adopting a positive attitude and
by truly enjoying the place you are in. The beautiful thing about being
open is that the earth will reveal its secrets, and your passion for the
journey will grow with each one. But know that no one will ever find
them all.

In a state of thrival in nature, a person achieves a feeling of pure bliss,
which is the second primary aspect. At that point, they are no longer an
alien on the planet. They are not separated from the earth by technology
and modern gadgetry, but are rather in a place where the earth becomes

their family and they find an indelible connection to it. We all need this in varying degrees, regardless of where we are.

This entry into thrival can be extended to any place or situation you find yourself in, not just in nature. It could be a new town, new job, or new house. Whatever the journey, being able to stay with it for a long period of time takes training, knowledge, and patience.

As a teacher, I've had the opportunity to work with hundreds, if not thousands, of students from all walks of modern and submodern life. I have watched them relate to the land, seen what sticks with them, and what they take away from the experience and apply to their everyday lives. From what I have seen, people on social levels and people in nature are very similar, but in a mental capacity, things are much different. We are striving too hard for technology and have become trapped in the dogma of scientific thinking. This has pulled us away from the land. There is also evidence, cited frequently by Paul Shepard and others, that shows that our brain capacity is shrinking now that we have moved away from the hunter-gatherer lifestyle.

There are different types of intelligence and smarts—social, business, mathematical, and hunter-gatherer. Each type of smarts functions well in a specific place. What I have learned is that in society everything is very linear. Society demands a routine where things consistently function the same. But in nature, that is not possible. Everything is constantly growing and changing in infinite shapes. Not one limb is the same in nature, but at a hardware store, every piece of lumber is the same. In our society, we have created a formula for putting everything into its slot. In nature, that cannot be done, and that is threatening to people. It takes a certain intellect to work with something that is not straight. I think we have lost that capacity.

Take a rocket scientist dropped into nature—literally. I actually had one as a student on a survival course. I gave him an intermediate task: setting a trap with natural materials. Nothing in nature grows straight so you must have the mental flexibility to piece together things in a natural way that allows the trap to stay in place and spring at the right moment. Teaching the rocket scientist to set that trap became a great test of my patience. It made me realize that someone who is a genius can be completely inept at understanding the natural world.

Students will often come to a course with the reference point that hunter-gatherers are our "primitive" past. But I teach them that from my experience of living in the wilderness for more than two decades, I can say with near certainty that those habitants of our primitive past had a greater mental capacity than we realize, or can even comprehend. Here's my point to them: someone cannot dominate the wilds because he or she has a strong modern intellect. In fact, to be a thrivalist they must adopt a completely different set of rules.

The first two virtues I impress upon students are patience and not alienating yourself. Being patient means the ability to stay open to possibilities and look and listen to what the land needs and what the person needs in return. That is often the reason Native Americans practiced vision quest circles. Sitting in a five-to-nine-foot circle of stone for four days and four nights without food and water makes you patient.

Not alienating oneself in nature means that a person often needs to leave technology—in the form of common, purchased survival tools— out of the wilderness experience. This may mean leaving behind a favorite knife and relying on rocks and other natural sharp edges in nature for cutting. By doing this, yes, survival is technically harder. However, what I tell my students they will find is that they will want to stay with the experience longer because they are doing it on their terms.

Everybody is born with different thought processes and patterns.

Some people are very good at concentration and laserlike focus, whereas others have more dispersed focus. Today, we label that as ADD and ADHD. Interestingly, people with ADD and ADHD actually excel in the wilderness because you need dispersed focus to be able to talk to somebody while hearing the sounds of nature at the same time. The ability to have your focus bounce around can be very helpful. There are many things happening at once in the wild. Focusing too closely on just one thing can cause you to miss something important.

If you think your job is complex—whether you are an electrical engineer, a tax lawyer, or a television executive—spending time in nature may be a stark and ultimately beautiful awakening to true complexity. It may help you simplify and streamline your everyday life. We often think of nature as simple, but it's quite the opposite. I have had students who were doctors, lawyers, hippies, jocks, naturalists, comedians, musicians, and belly dancers, and regardless of their professions and backgrounds, they have all benefited from being in nature.

Being an instructor, I have learned that I can take something I love more than myself, my passion for the land, and share it with others, and see how they respond. The best way to learn, I have concluded, is to teach.

People who return from the wilderness are never the same. Though at home they return to comfort foods, most of them eat far less highly processed foods and more natural foods. Their anxieties are softened, and they are easier to be around. Many tell me that they achieve a better balance in their lives. I find that when they do small tasks that help the environment, such as recycling, they feel that they know what they are saving.

Learning to survive in the wild even for short periods can translate to handling fear in the everyday world. Whenever I'm scared—say in a TV audition, which can be scarier than many places in the wild—I step back and ask myself what the bigger picture is. If I can see it, then I can isolate that one spot, and move forward without being consumed by feelings of fear. I believe the same is true in much scarier situations. If you can see that bigger picture—even if it is not the ultimate bigger picture of your life continuing—you can condition yourself to relax and push forward.

What I find from my students who establish a connection with the land is that they realize the potential of what they can be as a human being on a physical and even spiritual level, and it makes them want to return for more. After we are out for a month, they will develop themselves in ways they have never experienced. Sometimes they push themselves so hard that they feel enlightenment, but at the same time, they are craving a cheeseburger and an ice cream sandwich. They return to their city lives and get all those things they were dreaming about on the trail, but when they lose that wilderness boost and that feeling of peace and sanity, they almost always return for more.

There used to be a point where I thought people could find peace and sanity without nature. But from my students, I found that people who find peace and sanity do so because they are usually taking a moment to acknowledge a little bit of nature around them.

EPIC SURVIVAL LESSON:

PUT YOUR FEET IN THE DIRT

What I want most of all is for people to have a burning need to understand the ground we live on, and be willing to set aside their preconceived notions to learn from the smallest and grandest of all things. I want to inspire people to be closer to the earth in its rawest form. I am now convinced that the only way to access the human spirit is through creating your own intimate relationship with the land.

To me, it is about learning the wisdom of the wilderness at a time when far too many view it, at worst, as material to be plowed down if it gets in the way of urban development, or at best, something to gawk at through a car window or in a coffee table book. Most people have in some way felt a special connection to the wilderness at one time or another. This connection goes beyond the recesses of the mind. It's something that cries out: "Wake up! The landscape is right in front of you. It is real, and it is where your spirit can thrive."

It boils down to not isolating ourselves too much from nature. The more ways we can find to put our feet in the proverbial dirt, the better we will be in all aspects of our lives.

Chapter Twenty

A BAREFOOT
WILDERNESS RUN

The sun had not yet come up. I walked outside wearing shorts and a light shirt to go for a run. I didn't need shoes for this run. I was in a place where I had never been, and I wanted to connect with this new land. My legs have always taken me to the places I needed to go. Today, I hoped they would carry me somewhere special.

I started slowly. I didn't know the area. It felt slightly foreign, but as with all places I have been around the world, it was also somewhat familiar. The ground was rough but forgiving. The more comfortable I felt, the more I lengthened my stride. Soon, I forgot I was running. The ground was simply moving under me—the zazen of running—which allowed me to take notice of my surroundings.

The first thing I noticed was a large tree that forked in four directions. It seemed to be pointing to the right toward a trail, so I took it.

The trail narrowed. On the right, there was a massive area of ancient bedrock protruding from the ground. As I passed, I saw mica flecks and lines in the stone that had been shaped millions of years ago. I wondered how many different bands of tribes had walked over the rock. Nearby, on a sweet birch tree branch, two squirrels wrestled for space.

I pressed on to see what the immediate future held. The trail dipped down and then popped up. Eventually I passed a pond. With each step,

I felt small pebbles nudging between my toes. They weren't sharp but rather round and worn, oddly soft in way, likely from being walked on for generations.

The sun was mounting the sky. I spotted a smaller path and exited the trail to explore it. A near-perfect canopy of trees was overhead. I stayed close to the water. Water always leads to more life. I felt alone, but I wasn't.

From all sides, people were emerging from connecting trails. They were very much like me; they were out enjoying the natural world.

Following the water, I hooked to the left. The sun had crested the horizon, and it hit me squarely in the eyes. I was heading east. I ran another mile or so, and continued to hug the water. It moved me to the south, and the sun was no longer in my eyes.

I could hear more people joining me. I didn't look back, but they were there. They weren't chasing me, or threatening to pass. Even though I knew they were on their own, it felt like they were joining me.

With the sun fully hidden by trees, I looked up. I saw in front of me, not more than a half mile away, one of the most beautiful structures I had ever laid eyes on. It was yellow stone, so perfectly shaped that I slowed down for a better look. I held my breath at its beauty. I kept looking up, up, up. It was as tall as a mountain.

The structure was two massive rock towers rising from a common point with a plateau in the center. In the sunlight, the towers glowed. It was nothing like the rock formations in the western United States, as it was perfectly sculpted.

Marveling, I continued my run. I passed a line of oak cattails. On the other side, I heard dogs barking, but I couldn't see them.

I looked up again. In front of me was a row of shiny structures. Everything looked like pure glass, and each one reflected off the other. Just behind them, one towered above the rest. The structure rose straight up so high only man could have conceived it.

I stopped and moved off the trail. I looked around. Hundreds of people were there, running and walking. People were walking their dogs. Just past two guys tossing a Frisbee was a metal cart with a picture of a hot dog on top. I watched a few joggers pass. None of them looked down at my bare feet.

It was my first run in Central Park, and my first morning ever being in New York City.

Millions of people were waking up around me and going off to make their futures. But these few hundred in Central Park realized that the natural world was right in front of them. Central Park is the one piece of land in New York City that has been comparably less disrupted than the rest of the city.

The people in the park seemed to fall into two categories. Some were going places, with purpose. Others were simply enjoying their steps. They were breathing the air, taking in the beauty of nature, and perhaps wondering how a row of hundred-year-old magnolias could form such a perfect canopy. The consensus, however, seemed to be that if you can survive outside the park, you can survive in it.

When you realize what the wild can do for you, there is a desire to mesh it with the modern world, even if you are not going to quit your job and become a hunter-gatherer. We know what nature can do for people who live in urban environments. We know it can build their awareness, their physicality, and their senses, and all of these can be adapted into their everyday lives. In our society today, the earth underneath us has become a distant thing. For the most part, we live in a concrete world removed from the true ground. These people could easily trap themselves in a concrete existence, but chose to step off the concrete and onto the land.

I stood there, staring up at the most beautiful buildings I had ever seen. They were designed and built with architectural integrity. But in my mind, no man's artistic creation, no feat of architecture, will ever compare to the beauty of the earth.

As I looked around, I wondered, *How do I fit with these people? I live near a town of a couple hundred people. The people around me now live in a city of eight million. We have completely different baselines.*

My hunter-gatherer lifestyle has led me to struggle in this reality on occasion. Relationships with women, for one, have been a challenge. In a bar, I sometimes have trouble communicating in modern-speak with members of the opposite sex. Not long ago, a stunning young lady came up to me and told me I was strikingly handsome. I'm not sure what I said, but it wasn't the right thing because she smiled and walked away.

Even when a woman understands my lifestyle, there can be complications. Once a girlfriend was living with me in my primitive dwelling in Utah and a mountain lion that fascinated me would visit every night. She became freaked out, so I scared off the mountain lion. We soon broke up, but I never saw the mountain lion again.

The rules of everyday life can often be foreign to me in comical ways. For example, when I was in Los Angeles last year I checked into a nice hotel. I was pleasantly surprised to find a refrigerator stocked with snacks and drinks. I was even more surprised to find that after I consumed many of the items, the hotel restocked them like my mother refilling the fridge. Problem was, when I checked out I got hit with a $551 minibar bill. I had no idea the snacks weren't complimentary.

As I travel to different cities, there is not the same type of weight on me as I see on many other people. Being in survival situations has taught me what it takes to live. That connection is very powerful. It has devcloped all my senses. I have experienced heightened hearing, clearer eyesight, and cleaner smell. My sixth sense comes alive, and it makes me realize the potential of what I can be as a human being. All of that breeds contentment.

Still, I often look up at the sky and ask the creator where I fit into all of this. How wild am I supposed to be? How human? How much are these supposed to be intertwined? I always come to the same conclusion: it is going to take a lifetime to find the answers.

But what pushes me forward, what inspires me, is seeing people who want to connect to the earth in any way, as these people in Central Park were doing. I believe it's important not to wait for tomorrow, but to start today. If someone has an interest in nature, they should start by going out and getting in it. If that takes ten minutes out of the day, take the ten minutes. It can't be a matter of "I'll wait for the weekend." A weekend is an awesome time to get outside, but making it a daily practice is better. In our culture today, for most people, if they don't do something on a daily basis, it will slip out of their routine, which is just human nature.

For me, tapping into the natural laws of one environment is very applicable when you travel to a different environment, no matter how dissimilar to yours. I have lived off the land in a jungle of Kauai, in the California desert, in the Sierra mountains. I have explored the plains of Tanzania, the rain forests of Costa Rica, and the jungles of Vietnam, and now I had run in Central Park. What I have learned from traveling to all these places is that the earth is vast, but it's not such a big place.

Even though my reference point with the land was vastly different from that of everyone in Central Park that morning, I felt as if I was sharing my connection to a new place with the people who knew it well. It didn't matter that they weren't on a walkabout in Arizona, or on a mesa in the Kaiparowits, or at ten thousand feet on the Pacific Crest Trail. They were enjoying the wilderness that was right in front of them. And I felt that if I kept running through the trails of Central Park, some new secrets of the earth might reveal themselves to me.

ACKNOWLEDGMENTS

Matt would like to thank:

My good friend Dave Nessia, who has been a great coteacher, walkabout partner, and a mentor of the human spirit, and my good friend Breck Crystal, who has been my equal and more in so many aspects throughout our primitive-skills journey.

David Holladay, who shared his vast plant and primitive-skills wisdom of the Boulder, Utah, area.

Kirsten Rechnitz, who was my partner and a gifted woman who shared time with me living in a pit house and a wigwam, as well as on the trail.

The Native American people for showing me their earth wisdom and ancient roots.

Vikki Thorn, Hannah Oohwiler, Matt Thorn, Eric Scott, and Raymond Shurtz for being a huge inspiration in bringing heart and life to the journeys of music and voice.

My dad for taking me on those long hikes in the mountains, and my mom for indulging me with so much ocean time.

My grandma and grandfather for teaching me spirit and integrity.

Dave Wescott, who has been one of the greatest bridges for bringing primitive skills to mainstream lives.

Jamie Grossman Young, my manager, who has helped make possible so many new opportunities for me in TV and publishing.

Larry and Judy Davis for being like second parents when I moved to Boulder, Utah.

All my TV friends and coworkers, who have shared joys and hardships on our worldwide adventures to bring the world of survival to mainstream TV.

Friends and teachers at the Boulder Outdoor Survival School (BOSS) for our journeys and growth together.

Friends and family at the Winter Count gathering for being such a close tribe.

All my students, who have at times taught me more than I may have taught them.

The rock for teaching me how to "dance."

And the earth for teaching so much.

Matt and Josh would like to thank:

Andrew Stuart, our agent, for showing his enthusiasm for this project when it was an amorphous idea, guiding it through the proposal stage, and working his usual magic in finding us the best publisher.

Mitchell Ivers, who was our first choice to edit this book. He is the consummate writer's editor who knows how to bring out the best in any manuscript regardless of the subject matter. Not only are we thankful he stepped up and bought the book at the proposal stage, but we are even more grateful for his advice on constructing a narrative that delivered on all aspects of Matt's story.

The first-rate team at Gallery Books: Louise Burke (the boss who sets the professional tone), publisher Jen Bergstrom, director of publicity Jennifer Robinson, publicist Meagan Brown, art director Lisa Litwack, managing editor Susan Rella, senior production editor Jessica Chin, marketing director Liz Psaltis, and last but certainly not least, assistant editor Natasha Simons, who babysat the book at all stages, and Stephanie Evans Biggins for her sure-handed copyedit of the manuscript.

Jesse Perry, who wrote a wonderful diary that helped recall the events of his walk with Matt from Arizona to Utah.

Dr. Sam Parnia for his medical insights into what the body goes through in extreme conditions, and Ryan Koch for his thoughts on nutrition in the wild.

Bulletin of Primitive Technology (www.btprimitives.com) for granting permission to use Matt's original drawings that first appeared in the magazine.